CNS NEUROTRANSMITTERS
AND
NEUROMODULATORS

NEUROACTIVE STEROIDS

Edited by

Trevor W. Stone, Ph.D., D.Sc.
Professor of Pharmacology
Department of Pharmacology
University of Glasgow
Glasgow, Scotland

CRC Press
Boca Raton   New York

Library of Congress Cataloging-in-Publication Data

CNS neurotransmitters and neuromodulators: neuroactive steroids
    edited by Trevor W. Stone.
        p. cm.
    Includes bibliographical references and index.
    ISBN 0-8493-7633-5
    1. Steroid hormones—physiological effect. 2. Neurohormones.
    3. Neurotransmitters. I. Stone, T. W.
        (DNLM: 1. Steroids—physiology. 2. Neurotransmitters—physiology.
    3. Central Nervous System—physiology. 4. Steroids—pharmacology.
    5. Receptors, Steroids—physiology. 6. Receptors, Neurotransmitter—physiology.·
    WK 150 C651 1996)
    QP572.S7C67 1996
    612.8,042—dc20
    DNLM/DLC   95-46396
                                                                                 CIP

PREFACE

The rate at which neuroscience research is growing makes it increasingly difficult for active scientists to keep abreast of topics not in their immediate sphere of interest. Despite this, there is undoubtedly much to learn from workers in closely related fields. This series of review volumes on *CNS Neurotransmitters and Neuromodulators* is intended to provide the kind of overview of areas of neuroscience, covering aspects from molecular to behavioral, which should be valuable as reference and background material for anyone working in neuroscience, hopefully encouraging cross-talk between groups in those different areas which will facilitate the making of major advances in knowledge.

Trevor W. Stone

THE EDITOR

Trevor W. Stone, B. Pharm., Ph.D., D.Sc., is a Professor of Pharmacology at the University of Glasgow in Glasgow, Scotland. Professor Stone graduated in 1969 from the School of Pharmacy at London University and proceeded to the University at Aberdeen in Scotland to take a Ph.D degree under the supervision of Professor J. Laurence Malcolm. Professor Stone was appointed to a Lectureship in Physiology at Aberdeen where he remained until 1977 when he was appointed to a Senior Lectureship and subsequent Professorship in Neurosciences at St. Georges Medical School in London. Professor Stone is a member of the British Physiological and Pharmacological Societies, European and International Neuroscience Research Societies, The British Society of Medicine in London, and the New York Academy of Sciences.

Professor Stone has held research appointments at the National Institute of Mental Health in Washington, D.C. and has been a Visiting Professor of Pharmacology at the University of Auckland, New Zealand and at the Gulbenkian Institute of Science, Portugal.

Professor Stone has presented invited lectures at international meetings and has published more than 400 research papers and communications. In 1983, Professor Stone was awarded the degree of Doctor of Science by the University of London for his work on the physiology and pharmacology of the nervous system. His current research interests include the pharmacology of synaptic transmission in the nervous system particularly with respect to amino acids and purines, the interactions between synaptic transmitters, and the role of amino acids in neurological disorders.

CONTRIBUTORS

Gisela Barbany, Ph.D.
Medical Biochemistry and Biophysics
Karolinska Institute
Stockholm, Sweden

Darrell W. Brann, Ph.D.
Department of Physiology and Endocrinology
Medical College of Georgia
Augusta, Georgia

Peter C. Butera, Ph.D.
Department of Psychology
Niagara University
Niagara, New York

B. Dubrovsky, Ph.D.
Departments of Psychiatry
McGill University Medical School
and
Reddy Memorial Hospital
Montreal, Canada

David H. Farb, Ph.D.
Department of Pharmacology
Boston University School of Medicine
Boston, Massachusetts

L.M. Garcia-Segura, M.D.
Cajal Institute
C.S.I.C.
Madrid, Spain

Terrell T. Gibbs, Ph.D.
Department of Pharmacology
Boston University School of Medicine
Boston, Massachusetts

Antonio Guillamón, M.D.
Department of Psychobiology
UNED/Ciudad University
Madrid, Spain

J. Harris, Ph.D.
Department of Psychiatry
Reddy Memorial Hospital
McGill University Medical School
Montreal, Canada

Marian Joëls, Ph.D.
Department of Experimental Zoology
University of Amsterdam
Amsterdam, The Netherlands

Henk Karst, Ph.D.
Department of Experimental Zoology
University of Amsterdam
Amsterdam, The Netherlands

L. Leedom, M.D.
Department of Obstetrics and Gynecology
Center for Research in Reproductive Biology
Yale University School of Medicine
New Haven, Connecticut

Virendra B. Mahesh, Ph.D., D.Phil.
Department of Physiology and Endocrinology
Medical College of Georgia
Augusta, Georgia

Bruce S. McEwen, Ph.D.
Laboratory of Neuroendocrinology
The Rockefeller University
New York, New York

Christina R. McKittrick, Ph.D.
Laboratory of Neuroendocrinology
The Rockefeller University
New York, New York

F. Naftolin, M.D., D.Phil.
Department of Obstetrics and Gynecology
Yale University School of Medicine
New Haven, Connecticut

A.P. Payne, Ph.D.
Laboratory of Human Anatomy
University of Glasgow
Glasgow, Scotland

Jonathan R. Seckl, M.D., Ph.D.
Department of Medicine
University of Edinburgh
Edinburgh, Scotland

Santiago Segovia, Ph.D.
Department of Psychobiology
UNED/Ciudad University
Madrid, Spain

Trevor W. Stone, Ph.D., D.Sc.
Institute of Biomedical and Life Sciences
University of Glasgow
Glasgow, Scotland

A. Yoo, Ph.D.
Department of Psychiatry
Reddy Memorial Hospital
McGill University Medical School
Montreal, Canada

TABLE OF CONTENTS

Chapter 1

Expression and Properties of Corticosteroid Receptors in the CNS

Jonathan R. Seckl

CONTENTS

I. PREAMBLE

The pituitary gland has been described as the conductor of the endocrine orchestra, but it is clear that the brain determines most of the music that the glands play. Of course, the central nervous system (CNS) regulates neuroendocrine activity while receiving and integrating afferent hormonal and environmental information from the periphery, leading to the notion of a homeostatic relationship between the brain and neuroendocrine systems.[1,2] Perhaps the foremost example of this is the hypothalamic-pituitary-adrenal (HPA) axis and its limbic control.[3,4] When an external or internal stressor is perceived (by the brain), the HPA axis is stimulated, leading to adrenocortical secretion. Adrenal corticosteroids then exert negative feedback actions upon the limbic system (notably the hippocampus), the hypothalamus, and pituitary to inhibit further HPA activity. An alternative or additional principle to homeostasis, that of "allostasis" (the production of stability of a whole physiological system through change[5]) has been advocated to explain some of the long-term deleterious consequences of chronic HPA activation on the brain and other organs.[6] However, for either situation, the mechanism of corticosteroid interaction with the CNS is through specific receptors, and their biology and function is clearly crucial to the understanding of brain-HPA axis relationships. It is the molecular biology, biochemistry, and physiology of these receptors and some of their relationships with disease that form the topic of this review.

II. INTRODUCTION

A. GLUCOCORTICOIDS AND MINERALOCORTICOIDS

The adrenal cortex synthesizes a range of steroid hormones from cholesterol. Of greatest biological importance are the corticosteroids (glucocorticoids and mineralocorticoids), which play key roles in homeostasis and the response to stress. Additionally, the adrenal cortex produces progesterone, its derivatives, and adrenal androgens (indeed in humans the latter are the predominant adrenal products); their actions upon the brain are described elsewhere in this volume. Although production of steroids locally within the CNS has been mooted, levels of some key enzymes (21-hydroxylase, aldosterone synthase) are probably too low for significant biosynthesis of active corticosteroids, an assertion attested by the lack of occupancy of CNS receptors following adrenalectomy.[7,8] Synthesis of other steroids in the brain may be more significant.[9] Glucocorticoids (cortisol in humans and most mammals, corticosterone in rats, mice, and lower vertebrates) and mineralocorticoids (aldosterone, deoxycorticosterone) have been conserved through vertebrate evolution and influence many basic biological processes, maintaining metabolism, and protecting cells and organisms from external and internal challenge. Almost all tissues are targets for corticosteroid action and the CNS is no exception. Here, glucocorticoids influence brain activity and mediate stress-related functions, resulting in changes in cellular structure, biochemistry, electrical activity, and neurotransmission, thus affecting mood, behavior, cognition, neuroendocrine function, and cellular birth and death.

B. CORTICOSTEROID ACTION

Corticosteroid hormones have distinct tissue-specific effects indicating the presence of modulatory processes in target organs. In principle, steroid hormone action within tissues can be regulated at a variety of levels including (1) sequestration by plasma/extracellular binding proteins, (2) the presence of cell membrane uptake mechanisms, (3) prereceptor metabolism of ligand, (4) the concentration, affinities, and turnover of specific receptors, and finally (5) controls on postreceptor pathways (see Figure 1). For corticosteroids, almost all of these regulatory points play a role, though the importance of cell membrane uptake mechanisms remains unclear. Here we concentrate on the classical intracellular receptors and their regulation, but the other control points, although subject to less study, should not be ignored and will be mentioned where appropriate data are available. It should also be noted that not all corticosteroid actions are mediated by intracellular receptors and accumulating data suggest the presence of cell membrane binding sites which transduce some rapid corticosteroid effects.[10,11] These remain to be fully defined at biochemical and molecular levels and have been recently reviewed.[12]

III. TRANSCRIPTIONAL EFFECTS: GLUCOCORTICOID AND MINERALOCORTICOID RECEPTORS

The majority of steroid actions in the brain and elsewhere occur many minutes or even hours after application of steroid and require prior transcription and/or translation to become manifest. Such effects are mediated by intracellular corticosteroid receptors which are of two types, mineralocorticoid receptors (MR, type I) and glucocorticoid receptors (GR, type II).[4,13] Both receptors act as ligand-activated transcription factors, binding to specific DNA sequences in the control region of target genes and acting in "trans" to modulate (increasing or decreasing) the rate of transcription. The relationships between receptor structure and function have been elucidated in some detail, mostly in clonal cells or cell-free

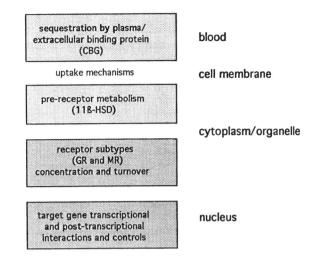

Figure 1 Regulatory points on the gluco-corticoid effector pathway.

systems.[14,15] However, the principle findings are likely to apply to receptors at all sites, including the brain, and are outlined below.

A. MODE OF ACTION OF GR AND MR

MR and GR have been categorized on the basis of their molecular structure as members of the steroid/thyroid hormone receptor gene superfamily.[15] In the absence of ligand, GR and MR are largely cytoplasmic,[16,17] existing in multimeric complexes with a variety of proteins, including heat shock protein-90 (HSP90), HSP70,[18] HSP56, a 27 kDa and smaller protein species (see Figure 2). These accessory proteins probably act as "molecular chaperones" and maintain receptors in a structure that allows the ligand to bind. Once steroid binding has occurred, the "activated" receptor dissociates from the HSP complex and migrates to the nucleus by an, as yet, ill-defined energy-requiring process that may require microtubule assembly, but does not apparently require HSPs.[19]

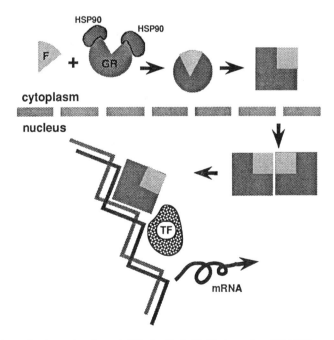

Figure 2 Classical mechanism of action of GR. Cortisol (F) binds to the GR-HSP complex, causing HSP dissociation, receptor transformation, nuclear translocation, and dimerization. GR homodimers bind to DNA sequences (GREs) and, often in association with other transcription factors (TF), alter the rate of transcription of a target gene.

In the classical mode of action, activated GR or MR then attach as homodimers[20] to specific DNA sequences in the regulatory regions of target genes; the point at which dimerization occurs is undefined, but may be the rate-limiting step.[21] The DNA target sequences, or glucocorticoid response elements (GREs), are typically, but not exclusively, approximate palindromes, comprising a 6 base pair (bp) dyad repeat separated by 3 bp (e.g., 5' AGAACAnnnTGTTCT 3'). The binding of activated GR homodimers to DNA (at least to naked DNA) may be affected by other proteins, which either increase binding (such as HSP70[22]) or decrease binding (e.g., calreticulin, which competes with GREs for attachment to the crucial residues at the base of the first zinc finger of the GR DNA binding domain[23]). *In vivo*, nucleosome structure also affects GR-DNA interactions, determining whether sites on DNA are accessible to receptors.[24,25] Once the GR-GRE complex is formed, other transcription factors may interact,[26-29] either cooperatively or antagonistically, as frequently noted for interactions between GR and AP-1.[30] Nevertheless, GR binding to the GRE may be very transient, acting perhaps to stabilize the transcription initiation complex, thus producing more prolonged effects than suggested merely by the duration of receptor-DNA interaction.[31] Environmental stimuli such as stress affect the efficiency of transcriptional induction by GR *after* binding to the GRE, presumably via induction of accessory factors which potentiate GR interactions with the RNA polymerase II complex.[32] Finally, a nonclassical mode of action of GR involves interaction with other transcription factors or the RNA polymerase II complex without direct DNA binding of the receptor.[33] Whether this mechanism also applies to MR, and its role if any in the CNS, remains to be determined.

B. DNA RESPONSE ELEMENTS

Intriguingly, both GR and MR (and indeed progesterone and androgen receptors) bind to the same DNA response elements, at least *in vitro*. This has suggested the current dogma that the obvious receptor-specificity of action upon target genes *in vivo* resides elsewhere than the sequence of the GRE itself[27,34,35] and may thus reflect the tissue-specific complement of other nuclear proteins. These proteins may interact directly with the receptors to determine their transcriptional effects, as recently demonstrated for some enhancer-binding proteins,[36] or may bind to other DNA sequences, particularly the regions flanking GREs, producing "composite elements"[26,27] which bind additional nuclear factors, such as AP1[27] and CREB.[37]

However, these studies have largely exploited synthetic response elements, and for natural promoters *in vivo* other possibile mechanisms may also produce the receptor-selectivity of effects upon target genes. Thus, *in vivo*, GR may regulate transcription by altering chromatin structure and nucleosome distribution,[25] thus revealing or concealing sites for other transcription factors, a function that the related progesterone receptor cannot mimic.[38] Of course, the concentration of receptors *in vivo* is likely to be important. Thus, it has been proposed that GR:MR heterodimers may form and bind cooperatively to GREs.[39] This might occur where the two receptors are co-localized in appropriate concentrations (perhaps in the hippocampus[40,41]). Finally, differences in the affinities of receptors for *natural* GREs, presumably determined by the particular base sequence of the dyad repeats, the triplet spacer, and the flanking sequences, may critically determine whether binding and transcriptional effects occur *in vivo*.[42] The detailed molecular events whereby GR and/or MR and their associated nuclear factors regulate physiological target genes in the CNS have yet to be elucidated.

IV. STRUCTURE OF GR AND MR AND THEIR ENCODING GENES

A. cDNAs AND GENES

MR and GR have been purified[43] and more recently the encoding cDNAs have been isolated and cloned.[44-48] The nucleic acid sequences encoding the two receptors show a high degree of homology, particularly within the domains responsible for binding to DNA and to hormones.[14,46] The genes encoding MR and GR are large and have not been completely defined. They comprise 9 coding exons[49] and a number of alternative 5' untranslated exons[50] with distinct promoters which presumably allow complex tissue-specific regulation. Thus, in the rat hippocampus, only the αMR promoter is sensitive to glucocorticoids,[50] whereas other promoters are glucocorticoid-insensitive, but are clearly regulated as they exhibit a distinct ontogeny.[51] The biological reasons for this transcriptional complexity is not known, but may relate to a requirement for a minimum basal level of expression of MR and GR (from constitutive promoters) to ensure essential cellular metabolic functions, with an overlay of regulated expression to mediate varying environmental and homeostatic (or allostatic) demands. Human GR, but not the rat or mouse homologs, exist in two distinct forms (GRα and GRβ) due to splicing of alternative last exons.[49]

Only GRα is active.[44] The β isoform may act as a negative regulator in the pituitary, although it is apparently absent from brain.[52]

B. PROTEIN STRUCTURE

GR and MR have been highly conserved in mammalian species and the DNA- and ligand-binding domains of the rat homologs[45,47,48] show ~90% identity to their human counterparts. The human GR comprises 777 amino acids[14] (Figure 3). The N terminal 421 amino acid (A/B or immunogenic) domains is not conserved between GR, MR, or other steroid/thyroid hormone receptors and contains many of the epitopes recognized by antisera raised against purified GR. This domain functions in transcriptional regulation or modulation,[53] and contains at least one (acidic or tau) transcriptional activator region.[34] The greater transactivation potency of GR compared to MR *in vitro* is due to its N terminal domain.[54] The DNA binding (C domain), by contrast, is very highly conserved indeed (94% identity between MR and GR), containing ~60 amino acid residues which form two "zinc finger" motifs, each with a Zn atom held covalently between four cysteine residues. Specific exposed residues of α-helical substructures of the zinc-fingers bind to the conserved bases of the GRE in the major groove of the DNA double helix.[20,55] Other residues in this domain function in transactivation of target genes.[53] The conserved (within the GR/MR subclass) hinge region or D domain comprises ~42 amino acids and is important for receptor translocation to the nucleus. The ligand binding region (E domain) represents the C terminal 249 residues which forms a hydrophobic steroid-binding pocket. This region is also involved in binding HSP90,[56] dimerization,[57] transactivation,[58] and nuclear translocation functions. Although not subjected to such detailed scrutiny, MR (984 amino acids in humans) can probably be similarly subdivided on a structural basis. Phosphorylation may be important for determining the function of some steroid receptors, and GR can be phosphorylated,[59] but its role, if any, in GR function is unclear.[60]

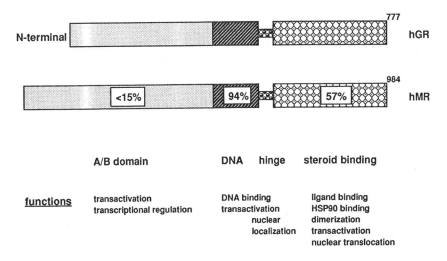

Figure 3 Domain structure of GR and MR. Showing percent identity between receptor types in various domains.

V. STEROID AFFINITIES OF GR AND MR

At the protein level, the two receptors can also be distinguished by their affinity for ligands.[7,46,47,61,62] Both rat and mouse GR show the highest affinity for synthetic glucocorticoids, such as dexamethasone and RU28362, lower affinity for physiological glucocorticoids (cortisol, corticosterone), and lower affinity still for mineralocorticoids. Other steroids bind very poorly, although the adrenal androgen dehydroepiandrosterone circulates in such high concentrations that it may bind GR under some circumstances *in vivo,* probably acting as a weak antagonist. MR, by contrast, shows the highest affinity for corticosterone and aldosterone, followed by cortisol. MR also binds some synthetic glucocorticoids such as dexamethasone,[62] although with lower affinity; any MR activation by synthetic glucocorticoids is less certain.

Some species differences in ligand affinities of GR and MR have been reported, though many can be predicted from the steroid chemistry involved. Thus in hamsters, which have cortisol as the major glucocorticoid, hippocampal MR and GR have a higher affinity for cortisol than the homologs in the rat[63] in which corticosterone is the glucocorticoid. Dogs and humans, which also have cortisol as the predominant glucocorticoid, have MR which, nevertheless, binds corticosterone better than cortisol and aldosterone.[46,64] The function in humans of corticosterone, which circulates in nanomolar concentrations,[65] remains obscure, and its affinity for MR has been exploited little experimentally. MR in dog brain binds dexamethasone well, although again the physiological relevance is obscure.[64] Dog brain GR, by contrast, binds dexamethasone poorly,[64] illustrating the potential pitfalls of pharmacological investigation when species differences in ligand affinities occur (human and rat GR have high affinity for dexamethasone which is frequently used for experimental and clinical investigation of central glucocorticoid feedback control[66]). The author's prejudice is that more attention should be paid to the effects of physiological glucocorticoids and the more selective receptor antagonists in exploring corticosteroid function in the CNS. Other species differences are less predictable. Thus, guinea pig GR has a markedly reduced affinity for physiological and synthetic glucocorticoids, but increased constitutive activity conferred, at least in large part, by the presence of tryptophan instead of cysteine at a crucial residue of the ligand binding domain.[67] This, and other residues that are not conserved in the guinea pig, normally form covalent bonds with steroids and are involved in HSP90 attachment, the latter perhaps explaining the constitutive activity of guinea pig GR. The entire complexity of guinea pig GR may have arisen secondary to a single amino acid-mutated ACTH in this species, which produces a superagonist at adrenal receptors and hence hypersecretion of glucocorticoids.[68] The detailed implications of these changes for guinea pig neuroendocrine and other central glucocorticoid functions are unexplored. Presumably, like new world monkeys, which show glucocorticoid resistance as part of more generalized steroid resistance, glucocorticoid feedback upon the HPA axis ensures that circulating steroid levels balance out relative receptor insensitivity to maintain an appropriate postreceptor "signal" to target genes.

A. OCCUPANCY

Brain MR, at least in the hippocampus, bind cortisol or corticosterone with high affinity but low capacity and are therefore largely, but not completely, occupied under basal conditions, whereas GR are lower affinity (but higher density) sites that are essentially unoccupied under basal conditions, but become progressively occupied during the diurnal maxima or after stress (see Figure 4).[7,8,61] This has some obvious implications. During the diurnal nadir when glucocorticoid levels are low, essentially only MR-mediated functions will be maintained.[4,69] These are likely to be stimulatory to crucial metabolic,[70] electrical,[71,72] and neurochemical effects of glucocorticoids. At higher glucocorticoid levels, GR will become occupied and may coordinate with or antagonize MR-mediated functions.[71,73-76] Examples of these complex interactions are well illustrated by the various electrophysiological effects of corticosteroids (addressed later in this volume).

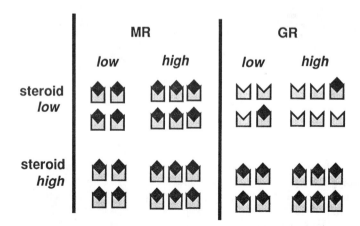

Figure 4 Receptor occupancy and signal in the hippocampus. MRs are already largely occupied by ligand under basal conditions, so the signal is partly dependent on receptor density. GRs are unoccupied by basal glucocorticoid levels and show a broad range of occupancy, so the signal is more sensitive to glucocorticoid levels than receptor density.

VI. ALDOSTERONE SELECTIVITY

The ligand affinities of receptors *in vitro* also reveal a paradox, since it is clear that MR in many peripheral target tissues (distal nephron, colon, salivary gland) selectively bind aldosterone *in vivo* in the face of a 100- to 1000-fold molar excess of cortisol or corticosterone, although these glucocorticoids are excellent ligands for purified or recombinant MR *in vitro*.[43,47] By contrast, most MR in the CNS, e.g., in the hippocampus, are nonselective and are therefore predominantly activated by corticosterone *in vivo,* reflecting the *in vitro* affinities.[7,13,61] Nevertheless, some central MR-mediated effects, such as actions upon salt appetite[77,78] and blood pressure,[75,79] are specific to aldosterone and are not mimicked by corticosterone. Moreover, the hypothalamus, in particular, concentrates endogenous aldosterone rather than corticosterone.[80,81] These discrepancies do not appear to reside in any intrinsic differences between MR at the various sites, since MR proteins[82] and the translated regions of cDNAs encoding MR in kidney and hippocampus[47,48] are identical.

A. CORTICOSTEROID-BINDING GLOBULIN

One suggested explanation involves the possible action of corticosteroid-binding globulin (CBG), a plasma and tissue protein with high affinity for cortisol and corticosterone.[43] The notion was that CBG sequestered glucocorticoids within cells before receptor binding could occur (reviewed in Reference 13). However, aldosterone selectivity of MR in the kidney is observed in neonatal rats which have very low levels of CBG.[83] Moreover, although some MR in the brain are aldosterone selective, CBG protein is almost absent from the adult rat brain.[13]

B. 11β-HYDROXYSTEROID DEHYDROGENASE

Any solution to this conundrum thus remained enigmatic until the recent discovery of the function of a hitherto obscure steroid-metabolizing enzyme, 11β-hydroxysteroid dehydrogenase (11β-HSD; recently reviewed in References 84 and 85). 11β-HSD catalyzes the conversion of physiological glucocorticoids to inert 11-keto derivatives (cortisol to cortisone, corticosterone to 11-dehydrocorticosterone). In aldo-sterone-selective target tissues, such as the distal nephron, abundant high-activity 11β-HSD excludes cortisol or corticosterone from MR, thus producing aldosterone selectivity *in vivo*[86,87] (aldosterone itself is protected from the enzyme by its 11 to 18 hemiacetal bridge structure[88]). In the congenital absence of 11β-HSD or when the enzyme is inhibited by liquorice or its derivatives, cortisol or corticosterone illicitly occupy and activate renal MR.[86,87,89,90] Thus, the specificity of MR *in vivo* is enzyme-, not receptor-mediated.

At least two isoforms of 11β-HSD exist.[85] 11β-HSD-1 is a reversible, NADP-dependent, lower affinity enzyme, initially characterized from liver.[91] In contrast, 11β-HSD-2 is a higher affinity, NAD-dependent enzyme, first identified in placenta and kidney,[92,93] which shows selective dehydrogenase (glucocorticoid inactivating) function. Many historical studies[94,95] and much recent data[96-98] clearly demonstrate 11β-HSD activity in many CNS subregions. The predominant isoform is 11β-HSD-1,[97,98] data in keeping with the lack of selectivity of most MRs in the CNS, which avidly bind corticosterone *in vivo.* 11β-HSD-2 has been provisionally reported in some subregions,[99] notably the amygdala and hypothalamus. Indeed this localization of 11β-HSD-2-like activity is in accordance with evidence suggesting that aldosterone-selective actions on salt appetite are mediated in the amygdala[100] and that the hypothalamus/preoptic area plays an important role in the central control of blood pressure.[101-103] Whether the sites of these aldosterone effects and functional cerebral 11β-HSD coincide anatomically is unknown, but administration of an 11β-HSD inhibitor to conscious unstressed rats alters cerebral function specifically in the preoptic area/periventricular hypothalamus.[104] It therefore seems reasonable to propose that cerebral 11β-HSD-2 may explain the aldosterone-selective MR sites in the hypothalamus and perhaps amygdala. Intriguingly, intracerebroventricular infusion of an 11β-HSD inhibitor increases blood pressure, suggesting this attenuates enzyme-mediated protection of aldosterone-selective MR, allowing corticosterone to activate these sites and elevate blood pressure.[105]

The function of the predominant and widespread 11β-HSD-1 isoform in the brain is obscure.[97,98] This enzyme is bi-directional and the reverse or reductase activity (glucocorticoid activation) may prevail *in vivo*.[98,106] This function would be anticipated to increase intracellular corticosterone levels from the circulating reservoir of inactive 11-dehydrocorticosterone. 11β-HSD reductase activity might maintain intracellular glucocorticoid levels and associated metabolic functions in cells sensitive to glucocorticoid depletion[107,108] during the diurnal nadir by reactivation of the more constant levels of otherwise inert

11-keto forms. However, it is this 11β-HSD isoform which is increased by chronically elevated gluco-corticoid levels and some stressors,[109] suggesting either that the reaction direction of 11β-HSD varies between cells or physiological states,[106,110] or that the enzyme acts further to expose already vulnerable neurons to glucocorticoid toxicity.[111] The ratio of corticosterone to its 11-keto metabolite in various brain subregions *in vivo* would be useful data to begin to address this question.

VII. MEASUREMENT OF GR AND MR IN THE BRAIN

A. RADIOLIGAND BINDING

Receptors were originally estimated by radioligand-binding techniques, either in cytosol/nuclear prepara-tions or by *in vivo* autoradiography, and these still form the mainstay of most studies.[13] However, the majority of MR and some GR sites are occupied by basal concentrations of glucocorticoids, and both receptors are largely occupied by stress-related glucocorticoid levels.[4,8,13,61] Moreover, the binding of steroid is essentially irreversible *in vitro,* as rebinding requires the receptor to dissociate from DNA, diffuse from the nucleus to the cytoplasm, and re-complex with HSP90 and associated proteins. This takes time (several hours, which allows some receptor degradation) and may be inefficient at the low temperatures conven-tionally employed for such assays. Thus, most studies have employed prior adrenalectomy to clear endog-enous steroids from receptors before estimation of their density, but of course this alters the state of the HPA axis, which itself exerts acute and, less certainly, chronic effects on receptor density.[7] A variety of exchange assays have been developed,[112,113] but the efficiency of rebinding is unclear,[114] and these approaches have been criticized.[111] *In vivo* autoradiography has also been applied,[115] but these techniques although yielding very useful anatomical detail, are capricious and very time consuming.

B. ANTISERA

Antisera to GR and, more recently, MR have been raised by a variety of means.[17,34] These have proved useful in some immunohistochemical studies for receptor localization,[116-119] but few groups have reported receptor estimation by techniques such as Western blotting, perhaps reflecting the problems of antibody specificity or recognition of particular epitopes in the different receptor activation states.

C. mRNA ESTIMATION

More recently, the cloning of cDNAs encoding GR and MR has allowed analysis of the expression and distribution of the respective mRNA species. Such approaches have intrinsic interest in the analysis of gene expression and are not hampered by problems of specificity in distinguishing GR and MR tran-scripts. Northern analysis and solution hybridization have provided useful quantitation of mRNA expres-sion in tissues[120,121] and *in situ* hybridization has produced semiquantitative data on receptor gene expression in brain sections[40,122] with excellent anatomical resolution at the level of individual cells and no apparent problems of cross hybridization between the two genes.[41] Of course, steady-state mRNA and receptor protein do not necessarily directly correlate, and indeed GR levels are regulated at the level of transcription,[123] mRNA stability,[124] and translation efficiency.[124] Nevertheless, in broad terms, ligand binding and mRNA studies in the brain produce coincident data and transcriptional control appears to predominate in many tissues.[125] The clear specificity, detailed anatomy, and quantitation afforded by molecular approaches have proved very useful indeed.

VIII. DISTRIBUTION

In the rat and mouse, GR are very widely expressed throughout the brain, not only in neurons, but also in glia[117,126] and vascular tissues. GR are abundant in the hippocampus, cerebral cortex, cerebellum, amygdala, some diencephalic nuclei, including the hypothalamic paraventricular nucleus where gluco-corticoids regulate corticotrophin-releasing factor-41 (CRF), and vasopressin synthesis, and in ascending monoaminergic neurons of the brainstem, including the raphé nuclei, locus ceruleus, and nucleus tractus solitarius.[61,127,128] Indeed, it is doubtful whether any CNS cell is entirely devoid of GR. By contrast, MR are at very low density in most brain regions, with high expression restricted to hippocampal neurons (both granular and pyramidal) and the septum.[61,127,128]

In general, the encoding mRNAs show similar distributions, but the higher level of resolution of *in situ* hybridization techniques have revealed more widespread MR mRNA expression,[47] including detectable levels of MR gene expression in areas of neuroendocrine interest, such as the hypothalamic paraventricular nucleus.[129,130] Recent immunohistochemical surveys (albeit using antisera raised against human MR on rat brain sections) also suggest MR-like immunoreactivity in many neurons outside the limbic system, but of course cannot quantitate protein levels.[119] Whether these low densities of MR exert physiologically significant effects remains unknown, but may be rather improbable. GR immunoreactivity broadly follows the distribution determined by ligand autoradiography and molecular approaches.[116,117]

The hippocampus has the highest density of both receptor types. Here corticosteroids alter electrical function[73] (discussed later in this volume), biochemical processes,[69] and neurotransmission,[77] thereby affecting mood, behavior, cognitive processes, neuroendocrine control,[131,132] and cell survival.[111,133,134] Both GR and MR are found throughout the hippocampal neuronal layers, with MR mRNA evenly expressed (perhaps highest in CA2 pyramidal neurons) and GR mRNA higher in dentate gyrus granule cells and pyramidal cells of CA1/2 than in CA3/4.[40,41,122,135] The encoded receptor proteins are similarly distributed.[128]

Other species show similar, although not identical distributions of MR and GR,[136] although the dog brain has more MR in extra-hippocampal sites than the rat.[64] Study of GR and MR in the human brain (e.g., hippocampus) has been greatly hampered by the paucity of surgical material,[137] the hazards of extrapolation from such pathological material to the normal situation, and the extreme instability of glucocorticoid binding sites in post-mortem specimens.[138] In contrast, mRNA is relatively stable in post-mortem brain[139] and recent studies have clearly shown high expression of genes encoding GR and MR in specific neuronal subregions of the human hippocampus,[140] suggesting that, at least with respect to the hippocampus, extrapolation from animal studies to humans may not be entirely inappropriate.

IX. REGULATION

Many groups have studied the regulation of corticosteroid receptors in the brain. Most data address the hippocampus and are briefly reviewed here.

A. CORTICOSTEROIDS

Glucocorticoid receptors can bind to the GR gene[141] and potently downregulate its transcription and translation in some peripheral cells.[142] However, at other loci glucocorticoids upregulate GR.[143,144] This tissue-specific control must therefore involve factors other than GR itself, which is ubiquitous. Similar site-specific effects are observed in the CNS. In the hippocampus, glucocorticoids autoregulate hippocampal GR mRNA in the short term, but not the longer term.[120-122,145] Moreover, this control is subregionally specific, with effects on GR mRNA expression confined to the dentate gyrus and CA1.[122,145] Similarly, little change in hippocampal GR sites is seen with chronic adrenal steroid manipulations.[7] Hippocampal expression of MR mRNA and protein are also altered in the short term in specific subfields by changes in adrenal steroid levels, effects mediated via GR, but more chronic manipulations are without effect.[61,120,122,145] The lack of persistence of effects in the hippocampus suggests that plastic processes occur, possibly affecting other regulatory pathways, such as the activity of afferent neural inputs (see below), which may compensate/correct direct glucocorticoid actions on receptor expression.

B. DIURNAL RHYTHM

Plasma glucocorticoids show a prominent diurnal variation, but data on circadian rhythms of MR are discrepant. Hippocampal MR binding sites at the corticosterone peak (dark phase in the rat) have been reported to be increased[146] or decreased.[121] At the transcriptional level the situation is equally confusing, with MR mRNA expression reported to show either no diurnal change,[121] a dark phase fall, specifically in the dentate gyrus and CA1,[147] or even a biphasic rhythm.[148] At least for GR there is some agreement, with two groups reporting decreased hippocampal neuronal GR mRNA expression in the dark phase, changes again limited to the dentate gyrus and CA1.[147,148] This dark phase fall in GR (and perhaps MR) mRNA expression is probably dependent on the rise in plasma corticosterone, as it is abolished by adrenalectomy.[148] The discrepancies between studies of diurnal effects upon hippocampal MR may reflect genetic (strain) or environmental (e.g., housing) differences, perhaps relating to the degree of GR activation by peak glucocorticoid levels[145] as GRs autoregulate their encoding gene transcription[122,145] and may also control co-localized MR.[145,149]

C. STRESS

Stress may also alter hippocampal MR and GR expression. Acute footshock has been reported to increase MR and GR in whole hippocampus after 14 days.[150] The effects may be stressor and time specific, however, as 2 h after acute restraint GR mRNA expression is unchanged,[148] while steady-state MR mRNA levels are increased, but MR gene transcription is decreased,[123] or MR and GR mRNA expression are unchanged (T. Olsson and J.R. Seckl, unpublished observations). More chronically, no changes in MR or GR mRNA levels were observed 6 h after surgical stress or after 2 weeks of chronic arthritis.[147] Interestingly, plasma corticosterone is persistently elevated in the arthritis model,[151] underlining the lack of chronic control by glucocorticoids of receptor expression in the hippocampus, in contrast to other sites.[142]

D. NEUROTRANSMITTERS
1. Monoamines
By contrast, neurotransmitters exert potent long-term control of hippocampal MR and GR.[131] Monoamines may be particularly important, as agents which deplete monoaminergic nerve terminals, such as reserpine,[152] methamphetamine,[153] and 3,4-methylenedioxymethamphetamine (MDMA),[154,155] markedly reduce corticosteroid-binding sites and MR and GR mRNA expression. Central serotonergic lesions in particular, decrease hippocampal corticosterone binding sites and GR and MR mRNA expression[41,155,156] and attenuate feedback sensitivity.[157] The mechanism is likely to be direct, since serotonin increases GR sites in primary fetal hippocampal and cortical neurones *in vitro*.[158,159] Noradrenergic lesions also decrease hippocampal corticosteroid receptors,[160] reducing MR, but not GR mRNA levels.[161] Other data confirm the specificity of noradrenergic lesions for hippocampal MR, but differ upon the direction of change.[112] This discrepancy may reflect methodological differences in the ligand-binding techniques used (total cell/cytosolic binding sites in intact rats vs. nuclear sites in adrenalectomized animals) and highlight the potential pitfalls of these approaches.

2. Acetylcholine
In contrast, lesions of the hippocampal cholinergic innervation, which lead to cognitive impairments, increase hippocampal MR and GR gene expression.[162] Combined cholinergic and serotonergic lesions, which produce more severe cognitive effects similar to those of age,[163] also elevate hippocampal GR, although not MR gene expression.[164] Indeed, the maintenance or elevation of GR expression in hippocampal neurons with cholinergic lesions or age may increase cellular vulnerability to glucocorticoid-mediated dysfunction and neurotoxicity.[111,162,165] Lesions affecting excitatory amino acid transmission also attenuate hippocampal GR and MR.[166] It should be noted that acute interruption of at least some hippocampal inputs (e.g., the glutamatergic perforant pathway from the entorhinal cortex) produce rapid, transient changes in GR expression in specific subregions.[167] These may have an adaptive value, with receptor downregulation perhaps promoting plastic responses normally sensitive to inhibition by glucocorticoids, such as axodendritic re-growth in the dentate gyrus.[168]

3. Neuropeptides
Neuropeptides may also control hippocampal GR and MR. The Brattleboro rat, which is deficient in vasopressin, shows decreased expression of GR in the hippocampus, presumably producing the documented hypersecretion of corticosterone and deficits in learning.[169] Treatment of Brattleboro rats with vasopressin, or even with analogs devoid of antidiuretic activity, restores hippocampal corticosteroid receptor levels, attenuates glucocorticoid hypersecretion, and improves learning.[169] Other studies have reported increased hippocampal corticosteroid receptor (MR) expression and improved cognition with ACTH analogs.[170-172] The site of action of peripherally administered peptides, which cross the blood-brain barrier poorly if at all, is unclear, as is any physiological relevance.

E. ANTIDEPRESSANTS AND OTHER DRUGS
The direct actions of serotonin on hippocampal GR expression[158] also suggest exploitation of agents that alter neurotransmission. Indeed several studies have demonstrated that antidepressant drugs, which potentiate monoaminergic neurotransmission, increase MR and, at least in the shorter term GR, mRNA expression and receptor protein in the hippocampus.[173-177] The effect of increased hippocampal receptor expression is to reduce HPA activity, presumably by enhancing sensitivity to glucocorticoid negative feedback.[174,175] In transgenic mice expressing a GR antisense construct that attenuates GR levels (leading to reduced sensitivity to glucocorticoid feedback and elevated corticosterone levels[178]), antidepressants increase brain GR expression (via increased gene transcription[179]) and reverse the neuroendocrine

deficit.[176] The mechanism whereby antidepressants increase GR gene expression has not been defined. Nevertheless, antidepressant induction of hippocampal GR may explain the amelioration of dexamethasone resistance and hypercortisolism with successful therapy of affective disorders.

Glucocorticoids (and stress) attenuate hippocampal spatial memory and learning in animals[74,180-182] and humans.[183] Stress-induced impairment of spatial memory is reversed by antidepressants,[184,185] and some agents even block the adverse effects of glucocorticoids upon hippocampal dendritic structure.[185] Recent data show a correlation between chronic antidepressant-mediated increases in hippocampal MR and improved cognitive performance in young rats,[186,187] possibly through the MR-associated increase in hippocampal excitability.[73,188] These data are particularly unexpected as the antidepressant used, amitriptyline, has marked sedative and anticholinergic properties that would be anticipated to attenuate cognitive processes rather than enhance them. Intriguingly, aged rats lose these plastic hippocampal responses to antidepressants. Although the mechanism is unclear, given the predominant control of MR and GR by neurotransmitter inputs it is tempting to speculate that loss of hippocampal innervations with age[189-193] may play a role in this maladaptive process.

F. ENVIRONMENTAL FACTORS

Several studies have documented that environmental manipulations during intrauterine life and the immediate postnatal period in rats potently influence stress responses, behavior, and hippocampal neuronal survival in later life (reviewed in Reference 194). Transient neonatal handling of rat pups leads to permanent increases in GR-binding sites in the hippocampus and frontal cortex, but not other brain regions,[195] effects secondary to increased GR mRNA expression in hippocampal neurons.[196,197] This causes more rapid suppression of stress responses and hence lower glucocorticoid exposure throughout the lifespan.[196] In consequence, neonatally handled rats largely avoid age-related memory impairments and hippocampal neuronal loss.[195,198,199] Chronic glucocorticoid treatment of handled rats, which downregulates hippocampal GR to levels indistinguishable from those in nonhandled rats, reverses many of the effects of handling, suggesting the protective effect is due to maintenance of low glucocorticoid levels.[114] Interestingly, the effect of handling on hippocampal GR may be mediated, at least in part, through the serotonergic system as it can be blocked by $5HT_2$ receptor antagonists.[194,200] Similar $5HT_2$ receptor-mediated effects of serotonin to increase GR expression are seen in primary fetal hippocampal neuron cultures,[201] suggesting direct action on hippocampal neurons. More severe perinatal manipulations in rodents permanently *reduce* hippocampal GR expression and increase basal and stress-induced glucocorticoid levels,[202,203] although the mechanisms are unclear. In adult rats, environmental conditions (i.e., conditions of housing) also affect hippocampal GR expression, albeit to a lesser extent,[204] and also alter adrenal activity[205] and cognitive function.[206] However, direct links between these parameters remain to be established.[207]

G. MOLECULAR MECHANISMS OF CORTICOSTEROID RECEPTOR
GENE REGULATION

Although accumulating data suggest that a number of steroids, peptides, and neurotransmitters affect GR and MR gene expression in the CNS, little work has been reported on the molecular mechanisms involved. Best documented is the control by serotonin, which appears to increase GR gene expression in hippocampal neurons *in vivo* and in primary cultures of fetal hippocampal neurons *in vitro*.[158] The latter system apparently acts via ketanserin-sensitive $5HT_2$-type receptors and can be mimicked by cyclic AMP analogs[201] (an unusual second messenger pathway for this receptor subgroup, which classically signals via inositol trisphosphate, implying the possible involvement of other 5HT receptors). A human GR gene promoter has been isolated and cloned,[49,208] and contains a DNA sequence compatible with a cyclic AMP response element (396 bp upstream of the major start site; Seckl, unpublished observations). In addition, the promoter has putative binding sites for other transcription factors, notably multiple sites for AP2 and SP1[49,208] and sites for NGFI-A (zif268, egr1, krox24) and NGFI-B (nur77), 662 and 1388 bp upstream of the start site, respectively (Seckl, unpublished observations). Intriguingly, expression of NGFI-A and NGFI-B varies in anatomical congruence with alterations in hippocampal GR mRNA levels following at least one manipulation (environmental enrichment),[204] suggesting these nuclear factors may be involved in GR gene control. Direct evidence of this assertion awaits confirmation by studies with GR-reporter gene constructs in transfected cells or transgenic models.

X. FUNCTION OF GR AND MR

Detailed examination of the myriad functions of GR and MR in the CNS is beyond the scope of this article, but this has been reviewed recently[4] and some examples are addressed in later chapters. One topic of relevance to this review is the thorny problem of determining the relative importance of circulating glucocorticoid levels or of receptor density. This is especially pertinent to regions of the CNS involved in negative feedback control of the HPA axis (e.g., the hippocampus).

A. GLUCOCORTICOIDS vs. RECEPTOR DENSITY

For GR, receptor density varies by less than 2-fold, whereas occupancy varies over a 10-fold or greater range. Therefore, it is probable that the signal associated with GR is more sensitive to glucocorticoid levels than receptor density (see Figure 4). The implications of this may be illustrated by recent data showing a direct correlation between hippocampal GR binding sites[209] or mRNA expression[186,187] and spatial memory in aged, but not young rats. This is unexpected as GR ligands *attenuate* "hippocampal-type" memory and its electrophysiological correlates.[74,180,188] Thus, reduced hippocampal GR would be predicted to mediate a less inhibitory signal and thus improve memory function, rather than correlate with cognitive impairment. A solution to this conundrum may lie in the low occupancy of GR under basal conditions. In aged cognitively impaired rats, basal plasma corticosterone levels are elevated and correlate negatively with both hippocampal GR mRNA expression and spatial memory.[186,187,209,210] Hippocampal GR levels, in part, determine feedback sensitivity and plasma glucocorticoid levels.[132,196] Thus, lower hippocampal GR will attenuate feedback sensitivity, producing elevated plasma corticosterone, which *itself* exerts the predominant negative effects upon memory, occupying otherwise vacant GR sites whatever their density. These observations accord with Sapolsky's hypothesis that reduced GR in the hippocampus impair glucocorticoid feedback on stress-induced HPA activation, eventually leading to elevated basal corticosterone levels, causing neuronal dysfunction and eventually loss.[211] Indeed, when there is age-related hippocampal cell loss, neurons with the highest GR density appear to be the most vulnerable,[111,212] presumably by transducing the greatest signal once glucocorticoid excess has occurred. For MR, by contrast, receptor occupation is high even at basal steroid levels and so any variation in receptor density (also around twofold) may be more crucial to the determination of the "effector" signal than glucocorticoid levels per se.

B. MOLECULAR CONSEQUENCES OF CORTICOSTEROID RECEPTOR ACTION

A few studies have addressed the molecular events occurring following receptor activation, either in normal neurons and glia or in pathology. Some genes regulated by GR or MR in hippocampal cells may be relevant to electrical function and perhaps memory.[213] Other potential target genes include neurotransmitter receptors[145,214,215] or have been implicated in neuronal endangerment, such as glucose transporters[216] and calcium-binding proteins, although effects on the latter appear slight.[217] However, the exact molecular mechanisms involved in the hippocampus (direct effects of GR/MR on the target gene promoters or indirect actions via other signaling systems) have not been defined. Some better defined examples in neuropeptidergic cells include the direct effects via GREs of GR upon proopiomelanocortin gene expression[218] and the possible indirect interaction of GR with other transcription factors to repress GnRH gene expression.[219] However, even for CRF, regulation by glucocorticoids is highly tissue and even subregion specific,[129] underlining the key importance of determining the nature and regulation of the factors responsible.

XI. GLUCOCORTICOIDS AND HUMAN HIPPOCAMPUS

In nonhuman primates the hippocampus is a major glucocorticoid target site;[111] glucocorticoid excess or severe social stress induces loss of hippocampal neurons,[220] and reduced hippocampal GR is specifically associated with attenuated glucocorticoid feedback sensitivity.[221] GR and MR mRNAs are also highly expressed in specific neuronal subregions of the human hippocampus,[140] suggesting that extrapolation from animal studies to humans may not be inappropriate. Moreover, some of the functional analyses in rodents may also apply to humans. Thus, electric stimulation of the hippocampus during neurosurgery inhibits adrenocortical secretion.[222] Hippocampal damage, for example in Alzheimer's disease, correlates with glucocorticoid feedback insensitivity.[223,224] Elevated glucocorticoid levels attenuate hippocampal memory in humans[225] and correlate with age-related cognitive decline.[183] In Cushing's disease and syndrome, in which cognitive abnormalities as well as mood changes are almost ubiquitous, neural atrophy and lesions in the limbic brain have been described[226-229] and hippocampal volume correlates positively with memory and negatively with cortisol levels.[230] Thus, many of the data on glucocorticoid receptors and their function derived from experimental models may also be applicable to the human brain. Indeed, some recent data have indicated that high expression of GR and MR mRNAs in hippocampal neurons is maintained in Alzheimer's disease in the face of marked hypercorticolemia.[165] These receptors may thus mediate deleterious effects of glucocorticoid excess upon cognitive function and neuronal survival. The exact parallels and dissimilarities between nonhuman mammalian species and the human CNS represent a key goal of future research.

ACKNOWLEDGMENTS

I am very grateful to Dr. Karen Chapman for a detailed critical review of the manuscript. Work from my laboratory discussed in this review was supported by the Wellcome Trust, the Medical Research Council, and the Scottish Hospital Endowments Research Trust.

REFERENCES

1. Harris, G. W., Neural control of the pituitary, *Physiol. Rev.,* 28, 139, 1948.
2. Gannon, W. F., Neurotransmitter mechanisms underlying stress responses, *Neuroendocrinology and Psychiatric Disorder*, Brown, G. M., Koslow, S. H., and Reichlin, S., Raven Press, New York, 1984, 133.
3. Selye, H., *The Physiology and Pathology of Exposure to Stress*, Acta Medica, Montreal, 1950.
4. de Kloet, E. R., Brain corticosteroid receptor balance and homeostatic control, *Frontiers in Neuroendocrinology*, 12, 95, 1991.
5. Sterling, P. and Eyer, J., Allostasis: a new paradigm to explain arousal pathology, *Handbook of Life Stress, Cognition and Health*, Fisher, S. and Reason, J., John Wiley & Sons, London, 1988.
6. McEwen, B. S. and Stellar, E., Stress and the individual: Mechanisms leading to disease, *Arch. Intern. Med.*, 153, 2093, 1993.
7. Reul, J. M. H. M., van den Bosch, F. R., and de Kloet, E. R., Relative occupation of type-I and type-II corticosteroid receptors in rat brain following stress and dexamethasone treatment: functional implications, *J. Endocrinol.*, 115, 459, 1987.
8. Spencer, R. L., Young, E. A., Choo, P. H., and McEwen, B. S., Adrenal steroid type I and type II receptor binding: estimates of *in vivo* receptor number, occupancy, and activation with varying level of steroid, *Brain Res.*, 514, 37, 1990.
9. Mellon, S. H. and Deschepper, C. F., Neurosteroid biosynthesis: genes for adrenal steroidogenic enzymes are expressed in the brain, *Brain Res.*, 629, 283, 1993.
10. Orchinik, M., Murray, T. F., and Moore, F. L., A corticosteroid receptor in neuronal membranes, *Science*, 252, 1848, 1991.
11. Wehling, M., Novel aldosterone receptors: specificity-conferring mechanism at the level of the cell membrane, *Steroids*, 59, 160, 1994.
12. McEwen, B. S., Non-genomic and genomic effects of steroids on neural activity, *Trends Pharmacol. Sci.*, 12, 141, 1991.
13. McEwen, B. S., de Kloet, E. R., and Rostene, W., Adrenal steroid receptors and action in the nervous system, *Physiol. Rev.*, 66, 1121, 1986.
14. Evans, R. M., The steroid and thyroid hormone receptor superfamily, *Science*, 240, 889, 1988.
15. Evans, R. M. and Arriza, J. L., A molecular mechanism for the actions of glucocorticoid hormones in the nervous system, *Neuron*, 2, 1105, 1989.
16. Picard, D., Kumar, V., Chambon, P., and Yamamoto, K. R., Signal transduction by steroid hormones: nuclear localization is differentially regulated in estrogen and glucocorticoid receptors, *Cell Regulation*, 1, 291, 1990.
17. Robertson, N. M., Schulman, G., Karnik, S., Alnemri, E., and Litwack, G., Demonstration of nuclear translocation of the mineralocorticoid receptor (MR) using an anti-MR antibody and confocal laser scanning microscopy, *Mol. Endocrinol.*, 7, 1226, 1993.
18. Hutchison, K. A., Dittmar, K. D., Czar, M. J., and Pratt, W. B., Proof that hsp70 is required for assembly of the glucocorticoid receptor into a heterocomplex with hsp90, *J. Biol. Chem.*, 269, 5043, 1994.
19. Yang, J. and DeFranco, D. B., Differential roles of heat shock protein 70 in the *in vitro* nuclear import of glucocorticoid receptor and simian virus 40 large tumor antigen, *Mol. Cell. Biol.*, 14, 5088, 1994.
20. Glass, C. K., Differential recognition of target genes by nuclear receptor monomers, dimers and heterodimers, *Endocrine Rev.*, 15, 391, 1994.
21. Drouin, J., Sun, Y. L., Tremblay, S., Lavender, P., Schmidt, T. J., de Léan, A., and Nemer, M., Homodimer formation is rate-limiting for high affinity DNA binding by glucocorticoid receptor, *Mol. Endocrinol.*, 6, 1299, 1992.

22. Srinivasan, G., Patel, N. T., and Thompson, E. B., Heat shock protein is tightly associated with the recombinant human glucocorticoid receptor: glucocorticoid response element complex, *Mol. Endocrinol.,* 8, 189, 1994.

23. Burns, K., Duggan, B., Atkinson, E. A., Famulski, K. S., Nemer, M., Bleackley, R. C., and Michalak, M., Modulation of gene expression by calreticulin binding to the glucocorticoid receptor, *Nature,* 367, 476, 1994.

24. Archer, T. K., Cordingley, M. G., Wolford, R. G., and Hager, G. L., Transcription factor access is mediated by accurately positioned nucleosomes on the mouse mammary tumor virus promoter, *Mol. Cell. Biol.,* 11, 688, 1991.

25. Hager, G. L. and Archer, T., Interaction of steroid receptors with chromatin, *Nuclear Hormone Receptors,* Parker, M. G., Academic Press, London, 1991, 217.

26. Diamond, M. I., Miner, J. N., Yoshinaga, S. K., and Yamamoto, K. R., Transcription factor interactions: selectors of positive or negative regulation from a single DNA element, *Science,* 249, 1266, 1990.

27. Pearce, D. and Yamamoto, K. R., Mineralocorticoid and glucocorticoid receptor activities distinguished by non-receptor factors at a composite response element, *Science,* 259, 1161, 1993.

28. Wheeler, T. T., Sadowski, H. B., and Young, D. A., Glucocorticoid and phorbol ester effects in 3T3-L1 fibroblasts suggest multiple and previously undescribed mechanisms of glucocorticoid receptor-AP-1 interaction, *Mol. Cell. Endocrinol.,* 104, 29, 1994.

29. Maroder, M., Farina, A. R., Vacca, A., Felli, M. P., Meco, D., Screpanti, I., Frati, L., and Gulino, A., Cell-specific bifunctional role of jun oncogene family members on glucocorticoid receptor-dependent transcription, *Mol. Endocrinol.,* 7, 570, 1993.

30. Pfahl, M., Nuclear receptor/AP-1 interactions, *Endocrine Rev.,* 14, 651, 1993.

31. Freedman, L. P., Yoshginaga, S. K., Vanderbilt, J. N., and Yamamoto, K. R., *In vitro* transcription enhancement by purified derivatives of the glucocorticoid receptor, *Science,* 245, 298, 1989.

32. Sanchez, E. R., Hu, J.-L., Zhong, S., Shen, P., Greene, M. J., and Housley, P. R., Potentiation of glucocorticoid receptor-mediated gene expression by heat and chemical shock, *Mol. Endocrinol.,* 8, 408, 1994.

33. König, H., Ponta, H., Rahmsdorf, H. J., and Herrlich, P., Interference between pathway-specific transcription factors: glucocorticoids antagonize phorbol ester-induced AP-1 activity without altering AP-1 site occupation *in vivo, EMBO J.,* 11, 2241, 1992.

34. Danielsen, M., Structure and function of the glucocorticoid receptor, *Nuclear Hormone Receptors,* Parker, M. G., Academic Press, London, 1991, 39.

35. Robins, D. M., Scheller, A., and Adler, A. J., Specific steroid response from a nonspecific DNA element, *J. Steroid Biochem. Mol. Biol.,* 49, 251, 1994.

36. Nordheim, A., CREB takes CBP to tango, *Nature,* 370, 177, 1994.

37. Stauber, C., Altschmied, J., Akerblom, I. E., Marron, J. L., and Mellon, P. L., Mutual cross-interference between the glucocorticoid receptor and CREB inhibit transactivation in placental cells, *New Biol.,* 4, 527, 1992.

38. Archer, T. K., Lee, H.-L., Cordingley, M. G., Mymryk, J. S., Fragoso, G., Berard, D. S., and Hager, G. L., Differential steroid hormone induction of transcription from the mouse mammary tumor virus promoter, *Mol. Endocrinol.,* 8, 568, 1994.

39. Trapp, T., Rupprecht, R., and Holsboer, F., Coexpression of the mineralocorticoid receptor and glucocorticoid receptor: characterization of transactivation and DNA binding properties, *J. Cell. Biochem.,* Suppl. 18B, 349, 1994.

40. Van Eekelen, J. A. M., Jiang, W., de Kloet, E. R., and Bohn, M. C., Distribution of the mineralocorticoid and the glucocorticoid receptor mRNAs in the rat hippocampus, *J. Neurosci. Res.,* 21, 88, 1988.

41. Seckl, J. R., Dickson, K. L., and Fink, G., Central 5,7-dihydroxytryptamine lesions decrease hippocampal glucocorticoid and mineralocorticoid receptor messenger ribonucleic acid expression., *J. Neuroendocrinol.,* 2, 911, 1990.

42. Chalepakis, G., Arnemann, J., Slater, E., Brüller, H.-J., Gross, B., and Beato, M., Differential gene activation by glucocorticoids and progestins through the hormone regulatory element of mouse mammary tumor virus, *Cell,* 53, 371, 1988.

43. Krozowski, Z. K. and Funder, J. W., Renal mineralocorticoid receptors and hippocampal corticosterone binding species have identical intrinsic steroid specificity, *Proc. Natl. Acad. Sci. U.S.A.,* 80, 6056, 1983.

44. Hollenberg, S. M., Weinberger, C., Ong, E. S., Cerelli, G., Oro, A., Lebo, R., Thompson, E. B., Rosenfeld, M. G., and Evans, R. M., Primary structure and expression of a functional human glucocorticoid receptor cDNA, *Nature,* 318, 635, 1985.

45. Miesfeld, R., Okret, S., Wilkstrom, A.-C., Wrange, O., Gustafsson, J.-A., and Yamamoto, K. R., Characterization of a steroid hormone receptor gene and mRNA in wild-type and mutant cells, *Nature,* 312, 779, 1984.

46. Arriza, J. L., Weinberger, C., Cerelli, G., Glaser, T. M., Handelin, B. L., Housman, D. E., and Evans, R. M., Cloning of human mineralocorticoid receptor complementary DNA: structural and functional kinship with the glucocorticoid receptor, *Science,* 237, 268, 1987.

47. Arriza, J. L., Simerly, R. B., Swanson, L. W., and Evans, R. M., The neuronal mineralocorticoid receptor as a mediator of glucocorticoid response, *Neuron,* 1, 887, 1988.

48. Patel, P. D., Sherman, T. G., Goldman, D. J., and Watson, S. J., Molecular cloning of a mineralocorticoid (type I) receptor complementary DNA from rat hippocampus, *Mol. Endocrinol.,* 3, 1877, 1990.

49. Encio, I. J. and Detera-Wadleigh, S. D., The genomic structure of the human glucocorticoid receptor, *J. Biol. Chem.,* 266, 7182, 1991.

50. Kwak, S. P., Patel, P. D., Thompson, R. C., Akil, H., and Watson, S. J., 5′-Heterogeneity of the mineralocorticoid receptor messenger ribonucleic acid: differential expression and regulation of splice variants within the rat hippocampus, *Endocrinology*, 133, 2344, 1993.
51. Vázquez, D. M., Kwak, S. P., López, J. F., Watson, S. J., and Akil, H., Localization of the mineralocorticoid receptor mRNA 5′ splice variants in the developing hippocampus, *Soc. Neurosci. Abstr.*, 19, 488, 1993.
52. Bamberger, C. M., Bamberger, A. M., and Chrousos, G. P., Glucocorticoid receptor beta, a potential endogenous antagonist of glucocorticoid action, *Endocrine Soc. 76th Annual Meeting*, 264, 1994.
53. Hollenberg, S. M., Giguere, V., Segui, P., and Evans, R. M., Co-localization of DNA binding and transcriptional activation functions in the human glucocorticoid receptor, *Cell*, 49, 39, 1987.
54. Rupprecht, R., Arriza, J. L., Sprengler, D., Reul, J. M. H. M., Evans, R. M., Hosboer, F., and Damm, K., Transactivation and synergistic properties of the mineralocorticoid receptor: relationship to the glucocorticoid receptor, *Mol. Endocrinol.*, 7, 597, 1993.
55. Luisi, B. F., Xu, W. X., Otwinowski, Z., Freedman, L. P., Yamamoto, K. R., and Sigler, P. B., Crystallographic analysis of the interaction of the glucocorticoid receptor with DNA, *Nature*, 352, 497, 1991.
56. Chakraborti, P. K. and Simons, Jr., S. S., Association of heat shock protein 90 with 16 kDa steroid binding core fragment of rat glucocorticoid receptors, *Biochem. Biophys. Res. Commun.*, 176, 1338, 1991.
57. Fawell, S. E., Lees, J. A., White, R., and Parker, M. G., Characterisation and colocalisation of steroid binding and dimerisation activities in the mouse estrogen receptor, *Cell*, 60, 953, 1990.
58. Godowski, P. J., Picard, D., and Yamamoto, K. R., Signal transduction and transcriptional regulation by glucocorticoid receptor-lex A fusion proteins, *Science*, 241, 812, 1988.
59. Bodwell, J. E., Ortí, E., Coull, J. M., Pappin, D. J. C., Smith, L. I., and Swift, F., Identification of phosphorylated sites in the mouse glucocorticoid receptor, *J. Biol. Chem.*, 266, 7549, 1991.
60. Moyer, M. L., Borror, K. C., Bona, B. J., DeFranco, D. B., and Nordeen, S. K., Modulation of cell signaling pathways can enhance or impair glucocorticoid-induced gene expression without altering the state of receptor phosphorylation, *J. Biol. Chem.*, 268, 22933, 1993.
61. Reul, J. M. H. M. and de Kloet, E. R., Two receptor systems for corticosterone in rat brain: microdissection and differential occupation, *Endocrinology*, 117, 2505, 1985.
62. Luttge, W. G., Davda, M. M., Rupp, M. E., and Kang, C. G., High affinity binding and regulatory actions of dexamethasone-type I receptor complexes in the mouse brain, *Endocrinology*, 125, 1194, 1989.
63. Sutanto, W., Reul, J. M. H. M., van Eekelen, J. A. M., and de Kloet, E. R., Corticosteroid receptor analysis in rat and hamster brains reveals species specificity in the type I and type II receptors, *J. Steroid Biochem.*, 30, 417, 1988.
64. Reul, J. M. H. M., de Kloet, E. R., van Sluijs, F. J., Rijnberk, A., and Rothuizen, J., Binding characteristics of mineralocorticoid and glucocorticoid receptors in dog brain and pituitary, *Endocrinology*, 127, 907, 1990.
65. Seckl, J. R., Campbell, J. C., Edwards, C. R. W., Christie, J. E., Whalley, L. J., Goodwin, G. M., and Fink, G., Diurnal variation of plasma corticosterone in depression, *Psychoneuroendocrinology*, 15, 485, 1990.
66. Carroll, B. J., Curtis, G. C., and Mendels, J., Neuroendocrine regulation in depression II: discrimination of depressed from non-depressed patients, *Arch. Gen. Psychiatry*, 138, 1218, 1976.
67. Keightley, M.-C. and Fuller, P. J., Unique sequences in the guinea pig glucocorticoid receptor induce constituitive transactivation and decrease steroid sensitivity, *Mol. Endocrinol.*, 8, 431, 1994.
68. Smith, A. I., Wallace, C. A., Moritz, R. L., Simpson, R. J., Schmauk-White, L. B., Woodcock, E. A., and Funder, J. W., Isolation, amino acid sequence and action of guinea pig ACTH on aldosterone production by glomerulosa cells, *J. Endocrinol.*, 115, R5, 1987.
69. de Kloet, E. R. and Reul, J. M. H. M., Feedback action and tonic influences of corticosteroids on brain function: a concept arising from heterogeneity of brain receptor systems, *Psychoneuroendocrinology*, 12, 83, 1987.
70. de Kloet, E. R., Kovács, G., Szabó, G., Telegdy, G., Bohus, B., and Versteeg, D., Decreased serotonin turnover in the dorsal hippocampus of the rat brain shortly after adrenalectomy: selective normalization after corticosterone substitution, *Brain Res.*, 239, 659, 1982.
71. Joels, M. and de Kloet, E. R., Effects of glucocorticoids and norepinephrine on the excitability in the hippocampus, *Science*, 245, 1502, 1989.
72. Joels, M. and de Kloet, E. R., Mineralocorticoid receptor-mediated changes in membrane properties of rat CA1 pyramidal neurons *in vitro*, *Proc. Natl. Acad. Sci. U.S.A.*, 87, 4495, 1990.
73. Joels, M. and de Kloet, E. R., Control of neuronal excitability by corticosteroid hormones, *Trends Neurosci.*, 15, 25, 1992.
74. Kerr, D. S., Huggett, A. M., and Abraham, W. C. C., Modulation of hippocampal long-term potentiation and long-term depression by corticosteroid receptor activation, *Psychobiology*, 22, 123, 1994.
75. Gomez-Sanchez, E. P., Fort, C. M., and Thwaites, D., Icv infusion of corticosterone antagonises icv-aldosterone hypertension, *Am. J. Physiol.*, E649-E653, 1990.
76. Bradbury, M. J., Akana, S. F., Casci, C. S., Levin, N., Jacobson, L., and Dallman, M. F., Regulation of basal ACTH secretion by corticosterone is mediated by both type I (MR) and type II (GR) receptors in rat brain, *J. Steroid Biochem.*, 40, 133, 1991.
77. McEwen, B. S., Lambdin, L. T., Rainbow, T. C., and De Nicola, A. F., Aldosterone effects on salt appetite in adrenalectomised rats, *Neuroendocrinol.*, 43, 38, 1986.

78. Sakai, R. R., Nicolaidis, S., and Epstein, A. N., Salt appetite is suppressed by interference with angiotensin II and aldosterone, *Am. J. Physiol.,* 251, R762, 1986.

79. Gomez-Sanchez, E. P., Intracerebroventricular infusion of aldosterone induces hypertension in rats, *Endocrinology,* 118, 819, 1986.

80. McEwen, B. S., Stephenson, B., and Krey, L. C., Radioimmunoassay of brain tissue and cell nuclear corticosterone, *J. Neurosci. Methods,* 3, 57, 1980.

81. Yongue, B. G. and Roy, E. J., Endogenous aldosterone and corticosterone in brain cell nuclei of adrenal-intact rats: regional distribution and effects of physiological variations in serum steroids, *Brain Res.,* 436, 49, 1987.

82. Wrange, O. and Yu, Z.-Y., Mineralocorticoid receptor in rat kidney and hippocampus: characterisation and quantitation by isoelectric focusing, *Endocrinology,* 113, 243, 1983.

83. Sheppard, K. and Funder, J. W., Mineralocorticoid specificity of renal type 1 receptors: *in-vivo* binding studies, *Am. J. Physiol.,* 252, E224, 1987.

84. Monder, C. and White, P. C., 11β-hydroxysteroid dehydrogenase, *Vitam. Horm.,* 47, 187, 1993.

85. Seckl, J. R., 11β-hydroxysteroid dehydrogenase isoforms and their implications for blood pressure regulation, *Eur. J. Clin. Invest.,* 23, 589, 1993.

86. Edwards, C. R. W., Stewart, P. M., Burt, D., Brett, L., McIntyre, M. A., Sutanto, W. S., de Kloet, E. R., and Monder, C., Localisation of 11β-hydroxysteroid dehydrogenase-tissue specific protector of the mineralocorticoid receptor, *Lancet,* ii, 986, 1988.

87. Funder, J. W., Pearce, P. T., Smith, R., and Smith, A. I., Mineralocorticoid action: target tissue specificity is enzyme, not receptor, mediated, *Science,* 242, 583, 1988.

88. Edwards, C. R. W. and Hayman, A., Enzyme protection of the mineralocorticoid receptor: evidence in favour of the hemi-acetal structure of aldosterone, *Aldosterone: Fundamental Aspects,* 215, Bonvalet, J.-P., Farman, N., Lombes, M., and Rafestin-Oblin, M. E., INSERM/John Libbey Eurotext Ltd., Paris, 1991, 67.

89. Stewart, P. M., Valentino, R., Wallace, A. M., Burt, D., Shackleton, C. H. L., and Edwards, C. R. W., Mineralocorticoid activity of liquorice: 11β-hydroxysteroid dehydrogenase deficiency comes of age, *Lancet,* ii, 821, 1987.

90. Stewart, P. M., Corrie, J. E. T., Shackleton, C. H. L., and Edwards, C. R. W., Syndrome of apparent mineralocorticoid excess: a defect in the cortisol-cortisone shuttle, *J. Clin. Invest.,* 82, 340, 1988.

91. Lakshmi, V. and Monder, C., Evidence for independent 11-oxidase and 11-reductase activities of 11β-hydroxysteroid dehydrogenase: enzyme latency, phase transition, and lipid requirements, *Endocrinology,* 116, 552, 1985.

92. Brown, R. W., Chapman, K. E., Edwards, C. R. W., and Seckl, J. R., Human placental 11β-hydroxysteroid dehydrogenase: partial purification of and evidence for a distinct NAD-dependent isoform, *Endocrinology,* 132, 2614, 1993.

93. Rusvai, E. and Náray-Fejes-Tóth, A., A new isoform of 11β-hydroxysteroid dehydrogenase in aldosterone target cells, *J. Biol. Chem.,* 268, 10717, 1993.

94. Peterson, N. A., Chaikoff, I. L., and Jones, C., The *in vitro* conversion of cortisol to cortisone by subcellular brain fractions of young and adult rats, *J. Neurochem.,* 12, 273, 1965.

95. Grosser, B. I., 11β-hydroxysteroid metabolism by mouse brain and glioma 261, *J. Neurochem.,* 13, 475, 1966.

96. Moisan, M.-P., Seckl, J. R., Brett, L. P., Monder, C., Agarwal, A. K., White, P. C., and Edwards, C. R. W., 11β-hydroxysteroid dehydrogenase messenger ribonucleic acid expression, bioactivity and immunoreactivity in rat cerebellum, *J. Neuroendocrinol.,* 2, 853, 1990.

97. Moisan, M.-P., Seckl, J. R., and Edwards, C. R. W., 11β-hydroxysteroid dehydrogenase bioactivity and messenger RNA expression in rat forebrain: localization in hypothalamus, hippocampus and cortex, *Endocrinology,* 127, 1450, 1990.

98. Lakshmi, V., Sakai, R. R., McEwen, B. S., and Monder, C., Regional distribution of 11β-hydroxysteroid dehydrogenase in rat brain, *Endocrinology,* 128, 1741, 1991.

99. Seckl, J. R., Brown, R. W., Rajan, V., Low, S. C., and Edwards, C. R. W., 11β-hydroxysteroid dehydrogenase and corticosteroid actions in the brain, *J. Endocrinol.,* 137 (Suppl.), S9, 1993.

100. Galaverna, O., de Luca, Jr., L. A., Schulkin, J., Yao, S.-Z., and Epstein, A. N., Deficits in NaCl ingestion after damage to the central nucleus of the amygdala in the rat, *Brain Res. Bull.,* 28, 89, 1991.

101. Berecek, K. H., Barron, K. W., Webb, R. L., and Brody, M. J., Vasopressin-central nervous system interactions in the development of DOCA hypertension, *Hypertension,* 4, 131, 1982.

102. Fink, G. D., Buggy, J., Johnson, A. K., and Brody, M. J., Prevention of steroid-salt hypertension in the rat by anterior forebrain lesions, *Circulation,* 56 (Suppl. III), III-242, 1977.

103. Sanders, B. J., Knardahl, S., and Johnson, A. K., Lesions of the anteroventral third ventricle and development of stress-induced hypertension in the borderline hypertensive rat, *Hypertension,* 13, 817, 1989.

104. Seckl, J. R., Kelly, P. A. T., and Sharkey, J., Glycyrrhetinic acid, an inhibitor of 11β-hydroxysteroid dehydrogenase, alters local cerebral glucose utilization *in vivo, J. Steroid Biochem. Mol. Biol.,* 39, 777, 1991.

105. Gomez-Sanchez, E. P. and Gomez-Sanchez, C. E., Central hypertensinogenic effects of glycyrrhizic acid and carbenoxolone, *Am. J. Physiol.,* 263, E1125, 1992.

106. Low, S. C., Chapman, K. E., Edwards, C. R. W., and Seckl, J. R., Liver-type 11β-hydroxysteroid dehydrogenase cDNA encodes reductase not dehydrogenase activity in intact mammalian COS-7 cells, *J. Mol. Endocrinol.,* 13, 167, 1994.

107. Gould, E., Woolley, C. S., and McEwen, B. S., Short-term glucocorticoid manipulations affect neuronal morphology and survival in the adult dentate gyrus, *Neuroscience*, 37, 367, 1990.
108. Sloviter, R. S., Valiquette, G., Abrams, G. M., Ronk, E. C., Sollas, A. L., Paul, L. A., and Neubort, S., Selective loss of hippocampal granule cells in the mature rat brain after adrenalectomy, *Science*, 243, 535, 1989.
109. Low, S. C., Moisan, M.-P., Edwards, C. R. W., and Seckl, J. R., Glucocorticoids and chronic stress up-regulate 11β-hydroxysteroid dehydrogenase activity and gene expression in the hippocampus, *J. Neuroendocrinol.*, 6, 285, 1994.
110. Agarwal, A. K., Monder, C., Eckstein, B., and White, P. C., Cloning and expression of rat cDNA encoding cortico-steroid 11dehydrogenase, *J. Biol. Chem.*, 264, 18939, 1989.
111. Sapolsky, R. M., *Stress, the Aging Brain and the Mechanisms of Neuron Death*, MIT, Cambridge, MA, 1992, 429.
112. Maccari, S., Le Moal, M., Angelucci, L., and Mormède, P., Influence of 6-OHDA lesion of central noradrenergic systems on corticosteroid receptors and neuroendocrine responses to stress, *Brain Res.*, 533, 60, 1990.
113. Eldridge, J. C., Brodish, A., Kute, T. E., and Landfield, P. W., Apparent age-related resistance of type II hippocampal corticosteroid receptors to down-regulation during chronic escape training, *J. Neurosci.*, 9, 3237, 1989.
114. Meaney, M. J., Aitken, D. H., Viau, V., Sharma, S., and Sarrieau, A., Neonatal handling alters adrenocortical negative feedback sensitivity and hippocampal type II glucocorticoid receptor binding in the rat, *Neuroendocrinology*, 50, 597, 1989.
115. de Kloet, E. R., Wallach, G., and McEwen, B. S., Differences in corticosterone and dexamethasone binding to rat brain and pituitary, *Endocrinology*, 96, 598, 1975.
116. Fuxe, K., Wikstrom, A.-C., Okret, S., Agnati, L. F., Harfstrand, A., Yu, Z.-Y., Granholm, L., Zoli, M., Vale, W., and Gustafsson, J.-A., Mapping of glucocorticoid receptor immunoreactive neurons in the rat tel- and diencephalon using a monoclonal antibody against rat liver glucocorticoid receptor, *Endocrinology*, 117, 1803, 1985.
117. Ahima, R. S. and Harlan, R. E., Charting of type II glucocorticoid receptor-like immunoreactivity in the rat central nervous system, *Neuroscience*, 39, 579, 1990.
118. Ahima, R. S. and Harlan, R. E., Differential corticosteroid regulation of type II glucocorticoid receptor-like immuno-reactivity in the rat central nervous system: topography and implications, *Endocrinology*, 129, 226, 1991.
119. Ahima, R. S., Krozowski, Z. S., and Harlan, R. E., Type I corticosteroid receptor-like immunoreactivity in the rat CNS: distribution and regulation by corticosteroids, *J. Com. Neurol.*, 313, 522, 1991.
120. Reul, J. M. H. M., Pearce, P. T., Funder, J. W., and Krozowski, Z. S., Type I and type II corticosteroid receptor gene expression in the rat: effect of adrenalectomy and dexamethasone administration, *Mol. Endocrinol.*, 3, 1680, 1989.
121. Chao, H. M., Choo, P. H., and McEwen, B. S., Glucocorticoid and mineralocorticoid receptor mRNA expression in rat brain, *Neuroendocrinology*, 50, 365, 1989.
122. Herman, J. P., Patel, P. D., Akil, H., and Watson, S. J., Localization and regulation of glucocorticoid and mineralo-corticoid receptor messenger RNAs in the hippocampal formation of the rat, *Mol. Endocrinol.*, 3, 1886, 1989.
123. Herman, J. P., Patel, P. D., and Watson, S. J., Rapid downregulation of mineralocorticoid receptor heteronuclear (hn) RNA by acute stress, *Soc. Neurosci. Abstr.*, 19, 1187, 1993.
124. Vedeckis, W. V., Ali, M., and Allen, H. R., Regulation of glucocorticoid receptor protein and mRNA levels, *Cancer Res.*, 49 (Suppl.), 2295, 1989.
125. Rosewicz, S., McDonald, A. R., Maddux, B. A., Goldfine, I. D., Miesfeld, R. L., and Logsdon, C. D., Mechanism of glucocorticoid receptor down-regulation by glucocorticoids, *J. Biol. Chem.*, 263, 2581, 1988.
126. De Vellis, J., McEwen, B. S., Cole, R., and Inglish, D., Relations between glucocorticoid nuclear binding, cytosol receptor activity and enzyme induction in a rat glial cell line, *J. Steroid Biochem.*, 5, 392, 1974.
127. Luttge, W. G., Rupp, M. E., and Davda, M. M., Aldosterone-stimulated down-regulation of both type I and type II adrenocorticosteroid receptors in mouse brain is mediated via type I receptors, *Endocrinology*, 125, 817, 1989.
128. Reul, J. M. H. M. and de Kloet, E. R., Anatomical resolution of two types of corticosterone receptor sites in rat brain with *in vivo* autoradiography and computerised image analysis, *J. Steroid Biochem.*, 24, 269, 1986.
129. Swanson, L. W. and Simmons, D. M., Differential steroid hormone and neural influences on peptide mRNA levels in CRH cells of the paraventricular nucleus: a hybridization histochemical study in the rat, *J. Com. Neurobiol.*, 285, 413, 1989.
130. Seckl, J. R., Dow, R. C., Low, S. C., Edwards, C. R. W., and Fink, G., The 11β-hydroxysteroid dehydrogenase inhibitor glycyrrhetinic acid affects corticosteroid feedback regulation of hypothalamic corticotrophin-releasing peptides, *J. Endocrinol.*, 136, 471, 1993.
131. McEwen, B. S., Glucocorticoid-biogenic amine interactions in relation to mood and behavior, *Biochem. Pharmacol.*, 36, 1755, 1987.
132. Jacobson, L. and Sapolsky, R., The role of the hippocampus in feedback regulation of the hypothalamic-pituitary-adrenal axis, *Endocrine Rev.*, 12, 118, 1991.
133. Sapolsky, R. M., Krey, L. C., and McEwen, B. S., Prolonged glucocorticoid exposure reduces hippocampal neuron number: implications for aging, *J. Neurosci.*, 5, 1221, 1985.
134. McEwen, B. S., Angulo, J., Cameron, H., Chao, H. M., Daniels, D., Gannon, M. N., Gould, E., Mendelson, S., Sakai, R., Spencer, R., and Woolley, C., Paradoxical effects of adrenal steroids on the brain: protection vs. degener-ation, *Biol. Psychiatry*, 31, 177, 1992.

135. Sousa, R. J., Tannery, N. H., and Lafer, E. M., *In situ* hybridization mapping of glucocorticoid receptor messenger ribonucleic acid in rat brain, *Mol. Endocrinol.*, 3, 481, 1989.

136. Sutanto, W., van Eekelen, J. A. M., Reul, J. M. H. M., and de Kloet, E. R., Species-specific topography of corticosteroid receptor types in the rat and hamster brain, *Neuroendocrinology*, 47, 398, 1988.

137. Sarrieau, A., Dussaillant, M., Sapolsky, R. M., Aitken, D. H., Olivier, A., Lal, S., Rostene, W. H., Quirion, R., and Meaney, M. J., Glucocorticoid binding sites in human temporal cortex, *Brain Res.*, 442, 157, 1988.

138. Sapolsky, R. M. and Meaney, M. J., Postmortem decay in glucocorticoid binding in human and primate brain, *Brain Res.*, 448, 182, 1988.

139. Leonard, S., Logel, J., Luthman, D., Casanova, M., Kirch, D., and Freedman, R., Parameters affecting the use of human postmortem brain for molecular biological studies, *Soc. Neurosci. Abstr.*, 17, 159, 1991.

140. Seckl, J. R., Dickson, K. L., Yates, C., and Fink, G., Distribution of glucocorticoid and mineralocorticoid receptor messenger RNA expression in human postmortem hippocampus, *Brain Res.*, 561, 332, 1991.

141. Burnstein, K. L., Bellingham, D. L., Jewell, C., Powell-Oliver, F. E., and Cidlowski, J. A., Autoregulation of glucocorticoid receptor gene expression, *Steroids*, 56, 52, 1991.

142. Silva, C. M., Powell-Oliver, F. E., Jewell, C. M., Sar, M., Allgood, V. E., and Cidlowski, J. A., Regulation of the human glucocorticoid receptor by long-term and chronic treatment with glucocorticoid, *Steroids*, 59, 436, 1994.

143. Eisen, L. P., Elsasser, M. S., and Harmon, J. M., Positive regulation of the glucocorticoid receptor in human T-cells sensitive to the cytolytic effects of glucocorticoids, *J. Biol. Chem.*, 263, 12044, 1988.

144. Denton, R. R., Eisen, L. P., Elsasser, M. S., and Harmon, J. M., Differential autoregulation of glucocorticoid receptor by glucocorticoids in human T- and B-cell lines, *Endocrinology*, 133, 248, 1993.

145. Holmes, M. C., Yau, J. L. W., French, K. L., and Seckl, J. R., Effect of adrenalectomy on 5-HT and corticosteroid receptor subtype mRNA expression in the rat hippocampus and dorsal raphe nucleus, *Neuroscience*, 64, 327, 1995.

146. Reul, J. M. H. M., van den Bosch, F. R., and de Kloet, E. R., Differential response of type I and type II corticosteroid receptors to changes in plasma steroid level and circadian rhythmicity, *Neuroendocrinology*, 45, 407, 1987.

147. Holmes, M. C., French, K. L., and Seckl, J. R., Modulation of serotonin and corticosteroid receptor gene expression in the rat hippocampus with circadian rhythm and stress, *Mol. Brain Res.*, 28, 186, 1995.

148. Herman, J. P., Watson, S. J., Chao, H. M., Coirini, H., and McEwen, B. S., Diurnal regulation of glucocorticoid receptor and mineralocorticoid receptor mRNAs in rat hippocampus, *Mol. Cell. Neurosci.*, 4, 181, 1993.

149. Brinton, R. E. and McEwen, B. S., Regional distinctions in the regulation of type I and type II adrenal steroid receptors in the central nervous system, *Neurosci. Res. Commun.*, 2, 37, 1988.

150. van Dijken, H. H., de Goeij, D. C. E., Sutanto, W., de Kloet, E. R., and Tilders, F. J. H., Short inescapable stress produces long-lasting changes in the brain-pituitary-adrenal axis of adult male rats, *Neuroendocrinology*, 58, 57, 1993.

151. Donaldson, L. F., McQueen, D. S., and Seckl, J. R., Endogenous glucocorticoids and the induction and spread of monoarthritis in the rat, *J. Neuroendocrinol.*, 6, in press.

152. Lowy, M. T., Reserpine-induced decrease in type I and II corticosteroid receptors in neuronal and lymphoid tissues of adrenalectomised rats, *Neuroendocrinology*, 51, 190, 1990.

153. Lowy, M. T. and Novotney, S., Methamphetamine-induced decrease in neural glucocorticoid receptors: relationship to monoamine levels, *Brain Res.*, 638, 175, 1994.

154. Lowy, M. T., Nash, J. J. F. and Meltzer, H. Y., Selective reduction of striatal type II glucocorticoid receptors in rats by 3,4-methylenedioxymethamphetamine (MDMA), *Eur. J. Pharmacol.*, 163, 157, 1989.

155. Yau, J. L. W., Kelly, P. A. T., Sharkey, J., and Seckl, J. R., Chronic 3,4-methylenedioxymethamphetamine (MDMA) administration decreases glucocorticoid and mineralocorticoid receptor, but increases 5-HT1C receptor gene expression in the rat hippocampus, *Neuroscience*, 61, 31, 1994.

156. Siegel, R. A., Weidenfeld, J., Chen, M., Feldman, S., Melamed, E., and Chowers, I., Hippocampal cell nuclear binding of corticosterone following 5,7-dihydroxytryptamine, *Mol. Cell. Endocrinol.*, 31, 253, 1983.

157. Vernikos-Daniellis, J., Kellar, K. J., Kent, D., Gonzales, C., Berger, P. A., and Barchas, J. D., Serotonin involvement in pituitary-adrenal function, *ACTH and Related Peptides: Structure Function and Actions*, 297, Krieger, D. T. and Ganong, W. F., NY Academy of Sciences, New York, 1977, 518.

158. Mitchell, J. B., Rowe, W., Boksa, P., and Meaney, M. J., Serotonin regulates type II corticosteroid receptor binding in hippocampal cell cultures, *J. Neurosci.*, 10, 1745, 1990.

159. Vedder, H., Weiß, I., Holsboer, F., and Reul, J. M. H. M., Glucocorticoid and mineralocorticoid receptors in rat neocortical and hippocampal brain cells in culture: characterization and regulatory studies, *Brain Res.*, 605, 18, 1993.

160. Weidenfeld, J., Siegel, R. A., Corcos, A. P., Chen, M., Feldman, S., and Chowers, I., Effect of 6-hydroxydopamine on *in vitro* hippocampal corticosterone binding capacity in the male rat, *Exp. Brain Res.*, 52, 121, 1983.

161. Yau, J. L. W. and Seckl, J. R., Central 6-hydroxydopamine lesions decrease mineralocorticoid, but not glucocorticoid receptor gene expression in rat hippocampus, *Neurosci. Lett.*, 142, 159, 1992.

162. Yau, J. L. W., Dow, R. C., Fink, G., and Seckl, J. R., Medial septal cholinergic lesions increase hippocampal mineralocorticoid and glucocorticoid receptor messenger RNA expression, *Brain Res.*, 577, 155, 1992.

163. Nilsson, O. G., Strecker, R. E., Daszuta, A., and Bjorklund, A., Combined cholinergic and serotonergic denervation of the forebrain produces severe deficits in a spatial learning task in the rat, *Brain Res.*, 453, 235, 1988.

164. Yau, J. L. W., Kelly, P. A. T., and Seckl, J. R., Increased glucocorticoid receptor gene expression in the rat hippocampus following combined serotonergic and medial septal cholinergic lesions, *Mol. Brain Res.*, 27, 174, 1994.

165. Seckl, J. R., French, K. L., O'Donnell, D., Meaney, M. J., Yates, C., and Fink, G., Glucocorticoid receptor gene expression is unaltered in hippocampal neurons in Alzheimer's disease, *Mol. Brain Res.*, 18, 239, 1993.

166. Lowy, M. T., Kainic acid-induced decrease in hippocampal corticosteroid receptors, *J. Neurochem.*, 58, 1561, 1992.

167. O'Donnell, D., Baccichet, A., Seckl, J. R., Meaney, M. J., and Poirier, J., Entorhinal cortex lesions alter glucocorticoid but not mineralocorticoid receptor gene expression in the rat hippocampus, *J. Neurochem.*, 61, 356, 1993.

168. Zhou, F. C. and Azmitia, E. C., The effect of adrenalectomy and corticosterone on homotypic collateral sprouting of serotonergic fibers in the hippocampus, *Neurosci. Lett.*, 54, 111, 1985.

169. Sapolsky, R. M., Krey, L. C., and McEwen, B. S., Glucocorticoid-sensitive hippocampal neurones are involved in termination the adrenocortical stress response, *Proc. Natl. Acad. Sci. U.S.A.*, 81, 6174, 1984.

170. Landfield, P. W., Modulation of brain aging correlates by long-term alterations of adrenal steroids and neurally-active peptides, *Neuropeptides and Brain Function: Progress in Brain Research*, 72, de Kloet, E. R., Wiegant, N., and De Wied, D., Elsevier, Amsterdam, 1987, 279.

171. Rigter, H., Veldhuis, H. D., and de Kloet, E. R., Spatial learning and the hippocampal corticosterone receptor system of old rats: effect of the ACTH4-9 analog ORG 2766, *Brain Res.*, 309, 393, 1984.

172. Reul, J. M. H. M., Tonnaer, J. A. D. M., and de Kloet, E. R., Neurotrophic ACTH analog promotes plasticity of type I corticosteroid receptor in brain of senescent male rats, *Neurobiol. Aging*, 9, 253, 1988.

173. Brady, L. S., Whitfield, Jr., H. J., Fox, R. J., Gold, P. W., and Herkenham, M., Long-term antidepressant administration alters corticotropin-releasing hormone, tyrosine hydroxylase and mineralocorticoid receptor gene expression in the rat brain, *J. Clin. Invest.*, 87, 831, 1991.

174. Seckl, J. R. and Fink, G., Antidepressants increase glucocorticoid and mineralocorticoid receptor mRNA expression in the rat hippocampus *in vivo*, *Neuroendocrinology*, 55, 621, 1992.

175. Reul, J. M. H. M., Stec, I., Söder, M., and Holsboer, F., Chronic treatment of rats with the antidepressant amitriptyline attenuates the activity of the hypothalamic-pituitary-adrenocortical system, *Endocrinology*, 133, 312, 1993.

176. Pepin, M.-C., Pothier, F., and Barden, N., Antidepressant drug action in a transgenic mouse model of the endocrine changes seen in depression, *Mol. Pharmacol.*, 42, 991, 1993.

177. Pepin, M.-C., Beaulieu, S., and Barden, N., Antidepressants regulate glucocorticoid receptor messenger RNA concentrations in primary neuronal cultures, *Mol. Brain Res.*, 6, 77, 1989.

178. Pepin, M.-C., Pothier, F., and Barden, N., Impaired glucocorticoid receptor function in transgenic mice expressing antisense RNA, *Nature*, 355, 725, 1992.

179. Pepin, M.-C., Govinda, M. V., and Barden, N., Increased glucocorticoid receptor gene promoter activity after antidepressant treatment, *Mol. Pharmacol.*, 41, 1013, 1992.

180. McGaugh, J. L., Involvement of hormonal and neuromodulatory systems in the regulation of memory storage, *Ann. Rev. Neurosci.*, 12, 255, 1989.

181. Oitzl, M. S. and de Kloet, E. R., Selective corticosteroid antagonists modulate specific aspects of spatial orientation learning, *Behav. Neurosci.*, 106, 62, 1992.

182. Kerr, D. S., Campbell, L. W., Applegate, M. D., Brodish, A., and Landfield, P. W., Chronic stress-induced acceleration of electrophysiologic and morphometric biomarkers of hippocampal aging, *J. Neurosci.*, 11, 1316, 1991.

183. Lupien, S., Lecours, A. R., Lussier, I., Schwartz, G., Nair, N. P. V., and Meaney, M. J., Basal cortisol levels and cognitive deficits in human aging, *J. Neurosci.*, 14, 2893, 1994.

184. Luine, V., Villegas, M., Martinez, C., and McEwen, B. S., Repeated stress causes reversible impairments of spatial memory performance, *Brain Res.*, 639, 167, 1994.

185. McEwen, B. S., Angulo, J., Gould, E., Mendelson, S., and Watanabe, Y., Antidepressant modulation of isolation and restraint stress effects on brain chemistry and morphology, *Eur. Psychiatry*, 8 (Suppl. 2), 41, 1993.

186. Yau, J. L. W., Olsson, T., Morris, R. G. M., Meaney, M. J., and Seckl, J. R., Chronic antidepressant treatment improves spatial memory performance in young but not aged rats, *Soc. Neurosci. Abstr.*, 19, 387, 1993.

187. Yau, J. L. W., Olsson, T., Morris, R. G., Meany, M. J., and Seckl, J. R., Glucocorticosterids, hippocampal corticosteroid receptor gene expression and antidepressant treatment: relationship with spatial learning in young and aged rats, *Neuroscience*, 66, 571, 1995.

188. Diamond, D. M., Bennett, M. C., Fleshner, M., and Rose, G. M., Inverted-U relationship between the level of peripheral corticosterone and the magnitude of hippocampal primed burst potentiation, *Hippocampus*, 2, 421, 1992.

189. Bowen, D. M., Allen, S. J., Benton, J. S., Goodhardt, M. J., Haan, E. A., Palmer, A. M., Sims, N. R., Smith, C. C. T., Spillane, J. R., Esiri, G. K., Neary, D., Snowden, J. S., Wilcock, G. K., and Davison, A. N., Biochemical assessment of serotonergic and cholinergic dysfunction and cerebral atrophy in Alzheimer's disease, *J. Neurochem.*, 41, 266, 1983.

190. Mann, D. M. A. and Yates, P. O., Serotonergic nerve cells in Alzheimer's disease, *J. Neurol., Neurosurg. Psychiatry*, 46, 96, 1983.

191. Whitford, G. M., Alzheimer's disease and serotonin: a review, *Neuropsychobiology*, 15, 1, 1986.

192. Collerton, D., Cholinergic function and intellectual decline in Alzheimer's disease, *Neuroscience*, 19, 1, 1986.

193. Perry, E. K., Curtis, M., Dick, D. J., Candy, J. M., Atack, J. R., Bloxham, C. A., Blessed, G., Fairbairn, A., Tomlinson, B. E., and Perry, R. H., Cholinergic correlates of cognitive impairment in Parkinson's disease: comparison with Alzheimer's disease, *J. Neurol., Neurosurg. Psychiatry*, 48, 413, 1985.

194. Meaney, M. J., O'Donnell, D., Viau, V., Bhatnagar, S., Sarrieau, A., Smythe, J., Shanks, N., and Walker, C.-D., Corticosteroid receptors in the rat brain and pituitary during development and hypothalamic-pituitary-adrenal function, *Growth Factors and Hormones*, Zagon, S. and McLaughlin, P. J., Eds., Chapman and Hall, 1993, 163.

195. Meaney, M. J., Aitken, D. H., van Berkel, C., Bhatnagar, S., and Sapolsky, R. M., Effect of neonatal handling on age-related impairments associated with the hippocampus, *Science*, 239, 766, 1988.

196. Meaney, M. J., Aitken, D. H., Sharma, S., and Viau, V., Basal ACTH, corticosterone and corticosterone-binding globulin levels over the diurnal cycle, and hippocampal corticosteroid receptors in young and aged, handled and non-handled rats, *Neuroendocrinology*, 55, 204, 1992.

197. O'Donnell, D., La Roque, S., Seckl, J. R., and Meaney, M., Postnatal handling alters glucocorticoid but not mineralocorticoid receptor mRNA expression in the hippocampus of adult rats, *Mol. Brain Res.*, 26, 242, 1994.

198. Levine, S., Infantile experience and resistance to physiological stress, *Science*, 126, 405, 1957.

199. Levine, S., Plasma-free corticosteroid response to electric shock in rats stimulated in infancy, *Science*, 135, 795, 1962.

200. Smythe, J. W., Rowe, W. B., and Meaney, M. J., Neonatal handling alters serotonin (5-HT) turnover and 5-HT2 receptor binding in selected brain regions: relationship to the handling effect on glucocorticoid receptor expression, *Dev. Brain Res.*, 80, 183, 1994.

201. Mitchell, J. B., Betito, K., Boksa, P., Rowe, W., and Meaney, M. J., Serotonergic regulation of type II corticosteroid receptor binding in cultured hippocampal cells: the role of serotonin-induced increases in cAMP levels, *Neuroscience*, 48, 631, 1992.

202. Reul, J. M. H. M., Stec, I., Wiegers, G. J., Labeur, M. S., Linthorst, A. C. E., Artz, E., and Holsboer, F., Prenatal immune challenge alters the hypothalamic-pituitary-adrenocortical axis in adult rats, *J. Clin. Invest.*, 93, 2600, 1994.

203. Plotsky, P. M. and Meaney, M. J., Early, postnatal experience alters hypothalamic corticotropin-releasing factor (CRF) mRNA, median eminence CRF content and stress-induced release in adult rats, *Mol. Brain Res.*, 18, 195, 1993.

204. Olsson, T., Mohammed, A. H., Donaldson, L. F., Henriksson, B. F., and Seckl, J. R., Glucocorticoid receptor and NGFI-A gene expression are induced in the hippocampus after environmental enrichment in adult rats, *Mol. Brain Res.*, 23, 349, 1994.

205. Uphouse, L., Reevaluation of mechanisms that mediate brain differences between enriched and impoverished animals, *Psychol. Bull.*, 88, 215, 1980.

206. Mohammed, A. K., Winblad, B., Ebendal, T., and Lärkfors, L., Environmental influence on behavior and nerve growth factor in the brain, *Brain Res.*, 528, 62, 1990.

207. Mohammed, A. K., Henriksson, B. G., Söderström, S., Ebendal, T., Olsson, T., and Seckl, J. R., Environmental influences on the central nervous system and their implications for the aging rat, *Behav. Brain Res.*, 57, 183, 1993.

208. Zong, J., Ashraf, J., and Thompson, E. B., The promoter and first untranslated exon of the human glucocorticoid receptor gene are GC-rich but lack consensus glucocorticoid receptor element sites, *Mol. Cell. Biol.*, 10, 5580, 1990.

209. Issa, A. M., Rowe, W., Gauthier, S., and Meaney, M. J., Hypothalamic-pituitary-adrenal activity in aged, cognitively impaired and cognitively unimpaired rats, *J. Neurosci.*, 10, 3247, 1990.

210. Meaney, M. J., Bodnoff, S. R., O'Donnell, D., Sarrieau, A., Nair, N. P. V., Diamond, D. M., Rose, G. M., Poirier, J., and Seckl, J. R., Adrenal glucocorticoids as modulators of hippocampal neuron survival, repair and function in the aged rat, Cuello, C., Ed., *Restorative Neurology*, 6, Elsevier, New York, 1993, 267.

211. Sapolsky, R. M., Krey, L. C., and McEwen, B. S., The neuroendocrinology of stress and ageing. The glucocorticoid cascade hypothesis, *Endocrine Rev.*, 7, 284, 1986.

212. Sapolsky, R. M., Krey, L. C., McEwen, B. S., and Rainbow, T. C., Do vasopressin-related peptides induce hippocampal corticosteroid receptors? Implications for aging, *J. Neurosci.*, 4, 1479, 1984.

213. Farman, N., Bonvalet, J.-P., and Seckl, J. R., Aldosterone selectively increases Na^+, K^+-ATPase a3 subunit mRNA expression in rat hippocampus, *Am. J. Physiol.*, 266, C423, 1994.

214. Chalmers, D., Kwak, S., Mansour, A., Akil, H., and Watson, S., Corticosteroids regulate brain Hippocampal $5-HT_{1A}$ receptor mRNA expression, *J. Neurosci.*, 13, 914, 1993.

215. Meijer, O. and deKloet, E., Mineralo- and glucocorticoid receptors synergistically mediate the action of corticosterone on 5HT1A mRNA expression in dentate gyrus, *Soc. Neurosci. Abstr.*, 23rd meeting, 564, 1993.

216. Horner, H. C., Packan, D. R., and Sapolsky, R. M., Glucocorticoids inhibit glucose transport in cultured hippocampal neurones and glia, *Neuroendocrinology*, 52, 57, 1990.

217. Gannon, M. N., Rai, A., and McEwen, B. S., Stress and dexamethasone effects on calmodulin mRNAs in rat brain and immune tissues, *Soc. Neurosci. Abstr.*, 19, 508, 1993.

218. Drouin, J., Sun, Y. L., Chamberland, M., Gautier, Y., de Léan, A., Nemer, M., and Schmidt, T. J., Novel glucocorticoid receptor complex with DNA element of the hormone-repressed POMC gene, *EMBO J.*, 12, 145, 1993.

219. Chandran, U. R., Attardi, B., Friedman, R., Dong, K.-W., Roberts, J. L., and DeFranco, D. B., Glucocorticoid receptor-mediated repression of gonadotrophin-releasing hormone promoter activity in GT1 hypothalamic cells lines, *Endocrinology*, 134, 1467, 1994.

220. Sapolsky, R. M., Zola-Morgan, S., and Squire, L. R., Inhibition of glucocorticoid secretion by the hippocampal formation in the primate, *J. Neurosci.*, 11, 3695, 1991.

221. Brooke, S. M., De Haas-Johnson, A. M., Kaplan, J. R., Manuck, S. B., and Sapolsky, R. M., Dexamethasone resistance among nonhuman primates associated with a selective decrease of glucocorticoid receptors in the hippocampus and a history of social instability, *Neuroendocrinology*, 60, 134, 1994.

222. Mandell, A., Chapman, L. F., Rand, R. W., and Walter, R. D., Plasma corticosteroids: changes in concentration after stimulation of hippocampus and amygdala, *Science,* 139, 1212, 1963.
223. Pasquier, F., Bail, L., Lebert, F., Pruvo, J. P., and Petit, H., Determination of medial temporal lobe atrophy in early Alzheimer's disease with computed tomography, *Lancet,* 343, 861, 1994.
224. de Leon, M. J., McRae, T., Tsai, J. R., George, A. E., Marcus, D. L., Freedman, M., Wolf, A. P., and McEwen, B. S., Abnormal cortisol response in Alzheimer's disease linked to hippocampal atrophy, *Lancet,* ii, 391, 1988.
225. Newcomer, J. W., Craft, S., Hershey, T., Askin, K., and Bardgett, M. E., Glucocorticoid-induced impairment in declarative memory performance in adult humans, *J. Neurosci.,* 14, 2047, 1994.
226. Starkman, M. N. and Schteingart, D. E., Neuropsychiatric manifestations of patients with Cushing's syndrome, *Arch. Intern. Med.,* 141, 215, 1981.
227. Cohen, S., Cushing's syndrome: a psychiatric study of 29 patients, *Br. J. Psychiatry,* 136, 120, 1980.
228. Mauri, M., Sinforiani, E., Bono, G. et al., Memory impairment in Cushing's disease, *Acta Neurol. Scand.,* 87, 52, 1993.
229. Trethowan, W. H. and Cobb, S., Neuropsychiatric aspects of Cushing's syndrome, *Arch. Neurol. Psychiatry,* 67, 283, 1952.
230. Starkman, M. N., Gebarski, S. S., Berent, S., and Schteingart, D. E., Hippocampal formation volume, memory dysfunction and cortisol levels in patients with Cushing's syndrome, *Biol. Psychiatry,* 32, 756, 1992.

Chapter 2

Steroids as Modulators of Amino Acid Receptor Function

David H. Farb and Terrell T. Gibbs

Considerable pharmacological evidence demonstrates that certain steroids exhibit a novel mode of action on nervous tissue. Such steroids can either inhibit or enhance amino acid receptor function by rapidly binding to amino acid receptors and allosterically altering their responsiveness to the neurotransmitter. The prevalence of glutamate-, GABA-, and glycine-mediated synapses in the central nervous system (CNS) and the pronounced rapid modulation of amino acid responses by steroids suggest that neuroactive steroids have the capacity for the global control of nervous system function. Abnormal activation of amino acid receptors has been implicated in the etiology of psychiatric disorders such as anxiety, depression, and schizophrenia, and it is hoped that an understanding of steroid interactions with excitatory and inhibitory amino acid receptors will lead to new strategies for the treatment of psychiatric disorders.

It has been known for some time that steroid hormones can profoundly influence CNS excitability.[1,2] Progesterone, when injected into animals at moderate doses, tends to act as a behavioral depressant, reducing excitatory transmission and promoting inhibitory neurotransmission in the CNS.[3,4] β-Estradiol, in contrast, has been described as a CNS "activator," and tends to enhance excitatory neurotransmission.[5] Traditionally, the effects of steroid hormones on the vertebrate CNS have been thought to be mediated by genomic steroid response elements.[6] Although this is undoubtedly the mechanism of action for many steroids, some endogenous and synthetic steroids rapidly induce sedation and anesthesia. The rapidity of such effects, as well as their resistance to protein synthesis inhibitors and conventional steroid antagonists, has led investigators to seek mechanisms whereby steroids may directly modulate neurotransmission. A large body of evidence has been accumulated that many of the physiological and behavioral effects of steroids are mediated via direct modulation of inhibitory and, possibly, excitatory synaptic transmission in the CNS.

Sulfated steroids have generally been thought to be biologically inactive, but recent evidence raises the possibility that biological sulfation and desulfation of steroids may play an important role in the control of nervous system function. This suggestion is derived from the observation that sulfation of certain steroids, such as pregnenolone, dehydroepiandrosterone (DHEA), pregnanolone, and allopregnanolone, can either activate or inactivate the steroids as modulators of amino acid receptor-mediated responses.[7,8] Sulfation may also serve to store neurosteroids in a reserve pool that could be drawn upon by sulfatase-mediated desulfation. It is interesting to note that the concentration of dehydroepiandrosterone sulfate (DHEAS) in the blood of males aged 40 (8.92 μM) is considerably higher than that of DHEA (0.01 μM), and that the concentrations of DHEAS and DHEA decline during aging.[9] This observation has led to the hypothesis that changes in DHEAS levels may somehow be involved with the decline in cognitive ability with age.

Sulfated steroids are formed from the corresponding sterols by a single-step conversion catalyzed by steroid sulfotransferases. PAPS (Figure 1) serves as the sulfate donor for all sulfotransferase reactions (Figure 2A) and free sterol can be formed by the action of a steroid sulfatase (Figure 2B). Following its formation from cholesterol, pregnenolone can give rise to progesterone, 17-hydroxypregnenolone, or pregnenolone sulfate (Figure 3). Progesterone and 17α-hydroxyprogesterone, from which corticosteroids are derived, are themselves formed from pregnenolone and 17-hydroxypregnenolone, which can in turn be formed by the action of steroid sulfatases on pregnenolone sulfate and 17-hydroxypregnenolone sulfate. Naturally, DHEAS could serve as a precursor pool for DHEA and, consequently, androst-4-ene-3,17-dione.

Progesterone and certain reduced metabolites of progesterone and deoxycorticosterone, such as 5α-pregnan-3α-ol-20-one (allopregnanolone) and 5α-pregnane-3α,21-diol-20-one (THDOC), have been shown to be potent positive modulators of the GABA$_A$ receptor, in that they potentiate the GABA-induced

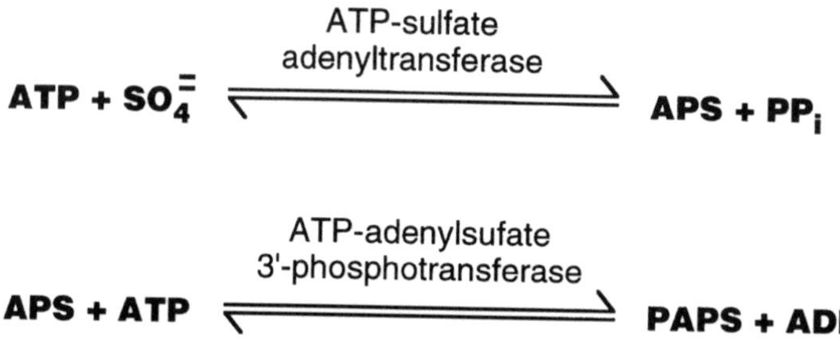

$$\text{ATP} + \text{SO}_4^= \underset{}{\overset{\text{ATP-sulfate}}{\rightleftharpoons}} \text{APS} + \text{PP}_i$$

ATP-sulfate adenyltransferase

$$\text{APS} + \text{ATP} \underset{}{\overset{\text{ATP-adenylsufate}}{\rightleftharpoons}} \text{PAPS} + \text{ADP}$$

ATP-adenylsufate 3'-phosphotransferase

Figure 1 The biosynthesis of PAPS (3'-phosphoadenosine-5'-phosphosulfate) from ATP and molecular sulfate. APS (adenosine-5'-phosphosulfate) is first synthesized from ATP and sulfate in a reaction catalyzed by ATP-sulfate adenylyltransferase. ATP-adenylylsulfate 3'-phosphotransferase then transfers a terminal phosphate of ATP to the 3' hydroxyl moiety of APS to form PAPS.

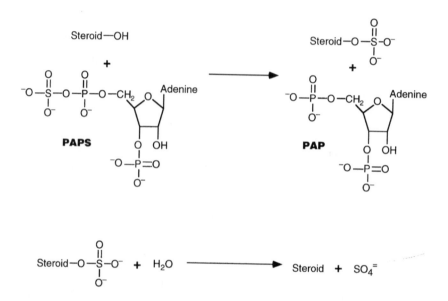

Figure 2 PAPS-dependent sulfation of steroids is catalyzed by steroid sulfotransferase.

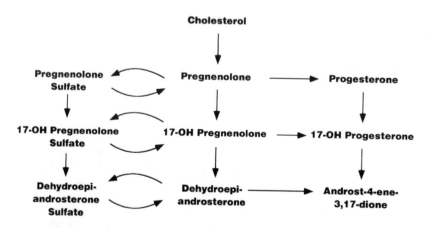

Figure 3 Metabolic pathway for sulfation of steroids.

Cl⁻ current,[7,10-16] stimulate both muscimol and benzodiazepine binding,[10,12] enhance GABA receptor-mediated Cl⁻ uptake,[10,17-19] and displace *t*-butylbicyclophosphorothionate (TBPS) binding.[10,20,21,17] The exact site of actions of these steroids on the GABA$_A$ receptor has not been clearly defined. Nevertheless, the remarkable potency of these steroids in potentiating the actions of GABA and their stereoselectivity suggests the existence of a specific site of interaction closely associated with the GABA$_A$ receptor.[13] Indeed, this idea is supported by recent electrophysiological studies that indicate that progesterone and allopregnanolone potentiate the GABA response by acting through a common, specific, saturable site at which progesterone is a partial positive modulator.[16] This steroid modulation was originally thought to be mediated via the barbiturate site on the GABA$_A$ receptor.[10,12] However, electrophysiological and receptor binding studies demonstrate that steroids and barbiturates act at different sites on the GABA$_A$ receptor.[21,14] There is evidence that the modulatory effects of steroids on the GABA$_A$ receptor may be regulated by the phosphorylation state of the receptor.[22]

In contrast, neurosteroids such as pregnenolone sulfate and DHEAS interact with the GABA$_A$ receptor in an antagonistic fashion.[23-26] The site of action of pregnenolone sulfate and DHEAS is not clear and whether these neurosteroids and allopregnanolone act through a common site on the GABA$_A$ receptor also remains to be elucidated. In the behavioral studies, THDOC has anxiolytic, anticonflict, hypnotic, and anti-aggressive effects on experimental animals.[27-29] Allopregnanolone, when injected intracerebroventricularly, produces analgesic effects in mice.[30] In contrast, pregnenolone sulfate injected either intracerebroventricularly or intraperitoneally into rats shortens pentobarbital-induced hypnosis.[31] These results are consistent with the hypothesis that modulation of the GABA$_A$ receptor by steroids plays an important physiological and pharmacological role in the regulation of CNS excitability.

Compared to the extensive studies of steroid effects on the GABA$_A$ receptor, interactions of steroids with the glycine receptor have been rather neglected. Progesterone itself decreases the glycine-induced Cl⁻ current but enhances the current induced by GABA.[16] Some, but not all, progesterone metabolites also display these effects. The site through which progesterone acts to inhibit the glycine response is distinct from the strychnine and glycine binding sites. Interestingly, allopregnanolone, one of the most potent modulators of the GABA$_A$ receptor, produces little effect on the glycine response. These findings not only provide an important distinction between Cl⁻-mediated GABA and glycine responses but also suggest that endogenous progesterone or its metabolites may differentially modulate the inhibitory actions of these two neurotransmitters. Like progesterone, the endogenous neurosteroid pregnenolone sulfate inhibits the glycine response, but with greater potency and efficacy. Thus far, no steroids have been identified as positive modulators of the glycine receptor.

While the role of β-alanine in fast synaptic inhibition is not yet clear, there is increasing evidence that β-alanine may act as an inhibitory neurotransmitter.[32,33] Moreover, whereas β-alanine evokes a Cl⁻ current, a major unanswered question remains: Does β-alanine act through a unique receptor[34,35] or through the distinct GABA$_A$ and/or glycine receptors?[36,37] As discussed above, progesterone potentiates the GABA-induced Cl⁻ current and antagonizes that induced by glycine, and its reduced metabolite allopregnanolone markedly enhances the GABA response, but has little effect on the glycine response.[16] This differential modulation of the GABA and glycine responses by progesterone and allopregnanolone provides a pharmacological tool to determine whether there is a distinct receptor for β-alanine. The data suggest that there is not a distinct receptor for β-alanine on spinal cord neurons, and that dual activation of GABA$_A$ and glycine receptors by β-alanine may explain the opposite effects of progesterone and allopregnanolone on the β-alanine response.[38] It is interesting to speculate that if β-alanine is a neurotransmitter, progesterone (which potentiates the GABA response and inhibits the glycine response) or allopregnanolone (which dramatically potentiates the GABA response and is without effect on the glycine response) could alter the flow of inhibition in the vertebrate CNS by selectively enhancing neurotransmission mediated by β-alanine acting on GABA$_A$ receptors.

L-Glutamate is known to be the major excitatory neurotransmitter in the CNS. This single neurotransmitter activates a broad class of ionotropic receptors, which can be pharmacologically divided into three broad classes on the basis of sensitivity to *N*-methyl-D-aspartate, kainate, and α-amino-3-hydroxy-5-methyl-4-isoxazolepropionate (AMPA). The NMDA receptor(s) derives its name from the observation that it responds preferentially to the synthetic analog of aspartate, *N*-methyl-D-aspartate. Like other glutamate-gated ionotropic receptors, the NMDA receptor activates ion channels that increase the membrane permeability to monovalent cations,[39-41] but the channel controlled by the NMDA receptor also exhibits a number of special properties, including high calcium permeability,[42] potentiation by glycine,[43] and susceptibility to blockade by magnesium in a voltage-dependent manner.[44,45] The NMDA receptor has been proposed to play a key function in long-term potentiation,[46-48] learning,[49] epilepsy,[50,51] hypoxic neuronal damage,[52] schizophrenia,[53,54] and excitotoxicity.[55]

Many of the depressant effects of pregnane steroids are probably explicable in terms of conversion into 3α-hydroxy-5α-pregnan-20-one (allopregnanolone), a highly potent positive modulator of the $GABA_A$ receptor.[10] The effects of steroids on excitatory neurotransmission are much less well understood. It does not seem likely that steroid effects on excitatory neurotransmission are entirely explicable in terms of modulation of $GABA_A$ neurotransmission, because attenuation of excitatory amino acid responses by progesterone persists in the presence of the $GABA_A$ antagonist, bicuculline,[4] suggesting that direct modulation of glutamate receptors by steroids may be involved.

Despite the important role of glutamate receptors in neurobiology and the profound actions of steroids on CNS excitability, relatively little is known about the effects of steroids on glutamate receptor-mediated responses. Although it is evident that steroids such as progesterone and 17β-estradiol can alter neuronal excitation mediated by glutamate,[3,5,56-58] the molecular mechanisms of their actions and the types of glutamate receptors involved are less clear.

The clearest example of direct steroid modulation of glutamate receptors is potentiation of the NMDA response by pregnenolone sulfate. In dissociated embryonic chick spinal cord[7] and adult[59] or embryonic rat hippocampal neurons,[60] pregnenolone sulfate enhances the NMDA-induced membrane current by up to 200 to 300%, while somewhat inhibiting responses to kainate and AMPA (Figure 4). Similarly, pregnenolone sulfate potentiates spontaneously occurring excitatory postsynaptic currents (EPSCs) in rat hippocampal neurons, and inhibits spontaneously occurring inhibitory postsynaptic potentials (IPSCs), indicating that pregnenolone sulfate is capable of potentiating the response to synaptically released glutamate and GABA. No potentiation is observed in the presence of the NMDA antagonist, APV, confirming that the potentiation of EPSCs by pregnenolone sulfate is mediated by NMDA receptors.[61] The rapidity of the onset of potentiation argues that pregnenolone sulfate modulates the NMDA receptor directly.[7] This conclusion is supported by the observation that pregnenolone sulfate potentiates NMDA-induced single channel activity in excised membrane patches.[59]

The EC_{50} for potentiation of the NMDA response by pregnenolone sulfate is 29 to 60 μM, but measurable potentiation can be observed at pregnenolone sulfate concentrations as low as 250 nM.[60] Pregnenolone sulfate has also been shown to potentiate NMDA-dependent Ca^{2+} transport in rat hippocampal neurons with similar potency, although the maximum percentage potentiation of Ca^{2+} uptake is greater than that observed for potentiation of the NMDA-induced current.[62]

Pregnenolone itself is without effect on the NMDA response, demonstrating that, if endogenous pregnenolone sulfate is biologically active, steroid sulfotransferases have the capacity to convert an inactive steroid into an active steroid. Similarly, DHEAS potentiates the NMDA response (by about 30% at 100 μM NMDA), whereas DHEA is without effect.[8] These results are consistent with the hypothesis that steroids such as pregnenolone sulfate and DHEAS may regulate the balance between excitation and inhibition on neurons derived from the vertebrate CNS by acting on excitatory and inhibitory amino acid receptors. For example, the net effect of pregnenolone sulfate on a synaptically active complex neural circuit involving NMDA receptors, $GABA_A$, and glycine receptors would be expected to be excitatory. If transmission through non-NMDA glutamate receptors were also important, the inhibitory effects of pregnenolone sulfate on these receptors would tend to reduce overall excitation.

Some limited structure-activity data is available for the interaction of steroids with the NMDA receptor (Table 1). In some cases, slight modifications of steroid structure result in large changes in activity, indicating that steroids produce their modulatory effects by specific interactions with the receptor, not by nonspecific effects on the lipid bilayer.

Reduction of the double bond between C-5 and C-6 results in complete loss of potentiation as measured either by potentiation of the NMDA-induced current[63] or by potentiation of NMDA-induced changes in $[Ca^{2+}]_i$.[64] Pregnanolone sulfate (3α-hydroxy-5β-pregnan-20-one sulfate), the 5β3α-reduced derivative of pregnenolone sulfate, inhibits the NMDA-induced current, as well as responses to kainate and AMPA (Figure 5).[8] The 5β3β-reduced derivative (3β-hydroxy-5β-pregnan-20-one sulfate; 5β3βS) is likewise inhibitory,[63] but the 5α3β-reduced derivative (3β-hydroxy-5α-pregnan-20-one sulfate) is without activity, suggesting that the inhibitory interaction with the NMDA receptor is stereospecific at C-5, but not at C-3. Interestingly, pregnanolone (3α-hydroxy-5β-pregnan-20-one), the nonsulfated form of the 5β3α-reduced derivative, is one of the most potent and efficacious positive modulators of the $GABA_A$ receptor, but is without activity at the NMDA receptor, indicating that replacement of the sulfate with a hydroxyl group results in loss of inhibitory activity.

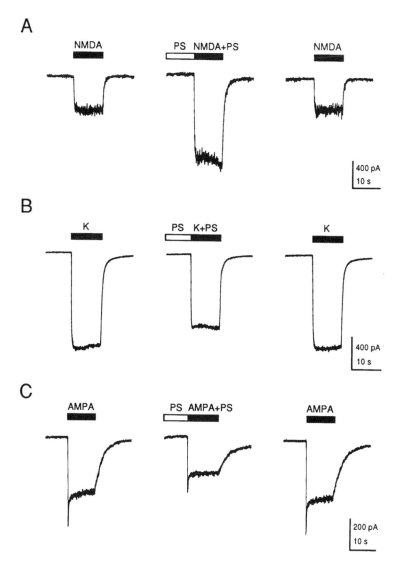

Figure 4 (A–C) Pregnenolone sulfate potentiates the NMDA response and inhibits kainate and AMPA responses of spinal cord neurons in tissue culture. (A) Pregnenolone sulfate (PS; 100 µM) dramatically potentiates the current response to 30 µM NMDA. (B) Pregnenolone sulfate (100 µM) inhibits response to 50 µM kainate. (C) Pregnenolone sulfate (100 µM) inhibits response to 25 µM AMPA. (From Wu et al., *Mol. Pharmacol.*, 40, 333–336, 1991. With permission.)

Replacement of the 3-sulfate with a glucosiduronate moiety greatly reduces potentiation of the NMDA response. In contrast, replacement of the sulfate with hemisuccinate does not reduce activity, but the hemisuccinate methyl ester is inactive, suggesting that the negative charge on the C-3 side chain is important for potentiation.

DHEAS has little if any activity, arguing that the D ring side chain is important for activity. In the presence of an acetate group at the C-21, potentiation is reduced but not eliminated, as measured electrophysiologically in spinal cord neurons,[63] whereas there is no decrease in potentiation of NMDA-induced Ca^{2+} uptake by hippocampal neurons.[64] Reduction of the ketone at C-20 reduces, but does not eliminate, potentiation. In contrast, potentiation is eliminated when a ketone or hydroxyl is introduced at the C-11 position, or when a ketone is present at the C-7 position.

Interpretation of structure-activity relationships for neurosteroid modulation of NMDA receptors is complicated by the fact that positive and negative modulation appear to be mediated by distinct sites on the receptor. This is evident from the observation that the percentage of inhibition of the NMDA response

Table 1 Effects of Steroids on the NMDA Response

Steroid	Percentage change in NMDA induced current	Percentage change in NMDA induced increase in $[Ca^{2+}]_i$ [d]
Pregnenolone sulfate	$+197 \pm 11 (6)$ [a]	$+310 \pm 38$
Pregnenolone hemisuccinate	$+204 \pm 18 (4)$ [b]	$+268 \pm 76$
Pregnenolone hemisuccinate methyl ester	$-0.5 \pm 1.9 (4)$ [b]	
Pregnenolone-3-D-glucosiduronate		47 ± 29
3β-Hydroxy-5β-pregnen-20-one sulfate	$-59 \pm 3 (4)$ [b]	-8 ± 4

Table 1 (continued)   Effects of Steroids on the NMDA Response

Steroid		Percentage change in NMDA induced current	Percentage change in NMDA induced increase in $[Ca^{2+}]_i{}^d$
3β-Hydroxy-5α-pregnen-20-one sulfate		$-1 \pm 3\ (6)^b$	$+32 \pm 19$
3α-Hydroxy-5β-pregnen-20-one sulfate (pregnenolone sulfate)		$-66 \pm 3\ (5)^c$	-76 ± 3
3α-Hydroxy-5β-pregnen-20-one (pregnenolone)		$+4 \pm 3^c$	
3α-Hydroxy-pregnen-20-one sulfate			-9 ± 8
3α-Hydroxypregnen-5,16-dien-20-one sulfate			26 ± 3
17-Hydroxypregnenolone 3-sulfate		$+178 \pm 15\ (4)^b$	$+163 \pm 69$

Table 1 (continued) Effects of Steroids on the NMDA Response

Steroid		Percentage change in NMDA induced current	Percentage change in NMDA induced increase in $[Ca^{2+}]_i$[d]
11β-OH-pregnenolone sulfate		-31 ± 2 (3)[b]	$+38 \pm 21$
11-Keto-pregnenolone sulfate		$+4 \pm 2$ (3)[b]	-1 ± 5
7-Keto-pregnenolone sulfate		-5 ± 5 (4)[b]	-12 ± 2
20β-Dihydropregnenolone sulfate		$+93 \pm 19$ (3)[b]	$+81 \pm 15$
21-Acetoxy-pregnenolone sulfate		$+103 \pm 34$ (3)[b]	$+316 \pm 71$
Dehydroepiandrosterone sulfate		$+28.8 \pm 8.8$ (4)[c]	$+46 \pm 13$
Androsterone sulfate		-18.1 ± 2.6 (3)[c]	

Table 1 (continued) Effects of Steroids on the NMDA Response

Steroid		Percentage change in NMDA induced current	Percentage change in NMDA induced increase in $[Ca^{2+}]_i$[d]
17β-Estradiol benzoate		24.3 ± 18 (3)[c]	

a	Results from Wu et al., 1991.[7]
b	Results from Wu et al., 1995, manuscript in preparation.
c	Results from Park-Chung et al., 1994.[8]
d	Results from Irwin et al., 1994.[64]
a,b,c	Potentiation (+)/inhibition (−) of NMDA-induced membrane currents of chick spinal cord neurons in tissue culture. Holding potential was −70 mV. NMDA concentration was 30 μM. Concentrations of steroids were 100 μM, except for β-estradiol benzoate, 20 μM and pregnanolone sulfate, 50 μM. Modulation of the NMDA response is expressed as the percentage of change of the NMDA response in the presence of steroid, as compared to control responses obtained from the same cell before application and after washout of steroid. Values are means ± standard errors. The number of neurons tested is indicated in parentheses.
d	Potentiation/inhibition of NMDA-induced increase in intracellular $[Ca^{2+}]$ of embryonic rat spinal cord neurons in culture, measured by fura-2 microspectrofluorimetry. Concentration of NMDA was 5 or 30 μM. Concentration of steroid was 50 or 100 μM. Results are means ± standard errors of 3 to 75 neurons.

by 5β3βS is not decreased in the presence of a high concentration of pregnenolone sulfate, and similarly, the percentage of potentiation by pregnenolone sulfate is not diminished in the presence of a high concentration of 5β3βS.[63]

The potentiating effect of pregnenolone sulfate appears to be predominantly due to an increase in the efficacy of NMDA,[60] which is evident even at saturating glycine concentrations. In addition, there is a leftward shift in the EC_{50} for NMDA, indicating that pregnenolone sulfate also increases the affinity of the receptor for NMDA.[7,60] At the single channel level, potentiation by pregnenolone sulfate manifests as an increase in channel open time and frequency of channel opening, with no change in single-channel conductance.[59,60] These results are consistent with a mechanism of action in which pregnenolone sulfate acts allosterically to enhance the potency and efficacy of NMDA and glycine as activators of the NMDA receptor.

The site of action of pregnenolone sulfate on the NMDA receptor is unclear, as pregnenolone sulfate potentiates the NMDA response regardless of whether it is added to the intra- or extracellular membrane surface.[59] Moreover, pregnenolone sulfate potentiates NMDA-induced single-channel activity when added to the outside of the neuron while recording from cell-attached patches,[60] indicating that pregnenolone sulfate can move through the membrane to its site of action.

An interesting question is whether neurosteroid modulation of NMDA receptor function is exerted through any of the other known modulatory sites of the NMDA receptor, such as the glycine,[43] polyamine,[65-68] redox,[69,70] or arachidonic acid[71,72] sites. Pregnenolone sulfate does not appear to act through the glycine modulatory site, as potentiation by pregnenolone sulfate can be demonstrated at saturating concentrations of glycine, and, conversely, potentiation by glycine can be observed in the presence of saturating pregnenolone sulfate.[7,60] Similarly, the potentiating effect of pregnenolone sulfate is not altered in the presence of dithiothreitol or high concentrations of spermidine or arachidonic acid.[63] These results argue that pregnenolone sulfate must be interacting with the NMDA receptor via a novel steroid modulatory site.

Inhibition of the NMDA-induced current by pregnenolone sulfate (5β3αS), the 5β3α-reduced derivative of pregnenolone sulfate, is primarily due to a decrease in the maximum response to NMDA, consistent with a noncompetitive or uncompetitive mechanism of action (Figure 6). Inhibition of the NMDA response by 5β3αS is not reversed in the presence of a saturating concentration of glycine,[8] indicating that 5β3αS does not act through the glycine modulatory site. The inhibitory effect of 5β3αS is not voltage dependent, suggesting that inhibition does not require penetration of the drug into the channel.[8]

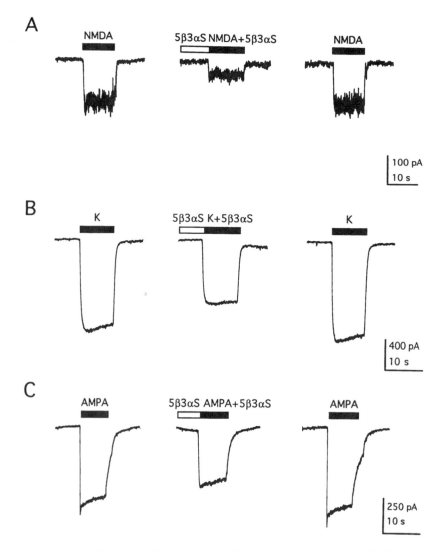

Figure 5 (A–C) 5β3αS inhibits the NMDA, kainate, and AMPA responses. $V_H = -70$ mV. (A) 5β3αS (100 μM) inhibits the current induced by 30 μM NMDA. (B) 5β3αS (100 μM) inhibits the current induced by 50 μM kainate (K). (C) 5β3αS (100 μM) inhibits the current induced by 25 μM AMPA. Horizontal bar above each trace represents a period of drug application. (From Park-Chung et al., *Mol. Pharmacol.*, 46, 146–150, 1994. With permission.)

Modulation of other types of glutamate receptors by steroids has not been studied in any detail, but modest inhibitory effects have been reported with sulfated steroids. In dissociated chick spinal cord neurons, pregnenolone sulfate (100 μM) produces a modest (25 to 29%) degree of inhibition of responses to kainate (50 μM) and AMPA (25 μM).[7] Pregnenolone sulfate (100 μM), in addition to inhibiting the NMDA response by 66%, also inhibited kainate and AMPA responses by 37 and 29%, respectively.[8] In rat hippocampal slices, 17β-estradiol (10 nM) has been reported to potentiate responses to iontophoretically applied quisqualate, AMPA, and kainate, but not NMDA, suggesting positive modulation of non-NMDA glutamate receptors.[73]

The discovery that steroids can modulate glutamate receptors raises the question of whether neurosteroids modulate excitatory neurotransmission *in vivo*. Although plasma levels of DHEAS are in the micromolar range,[74] neurosteroid levels in plasma, cerebrospinal fluid, or bulk brain tissue are more typically in the nanomolar range.[75-78] The EC_{50} for modulation of spontaneously occurring NMDA-receptor-mediated EPSCs by pregnenolone sulfate is about 5 μM,[61] which is 20- and 50-fold greater than the highest pregnenolone sulfate concentrations measured in nerve and brain, respectively.[77] It is possible, however, that elevated steroid levels may occur under certain conditions.

Figure 6 Antagonism of the NMDA response by 5β3αS is noncompetitive. Concentration-response curves for NMDA in the presence and absence of 50 μM 5β3αS. All responses are normalized to the peak current (*) induced by 100 μM NMDA. Data points represent normalized peak currents (mean of four to nine experiments). Error bars are standard errors. Error bars are not indicated when smaller than the size of the circle. Each set of data points is fitted by nonlinear regression to the logistic equation:

$$I/I_{max} = [NMDA]^n/([NMDA]^n + EC_{50}{}^n)$$

where I_{max} is the maximal normalized current, [NMDA] is the concentration of NMDA, and n is the Hill coefficient. In the absence of 5β3αS, $EC_{50} = 155.6$ μM, $I_{max} = 2.56$, and $n_H = 1.01$. In the presence of 5β3αS, $EC_{50} = 105.7$ μM, $I_{max} = 1.10$, and $n_H = 1.46$. (From Park-Chung et al., *Mol. Pharmacol.*, 46, 146–150, 1994. With permission.)

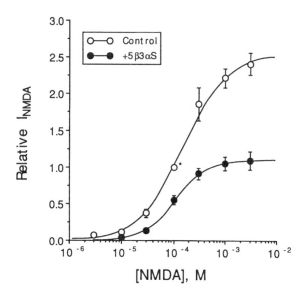

Changes in physiological state have been shown to alter brain neurosteroid levels. Surgical stress,[75,79] electroshock,[80] sexual encounters of male with female rats,[81] and the estrous cycle in female rats[82] all result in altered levels of neuroactive steroids. Aryl sulfotransferases have been detected in brain,[83] and mevalonnalactone can be used by primary cultures of astrocytes and oligodendrocytes for the biosynthesis of cholesterol and pregnenolone.[84] Moreover, isolated oligodendrocyte mitochondria can convert cholesterol into pregnenolone,[85] suggesting that, if oligodendrocytes contain steroid sulfotransferase, certain glial cells may synthesize pregnenolone sulfate and its metabolites. Synthesis of pregnenolone and pregnenolone sulfate has also been reported in isolated rat retina.[86] It is unknown whether pregnenolone sulfate is stored and concentrated in nervous tissue prior to release, such as might be expected if neurosteroids act as physiological neuromodulators. The fact that a sulfated steroid bears a full negative charge raises the possibility that such steroids could be stored within vesicular compartments in glial cells and released upon demand. Focal release could result in high local concentrations of neurosteroids, in which case measurements of pregnenolone sulfate levels in brain tissue would tend to underestimate the concentration at the receptor. Hence, the possibility of high local steroid or steroid sulfate concentrations cannot be dismissed.

It is also possible that experiments in cell culture underestimate the potency of neurosteroids *in vivo*. For example, rapid uptake by surrounding glial cells and/or degradation by sulfatases present in the cultures could reduce the effective concentration of pregnenolone sulfate at the cell membrane, resulting in an overestimate of the EC_{50}. Moreover, it is clear that glutamate receptors are heterogeneous. Three major families of glutamate receptors have been identified, with numerous subtypes, so it may be that some subtypes have greater sensitivity to steroids than those studied thus far. Finally, only a limited range of steroids have been studied for their effects on glutamate receptors, so it is possible that steroids other than pregnenolone sulfate may modulate glutamate receptors with greater affinity.

From a clinical point of view, an important question is whether neurosteroids can protect neurons from ischemic or anoxic damage, which is thought to be mediated at least in part by excitatory amino acids. Consistent with its ability to potentiate the NMDA response, pregnenolone sulfate potentiates neurotoxicity induced by chronic exposure of dissociated hippocampal cultures to NMDA[87] and increases the convulsant properties of NMDA in mice.[88] On the other hand, in another study, pregnenolone sulfate was protective against neurotoxicity induced by transient exposure of cortical neuronal cultures to NMDA.[89] This seems a surprising result, in view of pregnenolone sulfate's capacity to potentiate NMDA receptors and inhibit GABA_A and glycine receptors, but it is possible that pregnenolone sulfate's ability to inhibit non-NMDA glutamate receptors[7,62] could contribute to neuroprotection under some circumstances. Neurosteroids with inhibitory activity at glutamate receptors, such as 5β3αS[8] are more likely to have neuroprotective activity, but these have thus far received little attention. Pregnenolone sulfate has also been reported to enhance memory performance in the rat when injected directly into the nucleus basalis magnocellularis,[90] raising the interesting possibility that neurosteroids may have value as cognitive enhancers.

REFERENCES

1. Pfaff, D. W. and McEwen, B. S., Actions of estrogens and progestins on nerve cells, *Science (Washington, D.C.)*, 219, 808, 1983.
2. Majewska, M. D., Steroids and brain activity. Essential dialogue between body and mind, *Biochem. Pharmacol.*, 36, 3781, 1987.
3. Smith, S. S., Waterhouse, B. D., and Woodward, D. J., Sex steroid effects on extrahypothalamic CNS. II. Progesterone, alone and in combination with estrogen, modulates cerebellar responses to amino acid neurotransmitters, *Brain Res.*, 422, 52–62, 1987.
4. Smith, S. S., Progesterone administration attenuates excitatory amino acid responses of cerebellar Purkinje cells, *Neuroscience*, 42, 309–320, 1991.
5. Smith, S. S., Waterhouse, B. D., and Woodward, D. J., Sex steroid effects on extrahypothalamic CNS. I. Estrogen augments neuronal responsiveness to iontophoretically applied glutamate in the cerebellum, *Brain Res.*, 422, 40–51, 1987.
6. McEwen, B. S., Non-genomic and genomic effects of steroids on neural activity, *Trends Pharmacol. Sci.*, 12, 141, 1991.
7. Wu, F.-S., Gibbs, T. T., and Farb, D. H., Pregnenolone sulfate: a positive allosteric modulator at the NMDA receptor, *Mol. Pharmacol.*, 40, 333–336, 1991.
8. Park-Chung, M., Wu, F.-S., and Farb, D. H., 3α-Hydroxy-5β-pregnan-20-one sulfate: a negative modulator of the NMDA-induced current in cultured neurons, *Mol. Pharmacol.*, 46, 146–150, 1994.
9. Bélanger, A., Candas, B., Dupont, A., Cusan, L., Diamond, P., Gomez, J. L., and Labrie, F., Changes in serum concentrations of conjugated and unconjugated steroids in 40- to 80-year-old men, *J. Clin. Endocrinol. Metab.*, 79, 1086, 1994.
10. Majewska, M. D., Harrison, N. L., Schwartz, R. D., Barker, J. L., and Paul, S. M., Steroid hormone metabolites are barbiturate-like modulators of the GABA response, *Science (Washington, D.C.)*, 232, 1004, 1986.
11. Callachan, H., Cottrell, G. A., Hather, N. Y., Lambert, J. J., Nooney, J. M., and Peters, J. A., Modulation of the GABA$_A$ receptor by progesterone metabolites, *Proc. R. Soc. London B*, 231, 359, 1987.
12. Harrison, N. L., Majewska, M. D., Harrington, J. W., and Barker, J. L., Structure-activity relationships for steroid interaction with the γ-aminobutyric acid$_A$ receptor complex, *J. Pharmacol. Exp. Ther.*, 241, 346, 1987.
13. Lambert, J. J., Peters, J. A., and Cottrell, G. A., Actions of synthetic and endogenous steroids on the GABA$_A$ receptor, *Trends Pharmacol. Sci.*, 8, 224, 1987.
14. Peters, J., Kirkness, E., Callachan, H., Lambert, J., and Turner, A., Modulation of the GABA$_A$ receptor by depressant barbiturates and pregnane steroids, *Br. J. Pharmacol.*, 94, 1257–1269, 1988.
15. Puia, G., Santi, M., Vicini, S., Pritchett, D. B., Purdy, R. H., Paul, S. M., Seeburg, P. H., and Costa, E., Neurosteroids act on recombinant human GABA$_A$ receptors, *Neuron*, 4, 759, 1990.
16. Wu, F.-S., Gibbs, T. T., and Farb, D. H., Inverse modulation of γ-aminobutyric acid- and glycine-induced currents by progesterone, *Mol. Pharmacol.*, 37, 597, 1990.
17. Turner, D. M., Ransom, R. W., Yang, J. S.-J., and Olsen, R. W., Steroid anesthetics and naturally occurring analogs modulate the γ-aminobutyric acid receptor complex at a site distinct from barbiturates, *J. Pharmacol. Exp. Ther.*, 248, 960, 1989.
18. Im, W., Blakeman, D., Davis, J., and Ayer, D., Studies on the mechanism of interactions between anesthetic steroids and gamma-aminobutyric acid$_A$ receptors, *Mol. Pharmacol.*, 37, 429–434, 1990.
19. Morrow, A. L., Pace, J. R., Purdy, R. H., and Paul, S. M., Characterization of steroid interactions with γ-aminobutyric acid receptor-gated chloride ion channels: evidence for multiple steroid recognition sites, *Mol. Pharmacol.*, 37, 263, 1990.
20. Gee, K. W., Chang, W.-C., Brinton, R. E., and McEwen, B. S., GABA-dependent modulation of the Cl⁻ ionophore by steroids in rat brain, *Eur. J. Pharmacol.*, 136, 419, 1987.
21. Gee, K. W., Bolger, M. B., Brinton, R. E., Coirini, H., and McEwin, B. S., Steroid modulation of the chloride ionophore in rat brain: structure-activity requirements, regional dependence and mechanism of action, *J. Pharmacol. Exp. Ther.*, 246, 803, 1988.
22. Gyenes, M., Wang, Q., Gibbs, T. T., and Farb, D. H., Phosphorylation factors control neurotransmitter and neuromodulator action at the γ-aminobutyric acid type A receptor, *Mol. Pharmacol.*, 46, 542–549, 1994.
23. Majewska, M. D. and Schwartz, R. D., Pregnenolone-sulfate: An endogenous antagonist of the γ-aminobutyric acid receptor complex in brain?, *Brain Res.*, 404, 355, 1987.
24. Majewska, M. D., Mienville, J.-M., and Vicini, S., Neurosteroid pregnenolone sulfate antagonizes electrophysiological responses to GABA in neurons, *Neurosci. Lett.*, 90, 279, 1988.
25. Mienville, J. and Vicini, S., Pregnenolone sulfate antagonizes GABA$_A$ receptor-mediated currents via reduction of channel opening frequency, *Brain Res.*, 489, 190–194, 1989.
26. Majewska, M. D., Demirgören, S., Spivak, C. E., and London, E. D., The neurosteroid dihydroepiandrosterone sulfate is an allosteric antagonist of the GABA$_A$ receptor, *Brain Res.*, 526, 143, 1990.
27. Crawley, J., Glowa, J., Majewska, M., and Paul, S., Anxiolytic activity of an endogenous adrenal steroid, *Brain Res.*, 398, 382–385, 1986.
28. Mendelson, W., Martin, J., Perlis, M., Wagner, R., Majewska, M., and Paul, S., Sleep induction by an adrenal steroid in the rat, *Psychopharmacology*, 93, 226–229, 1987.

29. Kavaliers, M., Inhibitory influences of the adrenal steroid, 3 alpha, 5 alpha-tetrahydrodeoxycorticosterone on aggression and defeat-induced analgesia in mice, *Psychopharmacology,* 95, 488–492, 1988.

30. Kavaliers, M. and Wiebe, J., Analgesic effects of the progesterone metabolite, 3 alpha-hydroxy-5 alpha-pregnan-20-one, and possible modes of action in mice, *Brain Res.,* 415, 393–398, 1987.

31. Majewska, M. D., Bluet-Pajot, M.-T., Robel, P., and Baulieu, E.-E., Pregnenolone sulfate antagonizes barbiturate-induced hypnosis, *Pharmacol. Biochem. Behav.,* 33, 701, 1989.

32. Sandberg, M. and Jacobson, I., β-Alanine, a possible neurotransmitter in the visual system?, *J. Neurochem.,* 37, 1353, 1981.

33. Toggenburger, G., Felix, D., Cuenod, M., and Henke, H., *In vitro* release of endogenous β-alanine in pigeon optic tectum, *J. Neurochem.,* 39, 176, 1982.

34. DeFeudis, F. V. and Del Rio, M., Is β-alanine an inhibitory neurotransmitter?, *Gen. Pharmacol.,* 8, 177, 1977.

35. Parker, I., Sumikawa, K., and Miledi, R., Responses to GABA, glycine and beta-alanine induced in *Xenopus* oocytes by messenger RNA from chick and rat brain, *Proc. R. Soc. London B,* 233, 201–216, 1988.

36. Choquet, D. and Korn, H., Does beta-alanine activate more than one chloride channel associated receptor?, *Neurosci. Lett.,* 84, 329–334, 1988.

37. Horikoshi, T., Asanuma, A., Yanagisawa, K., Anzai, K., and Goto, S., Taurine and β-alanine act on both GABA and glycine receptors in *Xenopus* oocyte injected with mouse brain messenger RNA, *Mol. Brain Res.,* 4, 97–105, 1988.

38. Wu, F.-S., Gibbs, T. T., and Farb, D. H., Dual activation of GABA$_A$ and glycine receptors by β-alanine: inverse modulation by progesterone and 5α-pregnan-3α-ol-20-one, *Eur. J. Pharmacol.— Mol. Pharmacol. Section,* 246, 239–246, 1993.

39. Hablitz, J. and Langmoen, I., Excitation of hippocampal pyramidal cells by glutamate in the guinea-pig and rat, *J. Physiol. (London),* 325, 317–331, 1982.

40. Mayer, M. and Westbrook, G., The action of *N*-methyl-D-aspartic acid on mouse spinal neurones in culture, *J. Physiol. (London),* 361, 65–90, 1985.

41. Crunelli, V., Forda, S., and Kelly, J., The reversal potential of excitatory amino acid action on granule cells of the rat dentate gyrus, *J. Physiol. (London),* 351, 327–342, 1994.

42. Mac Dermott, A., Mayer, M., Westbrook, G., Smith, S., and Barker, J., NMDA-receptor activation increases cytoplasmic calcium concentration in cultured spinal cord neurones, *Nature (London),* 361, 65–90, 1986.

43. Johnson, J. W. and Ascher, P., Glycine potentiates the NMDA response in cultured mouse brain neurons, *Nature (London),* 325, 529–531, 1987.

44. Mayer, M., Westbrook, G., and Guthrie, P., Voltage-dependent block by Mg^{2+} of NMDA responses in spinal cord neurones, *Nature (London),* 309, 261–263, 1984.

45. Nowak, L., Bregestovski, P., Ascher, P., Herbet, A., and Prochiantz, A., Magnesium gates glutamate-activated channels in mouse central neurones, *Nature (London),* 307, 462–465, 1984.

46. Lynch, G., Gribkoff, V., and Deadwyler, S., Long term potentiation is accompanied by a reduction in dendritic responsiveness to glutamic acid, *Nature,* 263, 151–153, 1976.

47. Harris, E. W., Ganong, A. H., and Cotman, C. W., Long term potentiation in the hippocamppus involves activation of *N*-methyl-D-aspartate receptors, *Brain Res.,* 323, 132–137, 1984.

48. Wigstrom, H. and Gustafsson, B., A possible correlate of the postsynaptic condition for long-lasting potentiation in the guinea pig hippocampus *in vitro, Neurosci. Lett.,* 44, 327–332, 1984.

49. Rauschecker, J. and Hahn, S., Ketamine-xylazine anaesthesia blocks consolidation of ocular dominance changes in kitten visual cortex, *Nature (London),* 326, 183–185, 1987.

50. Herron, C., Lester, R., Coan, E., and Collingridge, G., Intracellular demonstration of an *N*-methyl-D-aspartate receptor mediated component of synaptic transmission in the rat hippocampus, *Neurosci. Lett.,* 60, 19–23, 1985.

51. Dingledine, R., Hynes, M., and King, G., Involvement of *N*-methyl-D-aspartate receptors in epileptiform bursting in the rat hippocampal slice, *J. Physiol. (London),* 380, 175–189, 1986.

52. Simon, R., Swan, J., Griffiths, T., and Meldrum, B., Blockade of *N*-methyl-D-aspartate receptors may protect against ischemic damage in the brain, *Science (Washington, D.C.),* 226, 850–852, 1984.

53. Carlsson, M. and Carlsson, A., Interactions between glutamatergic and monoaminergic systems within the basal ganglia — implications for schizophrenia and Parkinson's disease, *Trends Neurosci.,* 13, 272–276, 1990.

54. Wachtel, H. and Turski, L., Glutamate: a new target in schizophrenia?, *Trends Pharmacol. Sci.,* 11, 219–220, 1990.

55. Onley, J., Labruyere, J., Collins, J., and Curry, K., D-aminophosphonovalerate is 100-fold more powerful than D-α-aminoadipate in blocking *N*-methyl-aspartate neurotoxicity, *Brain Res.,* 221, 207–210, 1981.

56. Smith, S., Waterhouse, B., and Woodward, D., Locally applied progesterone metabolites alter neuronal responsiveness in the cerebellum, *Brain Res. Bull.,* 18, 739–747, 1987.

57. Smith, S. S., Waterhouse, B. D., Chapin, J. K., and Woodward, D. J., Progesterone alters GABA and glutamate responsiveness: a possible mechanism for its anxiolytic action, *Brain Res.,* 400, 353–359, 1987.

58. Smith, S. S., Waterhouse, B. D., and Woodward, D. J., Locally applied estrogens potentiate glutamate-evoked excitation of cerebellar Purkinje cells, *Brain Res.,* 475, 272–282, 1988.

59. Wong, M. and Moss, R. L., Patch-clamp analysis of direct steroidal modulation of glutamate receptor-channels, *J. Neuroendocrinol.,* 6, 347–355, 1994.

60. Bowlby, M., Pregnenolone sulfate potentiation of *N*-methyl-D-aspartate receptor channels in hippocampal neurons, *Mol. Pharmacol.,* 43, 813–819, 1993.

61. Park-Chung, M., Gibbs, T. T., and Farb, D. H., manuscript in preparation.

62. Irwin, R. P., Maragakis, N. J., Rogawski, M. A., Purdy, R. H., Farb, D. H., and Paul, S. M., Pregnenolone sulfate augments NMDA receptor medicated increases in intracellular Ca^{2+} in cultured rat hippocampal neurons, *Neurosci. Lett.*, 141, 30–34, 1992.

63. Wu, F.-S., Park-Chung, M., Purdy, R. H., Gibbs, T. T., and Farb, D. H., manuscript in preparation.

64. Irwin, R. P., Lin, S.-Z., Rogawski, M. A., Purdy, R. H., and Paul, S. M., Steroid potentiation and inhibition of *N*-methyl-D-aspartate receptor-mediated intracellular Ca^{++} responses: structure activity studies, *J. Pharmacol. Exp. Ther.*, 271, 677–682, 1994.

65. Ransom, R. W. and Stec, N. L., Cooperative modulation of [^{3}H]MK-801 binding to the *N*-methyl-D-aspartate receptor-ion channel complex by L-glutamate, glycine, and polyamines, *J. Neurochem.*, 51, 830–836, 1988.

66. Williams, K., Romano, C., and Molinoff, P., Effects of polyamines on the binding of [^{3}H]MK-801 to the *N*-methyl-D-aspartate receptor: pharmacological evidence for the existence of a polyamine site, *Mol. Pharmacol.*, 36, 575–581, 1989.

67. Sprosen, T. S. and Woodruff, G. N., Polyamines potentiate NMDA induced whole-cell currents in cultured striatal neurons, *Eur. J. Pharmacol.*, 179, 477–478, 1990.

68. Williams, K., Dawson, V. L., Romano, C., Dichter, M. A., and Molinoff, P. B., Characterization of polyamines having agonist, antagonist and inverse agonist effects at the polyamine recognition site of the NMDA receptor, *Neuron*, 5, 199–208, 1990.

69. Aizenman, E., Lipton, S. A., and Loring, R. H., Selective modulation of NMDA responses by reduction and oxidation, *Neuron*, 2, 1257–1263, 1989.

70. Tang, L.-H. and Aizenman, E., The modulation of *N*-methyl-D-aspartate receptors by redox and alkylating reagents in rat cortical neurones *in vitro*, *J. Physiol. (London)*, 465, 303–323, 1993.

71. Miller, B., Sarantis, M., Traynelis, S. F., and Attwell, D., Potentiation of NMDA receptor currents by arachidonic acid, *Nature*, 355, 722–725, 1992.

72. Nishikawa, M., Kimura, S., and Akaike, N., Facilitatory effect of docosahexaenoic acid on *N*-methyl-D-aspartate response in pyramidal neurones of rat cerebral cortex, *J. Physiol. (London)*, 475, 83–93, 1994.

73. Wong, M. and Moss, R. L., Long-term and short-term electrophysiological effects of estrogen on the synaptic properties of hippocampal CA1 neurons, *J. Neurosci.*, 12, 3217–3225, 1992.

74. Migeon, C. J., Androgens in human plasma, in *Hormones in Human Plasma: Nature and Transport*, Antoniades, H. N., Ed., Little, Brown, Boston, 1960.

75. Corpéchot, C., Robel, P., Axelson, M., Sjövall, J., and Baulieu, E.-E., Characterization and measurement of dehy-droepiandosterone sulfate in rat brain, *Biochemistry*, 78, 4704–4707, 1981.

76. Corpéchot, C., Synguelakis, M., Talha, S., Axelson, M., Sjöval, J., Vihko, R., Baulieu, E.-E., and Robel, P., Preg-nenolone and its sulfate ester in the rat brain, *Brain Res.*, 270, 119–125, 1983.

77. Lanthier, A. and Patwardhan, V. V., Sex steroids and 5-en-5β-hydroxysteroids in specific regions of the human brain and cranial nerves, *J. Steroid Biochem.*, 25, 445–449, 1986.

78. Robel, P., Bourreau, E., Corpéchot, C., Dang, D. C., Halberg, F., Clarke, C., Haug, M., Schlegel, M. L., Synguelakis, M., Vourch, C., and Baulieu, E.-E., Neuro-steroids: 3β-hydroxy-Δ5-derivatives in rat and monkey brain, *J. Steroid Biochem.*, 27, 649–655, 1987.

79. Purdy, R. H., Morrow, A. L., Jr., P. H. M., and Paul, S. M., Stress-induced elevations of GABA$_A$ receptor-active steroids in the rat brain., *Proc. Natl. Acad. Sci., U.S.A.*, 88, 4553–4557, 1991.

80. Korneyev, A., Guidotti, A., and Costa, E., Regional and Interspecies differences in brain progesterone metabolism, *J. Neurochem.*, 61, 2041–2047, 1993.

81. Corpéchot, C., Leclerc, P., Baulieu, E.-E., and Brazeau, P., Neurosteroids: Regulatory mechanisms in male rat brain during heterosexual exposure, *Steroids*, 45, 229–234, 1985.

82. Paul, S. M. and Purdy, R. H., Neuroactive steroids, *FASEB J.*, 6, 2311–2322, 1992.

83. Zhu, X., Veronese, M. E., Bernard, C. C. A., Sansom, L. N., and McManus, M. E., Identification of two human brain aryl sulfotransferase cDNAs, *Biochem. Biophys. Res. Commun.*, 195, 120–127, 1993.

84. Jung-Testas, I., Hu, Z. Y., Baulieu, E.-E., and Robel, P., Neurosteroids: biosynthesis of pregnenolone and progesterone in primary cultures of rat glial cells, *Endocrinology*, 125, 2083–2091, 1989.

85. Hu, Z. Y., Bourreau, E., Jung-Testas, I., Robel, P., and Baulieu, E.-E., Neurosteroids: oligodendrocyte mitochondria convert cholesterol to pregnenolone, *Proc. Natl. Acad. Sci. U.S.A.*, 84, 8215–8219, 1987.

86. Guarneri, P., Guarneri, R., Cascio, C., Pavasant, P., Piccoli, F., and Papadopoulos, V., Neurosteroidogenesis in rat retinas, *J. Neurochem.*, 63, 86–96, 1994.

87. Weaver, C. E., Jr., Wu, F.-S., and Farb, D. H., Pregnenolone sulfate potentiates NMDA-induced cell death in cultured neurons, *Soc. Neurosci. Abstr.*, 19, 1778, 1993.

88. Maione, S., Berrino, L., Vitagliano, S., Leyva, J., and Rossi, F., Pregnenolone sulfate increases the convulsant potency of *N*-methyl-D-aspartate in mice, *Eur. J. Pharmacol.*, 219, 477–479, 1992.

89. Wie, M. B., Tjan, C., and Choi, D. W., Pregnenolone sulfate attenuates excitotoxic neuronal injury in murine cortical cell cultures, *Soc. Neurosci. Abstr.*, 19, 1698, 1993.

90. Mayo, W., Dellu, F., Robel, P., Cherkaoui, J., Le Moal, M., Baulieu, E. E., and Simon, H., Infusion of neurosteroids into the nucleus basalis magnocellularis affects cognitive processes in the rat, *Brain Res.*, 607, 324–328, 1993.

Chapter 3

Regulation of Serotonergic Function in the CNS by Steroid Hormones and Stress

Christina R. McKittrick and Bruce S. McEwen

CONTENTS

I. INTRODUCTION

Hormone effects on brain development and adult brain function are among the most important means for the external world to coordinate and direct brain function and behavior; for example, hormones coordinate reproductive function with reproductive behavior and they mediate the effects of and adaptation to stressful events as well. Hormones of the gonads, adrenal cortex, and thyroid gland affect the brain during adult life as well as during development, and they do so at least in part via intracellular receptors that modulate gene expression.

Many neural systems respond to circulating hormones, and the serotonergic system is one of the oldest and most basic of the neurotransmitter systems of the brain. Its involvement in vegetative behaviors, affective state, and cognitive function makes its operation central to many aspects of behavior and pathophysiology. It is not surprising, therefore, that there have been fairly extensive investigations of the influence of circulating hormones on serotonergic function. However, the increasing recognition of the complexity of the serotonergic system in terms of the multiple 5HT receptor subtypes, neuroanatomical projections, and neurophysiological activity has increased the need for truly integrative studies that would link levels of analysis in an attempt to understand how serotonergic function is modified by circulating hormones, both developmentally and in adult life, and how these modifications are involved in behavioral changes. In the course of such studies, the pathophysiological aspects of serotonergic involvement in depressive illness and the mechanism of antidepressant action may become more accessible, particularly in relation to sex differences, actions of circulating gonadal hormones, stressful life events, and thyroid hormone balance. This review summarizes the main findings of studies on hormone effects on the serotonergic system and critically assesses what needs to be done to further develop an integrative approach.

II. OVERVIEW OF THE CENTRAL SEROTONIN SYSTEM

The interactions between steroid hormones on the central serotonergic system have been the focus of intense study for many decades. Gonadal steroids, adrenal steroids, and thyroid hormones all have profound effects on serotonin (5HT) function at many different levels. After a brief overview of the central 5HT system, each of these steroid systems will be discussed in terms of its relevant receptor systems and the effects each steroid has on different aspects of 5HT function, including 5HT levels and metabolism, binding and gene expression of 5HT receptor subtypes, 5HT-mediated electrophysiological responses, and 5HT-mediated behavior. Although the 5HT system itself has, in turn, been shown to be an important mediator of these hormonal systems, discussion of these effects is beyond the scope of this review.

A. SYNTHESIS AND METABOLISM

Serotonin is synthesized in the brain from tryptophan, which is transported into the brain by an active carrier of large neutral amino acids. Tryptophan is converted to 5-hydroxytryptophan (5HTP) by tryptophan hydroxylase (TPH), the rate-limiting enzyme in serotonin synthesis. 5HTP is then decarboxylated by amino acid decarboxylase to form 5-hydroxytryptamine (serotonin). Serotonin is stored primarily in vesicles at nerve terminals and is released by exocytotic mechanisms upon neuronal firing.

Serotonin transporters (5HTt) are located presynaptically on serotonin nerve terminals, where they remove serotonin in the synapse through a selective reuptake mechanism. Free 5HT is then catabolized by the enzyme monoamine oxidase (MAO) into 5-hydroxyindoleacetic acid (5HIAA), which is then excreted from the brain. The levels of 5HT and 5HIAA are routinely measured in extracts from tissue homogenates by high-performance liquid chromatography (HPLC). This technique measures both the intra- and extracellular content of these indoles; an increase in the levels of 5HIAA or in the 5HIAA/5HT ratio suggests an increase in the release and subsequent degradation of 5HT. In order to estimate 5HT turnover, a function of the rates of synthesis, release, and degradation, many investigators administer the MAO inhibitor pargyline and measure the accumulation of 5HT and 5HIAA absence of degradation; a higher accumulation rate suggests an increase in 5HT activity in those regions. In contrast, the newer techniques of *in vivo* voltammetry and microdialysis measure the extracellular content of transmitters, giving a more precise measure of 5HT release.

B. NEUROANATOMY

The majority of 5HT-containing cell bodies are located in the raphe nuclei in the brainstem; particularly in cell groups B4-B7 (dorsal raphe nucleus) and B8 (median raphe nucleus).[1] The two raphe nuclei send 5HT projections throughout the brain, with each nucleus having a distinct innervation pattern: the median raphe sends ascending projections to limbic regions such as hippocampus, septum, hypothalamus, and amygdala, while the dorsal raphe has a more diffuse projection throughout the striatum, basal ganglia, cerebral cortices, thalamus, and cerebellum.[2-8]

C. RECEPTOR SUBTYPES AND LOCALIZATION

The effects of 5HT in the brain are largely determined by the receptor subtypes present at their site of release. At the last count, 14 distinct receptor subtypes (including the 5HTt) have been identified in mammals, with an additional three cloned in *Drosophila* and one in snail (for an excellent review of 5HT receptors, see Reference 9). These subtypes can be divided into several families based on sequence similarities of the different receptors. The $5HT_1$ receptor family is the largest and comprises the $5HT_{1A}$, $5HT_{1B}$, $5HT_{1D}$, $5HT_{1E}$, $5HT_{1F}$, $5HT_{5A}$, $5HT_{5B}$, and $5HT_7$ receptors, as well as those found in *Drosophila* and snail. The receptors in the $5HT_1$ family are coupled to G proteins, as are the $5HT_2$ receptor family, which includes the $5HT_{2A}$, $5HT_{2B}$, and $5HT_{2C}$ receptors, and the one-membered $5HT_6$ receptor family. In contrast, the $5HT_3$ receptor is a ligand-gated ion channel, whereas the 5HTt is a member of a larger family of neurotransmitter transporters.

As a group, the $5HT_1$ receptors are characterized by their ability to inhibit adenylate cyclase; in addition, the $5HT_1$ subtypes and the $5HT_7$ receptor have a very high affinity for 5HT, although the affinity of $5HT_5$ receptors is somewhat lower. The $5HT_{1A}$ receptor is perhaps the best characterized of the 5HT receptors and is one of the few receptors to which several compounds bind selectively. The aminotetralin 8-OH-DPAT is the $5HT_{1A}$ agonist most commonly used in both binding and behavioral studies; 5CT also binds to $5HT_{1A}$ receptors with high affinity although it is less selective (Table 1). $5HT_{1A}$ receptor binding and mRNA are widely distributed throughout the brain, with particularly high levels in hippocampus, hypothalamus, septum, amygdala, entorhinal cortex, and dorsal raphe.[10,11] Most $5HT_{1A}$ receptors are believed to be postsynaptic, with the exception of those in the raphe, which serve as somatodendritic autoreceptors. The $5HT_{1B}$ receptors are far less abundant and are found predominantly in hippocampus, striatum, cerebellum, and raphe;[12] suggesting both pre- and postsynaptic localization of receptors. The rat $5HT_{1D}$ receptor has a relatively high degree of homology to the $5HT_{1B}$ receptor and is found in the raphe nuclei, olfactory tubercle, caudate nucleus, and accumbens.[13] Very little is known about the $5HT_{1E}$, $5HT_{1F}$, $5HT_{5A}$, $5HT_{5B}$, and $5HT_7$ receptors, although their cloning has opened several new avenues of investigation, limited primarily by a lack of receptor-selective drugs.

In contrast to the $5HT_1$ family, the $5HT_2$ receptors stimulate phosphoinositol hydrolysis through the activation of phospholipase C, leading to the release of Ca^{2+} from intracellular stores. Several $5HT_{2A}$ agonists are available for pharmacological studies, such as the agonists DOI and MK 212 and the antagonists ketanserin, mianserin, and cyproheptadine; spiroperidol and spiperone are also commonly used for receptor-binding studies, although these compounds also have a high affinity for the dopamine D_2 receptor (Table 1). $5HT_{2A}$ receptor binding and mRNA are found primarily in layers IV and V of cerebral cortex, claustrum, CA3 of ventral hippocampus, olfactory nuclei, pontine nuclei, and several brainstem nuclei.[14,15] The $5HT_{2B}$ receptor is located primarily in the stomach fundus and will not be discussed here. The $5HT_{2C}$ receptor was originally called the $5HT_{1C}$ receptor, due to its high affinity for 5HT, but subsequent studies have shown that it is similar to the $5HT_{2A}$ receptor in sequence, binding properties, and second messenger coupling. Many ligands do not differentiate well between $5HT_{2C}$ and $5HT_{2A}$ receptors, although the $5HT_{2C}$ receptor does have a higher affinity for mCPP and 5HT itself. $5HT_{2C}$ receptors are found throughout the brain with high densities of mRNA found in choroid plexus, basal ganglia, ventral CA3, amygdala, hypothalamus, thalamus, olfactory nuclei, and brainstem.[15,16]

The $5HT_3$ receptor is a member of a family of ligand-gated ion channels which also includes GABA, glutamate, glycine, and nicotinic acetylcholine receptors. Binding studies have found high levels of this receptor in cortex, amygdala, hippocampus, and accumbens.[17,18] The 5HT transporter structurally resembles several other Na^+ and Cl^- coupled neurotransmitter transporters.[19,20] 5HT transporters are localized presynaptically on 5HT nerve terminals, so binding sites are located throughout the brain, while most 5HTt mRNA expression is in the 5HT neurons of the raphe nuclei.[21,22] Several clinically active antidepressants, such as fluoxetine, paroxetine, and citalopram bind the 5HTt selectively and with high affinity,

Table 1 Commonly Used 5HT Drugs

Compound	Function
Cyproheptadine	$5HT_{2A/2C}$ antagonist
Cyanoimipramine	Selective 5HT uptake inhibitor; binds very selectively to the 5HTt
DOI: 1-(2,5-dimethoxy-4-iodophenyl)-2-aminopropane	$5HT_{2A}$ agonist
Fenfluramine	5HT releaser; depletes 5HT stores
Fluoxetine	Selective 5HT uptake inhibitor; binds very selectively to the 5HTt
Imipramine (IMI)	Monoamine uptake inhibitor; binds relatively selectively to the 5HT
L-Trp: L-tryptophan	5HT precursor
Ketanserin	$5HT_{2A}$ antagonist
mCPP: 1-(m-chlorophenyl)piperazine	$5HT_{2C}$ agonist; may also have $5HT_3$ antagonist activity
Methysergide	$5HT_{2A/2C}$ antagonist
MK 212: 6-chloro-2-(1-piperazinyl)-pyrazine	$5HT_{2A}$ agonist
Pargyline	Monoamine oxidase inhibitor
Paroxetine	Selective 5HT uptake inhibitor; binds very selectively to the 5HTt
PCA: p-chloroamphetamine	5HT releaser; depletes 5HT stores
PCPA: p-chlorophenylalanine methyl ester	5HT synthesis inhibitor
Quipazine	$5HT_3$ agonist
Ritanserin	$5HT_{2A}$ antagonist
Spiperone	$5HT_{2A}$ antagonist; D_2 antagonist
TFMPP: 1-(m-trifluoromethylphenyl)-piperazine	$5HT_{1B}$ and $5HT_{2C}$ agonist
Tranylcypromine	Monoamine oxidase inhibitor
5CT: 5-carboxamidotryptamine	$5HT_{1A}$ and $5HT_{1B}$ agonist
5HTP: 5-hydroxytryptophan	5HT precursor
5MeODMT: 5-methoxy-N,N-dimethyltryptamine	Nonselective 5HT agonist
5,7DHT: 5,7-dihydroxytryptamine	5HT-selective neurotoxin
8-OH-DPAT: 8-hydroxy-2-(di-n-propylamino)tetralin	$5HT_{1A}$ agonist

although several older studies have used the less selective compound imipramine to label 5HT transporters (Table 1).

D. ELECTROPHYSIOLOGY

In the hippocampus and dentate gyrus, 5HT has been shown to inhibit both population responses and those of individual pyramidal and granule cells. The most pronounced effect in these cells is an increase in K^+ conductance, which leads to hyperpolarization; $5HT_{1A}$ receptors have been shown to mediate this response.[23-27] Following the initial inhibitory hyperpolarization, 5HT leads to a decrease in the slow afterhyperpolarization mediated by a slow Ca^{2+}-dependent K^+ current, which has a net excitatory effect.[25,28,29] Studies in the presence of $5HT_{1A}$ blockade suggest that the 5HT effect on the slow afterhyperpolarization may be mediated by the $5HT_{2C}$ receptor subtype.[30]

In regions other than the hippocampus, 5HT and 8-OH-DPAT were shown to inhibit neuronal activity in both the paraventricular nucleus (PVN) and the ventromedial nucleus (VMN) of hypothalamus.[31,32] In contrast, however, electrical stimulation of the raphe nuclei led to excitation of neurons in the PVN, an effect that was blocked by the $5HT_{2A/2C}$ antagonists methysergide and cyproheptadine.[33] This suggests that the effects of 5HT on neuronal excitability are quite complex and depend on the complement of receptor subtypes present.

E. BEHAVIOR

Administration of 5HT precursors or agonists has been shown to produce a constellation of stereotypical behaviors; this response is generally referred to as the "5HT syndrome" and includes such responses as headweaving, flat body posture, hind limb abduction, tremor, forepaw treading, Straub tail, and generally increased activity.[34] This syndrome can be elicited by combined treatment with 5HT precursors and MAO inhibitors, such as L-tryptophan and pargyline, by nonspecific 5HT agonists, such as 5MeODMT, or by more selective compounds, such as 8-OH-DPAT. The behavioral response to these pharmacological agents is often used to determine the functional state of the 5HT system following other experimental manipulations. In addition to the 5HT syndrome, 8-OH-DPAT induces other quantifiable physiological changes; these include hypothermia, which appears to be mediated by postsynaptic $5HT_{1A}$ receptors,[35] and hyperphagia, which is mediated by somatodendritic $5HT_{1A}$ autoreceptors.[36,37] Head twitches and "wet-dog shakes" elicited by 5HTP are similar to those induced by DOI, suggesting that these responses are mediated by $5HT_{2A}$ receptors,[38,39] while the $5HT_{2C}$ receptor agonist mCPP has been shown to reduce locomotor activity.[40] There is, however, considerable evidence for functional interactions between 5HT receptors, suggesting that a given receptor subtype may oppose or enhance responses mediated by selective activation of another receptor.[41-46]

Activation of various 5HT receptor subtypes has also been shown to regulate several neuroendocrine systems, leading to increased secretion of ACTH (and subsequent increases in glucocorticoids), β-endorphin, prolactin, renin, vasopressin, and oxytocin (reviewed in Reference 47). These endocrine responses are used as physiological endpoints in measuring the responsiveness of the 5HT system. 5HT has been shown to modulate anxiety, sex behavior, feeding behavior, sleep, and nociception, in addition to playing a regulatory role in the maintenance of circadian rhythmicity (see References 48 and 49). The effects of steroid manipulations on several of these behaviors are discussed below.

III. GONADAL HORMONES

A. RECEPTORS AND THEIR DISTRIBUTION

Like other endogenous steroids, gonadal hormones are believed to exert their effects by binding to cytoplasmic or nuclear receptors, which then bind to specific target sequences on DNA and activate transcription. Androgen receptors (AR), estrogen receptors (ER), and progestin receptors (PR) have characteristic, if somewhat overlapping, patterns of distribution throughout the brain. All are found in relatively high density in areas associated with sexual behavior, such as the medial preoptic area (mPOA), the VMN, arcuate nucleus (ARC), and PVN. High levels of AR and ER binding and mRNA are also found in bed nucleus of stria terminalis (BNST) and septum. In addition, high levels of gonadal steroid receptors are also found throughout the amygdala, particularly the medial and cortical amygdaloid nuclei. Moderate levels of both AR and ER binding and mRNA can be found in the raphe nuclei and in midbrain central gray. The pontine nuclei, the hippocampal formation, and brainstem also contain moderate amounts of AR, but few ER, while ER, but few AR, can be found in dorsolateral hypothalamic nucleus, superchiasmatic nuclei of POA (SCN-POA), diagonal band of Broca, basolateral amygdaloid nuclei, and the parabrachial nuclei.[50-52] Sex differences in gonadectomized animals have also been noted, with males having fewer ER in mPOA and fewer PR in periventricular POA and VMN.[53]

High densities of PR are found in pituitary, hypothalamus, POA, cerebral cortex, and amygdala, with less in midbrain and cerebellum.[54-57] Serotonergic neurons in dorsal and median raphe and raphe magnus are also immunopositive for PR in macaque brain.[58] Progestin receptors can be induced by estrogen in ovariectomized and adrenalectomized females to a large degree in mediobasal hypothalamus, POA, PVN, SCN-POA, VMN, ARC-median eminence, and to a lesser extent in anterior and lateral hypothalamus, BNST, cingulate cortex, and CA1 of hippocampus.[55,56] Estrogen receptors are abundant in most of the regions in which PR binding is increased; however, no PR induction is seen in septum or amygdala, which also contains high densities of ER. Estrogen also has no effect on immunoreactive PR in the supraoptic nucleus, suggesting that estrogen responsiveness varies with individual neuronal populations.[57]

B. ANDROGENS
1. 5HT Levels and Metabolism
a. Adult

Relatively few studies have examined the effects of androgens on the levels and metabolism of 5HT. Studies looking at testosterone (T) administration to castrated (CAST) and intact adult male rats suggest a regional specificity in androgen effects. Bitar and co-workers showed that castration increased 5HIAA levels in hypothalamus and striatum while decreasing 5HT in hypothalamus and hippocampus; chronic treatment with estradiol benzoate (EB) reversed the effects of gonadectomy in hypothalamus, while either T or EB decreased 5HIAA in striatum to the levels of intact animals.[59] In contrast, another study found no differences in 5HT or 5HIAA content in medial basal hypothalamus or POA among CAST, CAST plus T, or intact animals.[60] In intact male rats receiving chronic (14-day) T treatment, however, 5HT and 5HIAA levels were decreased in hippocampus, but not in striatum or frontal cortex.[61] It is probable that age may play a role in regulating sensitivity of the 5HT system to androgens, as exogenous

T given to intact rats increased the 5HIAA/5HT ratio in locus coeruleus and hippocampus and decreased it in substantia nigra in young, but not old, rats.[62]

Most reports of sex differences in 5HT content and metabolism in the brain suggest that males have a less active 5HT system than females. Several authors have reported higher 5HT and 5HIAA levels in whole brain, hippocampus, and frontal cortex, as well as increased 5HT synthesis and turnover rates, in females relative to males.[63-70] The $5HT_{1A}$ agonist, 8-OH-DPAT, also inhibited 5HT synthesis to a greater extent in females.[67] Although Gonzales and Leret have shown a testosterone-dependent increase in the 5HIAA/5HT ratio in striatum and limbic regions of males compared to females (see below), any sex differences in the absolute values of 5HT and 5HIAA are not reported.[71]

b. Developmental Effects

Androgens appear to play a role in the development of 5HT innervation of the sexually dimorphic medial preoptic nucleus (MPN). Serotonin-immunoreactive (5HT-IR) fibers are found throughout the MPN, with relatively low densities in the medial and central MPN and much heavier innervation found in the lateral MPN. The sex differences in 5HT innervation of these subnuclei are related to the sexual dimorphism in their relative sizes; males have a larger medial and central MPN and thus have a larger area of low 5HT-IR density, while females have a relatively larger lateral MPN and proportionately greater region of high fiber density.[72] The 5HT innervation and cytoarchitecture of the MPN can be masculinized in female rats by perinatal treatment with testosterone (from day 16 of gestation to postnatal day 10).[73] There seems to be a critical period for the effects of testosterone, however, as administration of T on postnatal day 5 had no effect on 5HT-IR fiber distribution in the MPN.

Testosterone may have other organizational effects on the 5HT system during development as well. Hardin found higher 5HT synthesis rates and 5HT levels in brains from 2-day-old female rats compared to males, with these differences disappearing by day 5, while testosterone administration to 5-day-old females increased 5HT levels in some animals and decreased them in others.[74,75] However, female rats injected with T during the first 24 h of life had masculinization of proceptive sex behavior (increased mounting) and a higher 5HIAA/5HT ratio in striatum and limbic regions, similar to that observed by these investigators in adult males.[71] Another group saw a higher 5HIAA/5HT ratio in hypothalamus and POA in 40-day-old males compared to females, although when the animals were 80 days old, the females had higher 5HIAA/5HT ratios and higher rates of 5HT synthesis in the same regions; these sex differences were not apparent at earlier time points, suggesting that endogenous sex steroids do not modulate 5HT activity in these regions until relatively late in development.[76]

2. 5HT Receptors

In addition to the cytoarchitectonic effects on the medial POA, testosterone also alters 5HT receptor binding in this region. Our laboratory reported sex differences in gonadectomized animals in high affinity [³H]5HT receptor binding (presumably $5HT_1$ receptors) in POA, with higher binding in males, although castration itself decreased receptor number in septum, hypothalamus, and midbrain, and decreased affinity in POA.[77] However, recent studies with more selective ligands have failed to find a sex difference in intact or gonadectomized animals in either $5HT_{1A}$ or $5HT_{1B}$ binding in POA or any other region,[78] although castration did decrease 8-OH-DPAT binding to $5HT_{1A}$ receptors in POA, whereas T administration to castrated rats increased binding (Figure 1).[78,79] Chronic (14-day) T administration to intact males has also been shown to increase the affinity, but not the number, of $5HT_{1A}$ binding sites in hippocampus, while leaving $5HT_{2A}$ receptor binding unaltered.[61]

Despite recent studies linking $5HT_{1B}$ receptors to aggression in mice,[80] no differences in $5HT_{1B}$ binding were observed between CAST and intact males, or between males and females.[78] However, testosterone given to castrated rats decreased $5HT_3$ receptor binding specifically in the lateral and basolateral amygdaloid nuclei, while having no effect on binding in the posterolateral and posteromedial amygdala.[81]

Castration also induced a progressive decrease in [³H]imipramine ([³H]IMI) binding to cortical and hypothalamic membranes, while increasing binding to hippocampal membranes; testosterone blocked the castration-induced effects but had no effect on its own.[82] Quantitative autoradiography of [³H]paroxetine binding to the 5HT transporter (5HTt) in hippocampus also showed a sex difference in gonadectomized rats, with higher levels of 5HTt binding in castrated males in CA2 and CA4, as well as in the suprapyramidal blade of dentate gyrus as compared to ovariectomized females.[83]

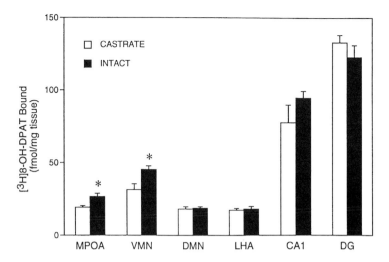

Figure 1 Effect of castration on 5HT$_{1A}$ receptor binding. Binding of [^{3}H]8-OH-DPAT in male rats castrated 7 days prior to sacrifice or left intact. Data expressed as mean ± SEM; n = 8; *p <0.05. MPOA = medial preoptic area; VMN = ventromedial hypothalamic nucleus; DMN = dorsomedial hypothalamic nucleus; LHA = lateral hypothalamic area; CA1 = CA1 subfield of hippocampus; and DG = dentate gyrus. (Adapted from Frankfurt et al., *Neuroendocrinology*, 59, 245, 1994.)

3. 5HT-Mediated Behavior

Several 5HT-mediated behaviors appear to be modulated by gonadal hormones. A sex difference has been observed in the hypothermic and the serum corticosterone (CORT) response to the 5HT$_{1A}$ agonist 8-OH-DPAT with female rats and mice having a greater response than males.[64,84-86] In males, testosterone seems to contain the hypothermic and CORT response to 8-OH-DPAT, since castration potentiated the responses, making them similar in magnitude to the female responses, although the effect on CORT secretion was not seen until five weeks after castration.[84,85] However, another study using animals matched by weight instead of age found no sex differences in 8-OH-DPAT-induced hypothermia, hyperphagia, or 5HT syndrome, although females still had a greater increase in plasma CORT following 8-OH-DPAT.[87]

Other sex differences vary with the experimental conditions used. Although females had greater hypothermic responses and a more pronounced 5HT behavioral syndrome in response to pargyline pretreatment and exogenously administered L-tryptophan, there were no sex differences in the 5HT syndrome induced by 8-OH-DPAT, the 5HT releaser PCA, or the nonspecific 5HT agonist 5MeODMT.[64,65] The sex difference in the pargyline/L-tryptophan response appears to be due to testosterone activation of androgen receptors, as castration eliminates the sex difference in males, and dihydroxytestosterone administration to CAST rats suppresses the behavioral response and restores the sex difference (Figure 2).[88] In contrast, ovariectomy of females and estrogen administration to either sex has no effect.[88]

There also appear to be interactions between gonadal hormones, 5HT, and sexual behavior. The 5HT$_{1B/2C}$ agonists TFMPP and mCPP inhibit copulatory behavior in intact male rats,[89] while having a limited facilitative effect in females (see below). Other receptor subtypes may also play a role in mediating male sexual behavior, as 5HT$_{2A}$ activation inhibits testosterone-dependent sex behavior, while the 5HT$_{2A}$ antagonist, ritanserin, stimulates it.[90] In addition, the agonist DOI inhibits mounting behavior in both males and in neonatally androgenized females.[90] Interestingly, ritanserin also selectively increases anxiety in males, while decreasing it in females, independent of the presence of testosterone in either sex.[90]

Testosterone and serotonin have both been shown to play a role in aggression and the maintenance of dominance in social groups. In dominance induced by chronic T treatment in rats, agonists to 5HT$_{1A}$, 5HT$_{1B}$, or 5HT$_{2A/2C}$ receptors produced a dose-dependent decrease in behaviors associated with dominance, suggesting a negative correlation between aggressive dominance and serotonergic activity.[61] Similarly, dominant male rats housed in mixed-sex groups had higher plasma testosterone concentrations and lower 5HIAA levels and 5HIAA/5HT ratios in several brain regions, as compared to subordinate males.[91] A negative correlation between aggression and CSF 5HIAA levels has also been found in rhesus monkeys,[92] although plasma 5HT levels are higher in dominant male vervet monkeys as compared to

A

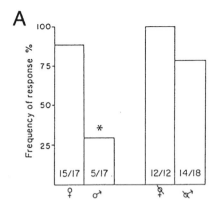

15/17 | 5/17 | | 12/12 | 14/18

♀ ♂ ♀ ♂

B

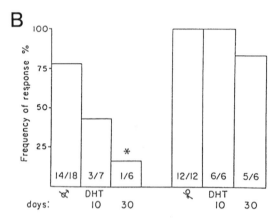

14/18 | 3/7 | 1/6 | 12/12 | 6/6 | 5/6

♂ DHT ♀ DHT

days: 10 30 10 30

Figure 2 Effect of gonadal steroids on the 5HT behavioral syndrome. The 5HT behavioral syndrome was elicited in male and female rats by pargyline (50 mg/kg) and L-tryptophan (50 mg/kg). A sex difference in frequency of response and the effect of gonadectomy are shown in (A) Administration of 30 mg/kg dihydroxytestosterone (DHT) restored the sex difference in male rats, while having no effect on ovariectomized females (B) *p <0.02. (From Fischette et al., *Life Sci.,* 35, 1197, 1984. With permission.)

subordinates.[93] Raleigh and co-workers also showed that dominant monkeys exhibited an enhanced behavioral response to fluoxetine, quipazine, and L-tryptophan.[94] Furthermore, treatment with fluoxetine and L-tryptophan facilitated dominance acquisition in unstable social groupings, but decreased aggression, while the 5HT depleter fenfluramine and the 5HT$_{2A}$ antagonist cyproheptadine increased aggressive behavior but impaired dominance acquisition.[95] These results suggest that, while 5HT activity may be negatively correlated with aggressive behavior, dominance in primate groups is determined by behaviors other than aggression, and 5HT may facilitate these behaviors and promote dominance acquisition under unstable social conditions.

C. ESTROGENS AND PROGESTINS
1. 5HT Levels and Metabolism
a. Adult

As mentioned above, females generally appear to have a higher brain content and synthesis rate of 5HT. The role of female sex steroids in mediating this relatively higher level of 5HT metabolism has been extensively studied; the results of experiments using exogenously administered EB and progesterone (P) are summarized in Table 2. Overall, administration of these hormones appears to increase 5HT activity in most regions, with the exception of the SCN, where metabolism is decreased by estrogen.[99,109]

In the SCN, as well as other hypothalamic regions, 5HT turnover has been shown to follow a diurnal rhythm that may be regulated, in part, by estrogens. In ovariectomized (OVX) rats, a diurnal pattern of 5HT accumulation was found in the SCN, MPN, and ARC, with a higher rate of turnover found during the light phase; no such rhythmicity was apparent in median eminence (ME).[108] Two days of EB induced a diurnal rhythm in the ME, abolished the rhythm in ARC and MPN, and reversed the rhythm in SCN such that turnover became higher during the dark phase.[108] In contrast, no diurnal rhythms were present in either OVX or EB-treated animals in globus pallidus.

Table 2 Effects of Exogenous Estrogens and Progestins on 5HT Levels and Metabolism

Condition	Regions examined	Method	Effects[a]	Ref.
OVX + low P (acute)	SPT, HYP, THAL, HC, CTX, raphe	Fluorescence assay (+/- pargyline)	↓5HT in THAL, CTX; ↑5HT accumulation in SPT, HYP, raphe	96
OVX + high P (acute)			↓5HT in THAL, HC; ↓5HIAA in SPT, HC, CTX, raphe; ↑5HT accumulation in SPT, HYP, raphe (larger than low P)	
OVX + EB (acute)	CTX, POA + SPT, HYP, THAL, STR, PIT, MBR	Synaptosomal uptake of [³H]5HT	NC	97
OVX + P (acute)			↑Uptake in POA + SPT	
OVX + EB + P (acute)			NC	
OVX + EB (acute)	HYP, AMY, CTX, BST	HPLC (+/- pargyline)	↑5HT in raphe only NC in accumulation	98
OVX + EB + P (acute)			↑5HT in raphe only NC in accumulation	
		HPLC: micropunches	↑5HT levels and accumulation in DR	
OVX + 7 day EB	SCN, ME, SON	Column chromatography after [³H]Trp loading	↑5HT levels and ↓5HIAA/5HT ratio in SCN (15 vs. 9 h)	99
OVX + 14 day EB	CTX	HPLC	↓5HT; NC in 5HIAA	100
OVX + 14 day P			NC in 5HT or 5HIAA	
OVX + 14 day (EB + P)			↓5HT; NC in 5HIAA	
OVX + EB (acute	Nuclei from STR, HYP, THAL, HC, AMY, SN, BST	HPLC: micropunches	↑5HT levels in SN and DR NC in 5HIAA	101
OVX + 28 day EB			NC in 5HT or 5HIAA	
OVX + EB (acute)	Nuclei from HYP, AMY, other	HPLC: micropunches (+/- pargyline)	↑5HT and 5HIAA accumulation in MPN, VMN, ACo	102
OVX + P (acute)	STR	HPLC	↑5HIAA; NC in 5HT	103
Intact male + P (acute)			NC	
OVX + EB (acute)	MBH, ME, AHA-POA	HPLC (+/- AADC inhibition)	NC in 5HT synthesis	104
OVX + P (acute)			↑5HT synthesis in AH-POA	
OVX + EB + P (acute)			↑5HT synthesis in AH-POA	
OVX + EB (acute)	Discrete nuclei from HYP, BST, MBR, etc.	HPLC: micropunches (+/- pargyline)	↑5HIAA in DR; NC in 5HT levels or accumulation	105
OVX + varying doses of EB (acute)	Discrete nuclei from HYP, BST, MBR, etc.	HPLC: micropunches	Dose-dependent ↑ in 5HIAA in VMN; NC in 5HT	106
OVX + EB + P (acute)	Discrete nuclei from HYP, BST, MBR, etc.	HPLC: micropunches (+/- pargyline)	↓5HT accumulation in VMN and MCG; NC in 5HT or 5HIAA (vs. OVX + EB)	107
OVX + 2 day EB	ME, SCN, MPN, ARC, GP	HPLC: micropunches	Altered diurnal rhythm of 5HT turnover in ME, SCN, ARC and MPN (see text for details)	108
OVX + EB (acute)	POA, ARC, VMN, SCN, ZI	HPLC: micropunches	↓5HT and 5HIAA in SCN; ↑5HIAA/5HT ratio in ZI	109
OVX + EB + P (acute)			↓5HT and 5HIAA in ZI; ↑5HIAA/5HT ratio in ME and ↓5HIAA/5HT ratio in VMN (vs. OVX); ↓5HT in ME; ↑5HT and 5HIAA in SCN; ↓5HIAA/5HT ratio in ARC (vs. OVX + EB)	
OVX + EB + P (acute)	STR	HPLC	Transient ↑ in 5HT and 5HIAA	110
Intact male + EB + P (acute)			Transient ↓ in 5HT and 5HIAA	

Table 2 (continued) Effects of Exogenous Estrogens and Progestins on 5HT Levels and Metabolism

Condition	Regions examined	Method	Effects[a]	Ref.
OVX + 45 day EB CAST + 45 day EB	HYP, STR, BST, HC	HPLC	↓5HIAA in HYP, BST ↓5HT in HYP ↑5HIAA in HYP, STR	59
OVX + 2 day EB implants in VMN + acute P	Discrete nuclei from AMY, HYP, MBR, etc.	HPLC: micropunches (+/- pargyline)	↓5HIAA/5HT ratio in VMN and PVN; ↓5HT accumulation in MCG, VMN and PVN; NC in 5HT or 5HIAA levels (vs. OVX + EB)	111
OVX + P (acute)	MCG, PVN, VMN		NC in levels or accumulation (vs OVX)	

Note: Regions: ABl = basolateral amygdaloid nucleus; Acc = nucleus accumbens; ACe = central amygdaloid nucleus; ACo = cortical amygdaloid nucleus; AH-POA = anterior hypothalamus-preoptic area; AMY = amygdala; ARC = arcuate nucleus; BST = brainstem; CB = cerebellum; CN = caudate nucleus; CTX = cerebral cortex; DR = dorsal raphe; GP = globus pallidus; HC = hippocampus; HYP = hypothalamus; LHA = lateral hypothalamic area; MBH = mediobasal hypothalamus; MBR = midbrain; MCG = midbrain central gray; ME = median eminence; MPN = medial preoptic nucleus; MR = median raphe; OLF = olfactory tubercle; PIT = pituitary; POA = preoptic area; PVN = periventricular hypothalamic nucleus; SCN = suprachiasmatic nucleus; SN = substantia nigra; SON = supraoptic nucleus; SPT = septum; STR = striatum; THAL = thalamus; VMN = ventromedial hypothalamic nucleus; and ZI = zona incerta.

Other abbreviations: AADC = amino acid decarboxylase; ADX = adrenalectomy; CAST = castrated males; EB = estradiol; F = female; HPLC = high performance liquid chromatography; Ile = isoleucine; Leu = leucine; M = male; NC = no change; OVX = ovariectomized females; P = progesterone; Phe = phenylalanine; TPH = tryptophan hydroxylase; Trp = tryptophan; Tyr = tyrosine; Val = valine; 5HT = serotonin; 5HIAA = 5-hydroxy-indoleacetic acid.

[a] Unless specified, comparisons are vs. animals receiving vehicle only.

Changes in 5HT content have also been examined during natural fluctuations in hormone levels, such as during the estrous cycle and pregnancy. In cerebral cortex, hippocampus, and cerebellum, 5HT levels fell in cortex during pregnancy and remained low, while they rose in hippocampus postpartum; levels of 5HIAA in all three regions also rose postpartum.[112] No changes in 5HT and 5HIAA content were found across the estrous cycle in these regions, however. In contrast, 5HT and 5HIAA levels have been shown to fluctuate across the estrous cycle in discrete nuclei specifically associated with female sexual behavior (Figure 3).[113] The author suggests that 5HT levels decrease during proestrus in regions where 5HT has been shown to inhibit lordosis, such as the VMN, whereas levels increase in regions where 5HT may play a facilitative role, such as midbrain central gray and mPOA. In support of this idea, James et al.[109] found a positive correlation between the 5HIAA/5HT ratio in ME and sexual receptivity, while 5HT and 5HIAA levels in VMN and ARC were negatively correlated with receptivity. In homogenates of whole hypothalamus, however, the 5HIAA/5HT ratio was found to be higher in sexually receptive rats.[114]

Variations in synaptosomal uptake of [³H]5HT across the estrous cycle have also been observed, with a transient increase in uptake in the SCN, POA, ME, and amygdala at 12 h of proestrus, just prior to the luteinizing hormone (LH) surge.[115] In addition, during diestrus II, 5HT uptake was elevated in amygdala at 12 h and in POA at both 12 and 19 h. Lesions with 5,7-DHT in all four brain regions disrupted cyclicity, suggesting that 5HT is an important mediator of the estrous cycle in rats.

b. Developmental Effects
The effect of estrogen and progesterone on the 5HT system also varies during development. In pubertal rats, the concentration of 5HT and the 5HIAA/5HT ratio are higher in the ARC and ME than in adults.[116] In addition, administration of EB and P to OVX prepubertal (16-day-old) rats induced a decrease in 5HT and 5HIAA levels in the anterior hypothalamus preoptic area, but increased levels in pubertal (30-day-old) rats, suggesting that the responsiveness to gonadal hormones changes with development and sexual maturation.[117]

2. 5HT Receptors
In addition to altering 5HT metabolism directly, sex steroids also modulate neuronal sensitivity to 5HT via changes in 5HT receptors. Early studies in our laboratory showed an effect of estrogen on high-affinity [³H]5HT binding in gonadectomized male and female rats, with short-term administration of

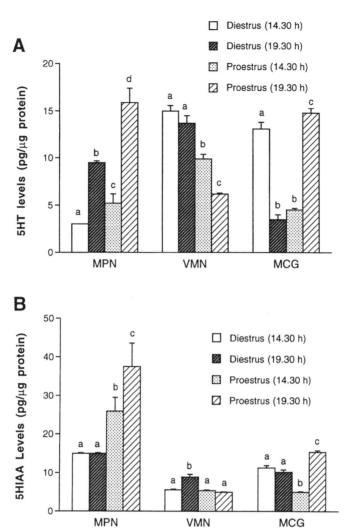

Figure 3 Serotonin levels during the estrous cycle. Levels of 5HT (A) and 5HIAA (B) were measured by HPLC in discrete brain regions involved in lordosis behavior. Values are pg/μg protein ± SEM; n = 8 to 9. Differences between groups tested by one-way ANOVA followed by Newman-Keuls analysis. Entries labeled with different letters are different from each other; $p < 0.05$. MPN = medial preoptic nucleus; VMN = ventromedial hypothalamic nucleus; and MCG = midbrain central gray. (Adapted from Luine, V. N., *Neuroendocrinology*, 57, 946, 1993.)

estrogen increasing binding in females in lateral POA, anterior hypothalamus, lateral septum, and amygdala, while in castrated males, estrogen increased receptor numbers in anterior hypothalamus and dorsal raphe and decreased them in mPOA and midbrain central gray, with no change in affinity.[77,118,119] The effect of estradiol of OVX female was biphasic, however, with a transient decrease in $5HT_1$ binding throughout the brain followed by a selective increase in brain regions rich in estrogen receptors.[118] In contrast, Williams and Uphouse found a decrease in $5HT_1$ receptor number and K_d in hypothalamic and cortical membranes following acute estrogen treatment[120] while others report no effects at all.[121] A recent study in our laboratory using quantitative autoradiography also found no effects of either estrogen or progesterone on $5HT_{1A}$ receptor binding, although EB and P administration to OVX rats did increase $5HT_{1B}$ binding in VMN as compared to OVX plus EB alone.[78] A lack of effects in binding does not preclude other functional alterations, however, as Clarke and Maayani found that although four-day estrogen treatment did not alter the number or affinity of $5HT_{1A}$ receptors on hippocampal membranes, it enhanced the ability of 5HT to inhibit adenylate cyclase activity.[122]

The effects of ovarian steroids on 5HT receptors also appear to vary with duration of treatment. Treatment of OVX females with estrogen and progesterone, individually or concurrently administered for one week, decreased $5HT_1$ receptor number in cerebral cortex to a similar extent.[100] The effects on [³H]spiroperidol binding to $5HT_{2A}$ receptors were more complex, as either EB or P treatment increased $5HT_{2A}$ receptor number, while they had no effect when administered together.[100] In contrast, 28 days of estrogen administration to intact rats decreased $5HT_1$ receptor number in hippocampus, but not cortex, while either 10 or 28 days of estrogen had no effects on $5HT_{2A}$ binding.[123,124]

In intact, cycling rats, endogenous ovarian hormones also modulate 5HT receptor binding. Binding sites for [³H]5HT were decreased during proestrus and estrus in basal forebrain membranes, although no effects were seen in hippocampus, cortex, or caudate.[125] In contrast, other investigators found the highest binding in frontal cortex during estrus, with the lowest levels of binding occurring during proestrus and intermediate binding seen in diestrus.[126] However, the time of sacrifice is also important, because during proestrus itself, [³H]5HT binding in hypothalamus and cortex increases from 14 to 24 h, then decreases to early proestrus levels by 14 h of estrus in hypothalamus, but not cortex.[120,126] Individual receptor subtypes are differentially regulated during the estrous cycle, as no variations in [³H]ketanserin or [³H]spiroperidol binding to $5HT_{2A}$ receptors were observed.[126]

In addition to effects on post- and presynaptic 5HT receptors, estrogen and progesterone affect binding to the 5HT transporter, which may reflect increases in presynaptic reuptake of the transmitter. While one study reported a large ovariectomy-induced increase in [³H]IMI binding in striatum, hippocampus, and hypothalamus, with no effects in cortex or brainstem,[127] others reported no effects of OVX in either cortex or hypothalamus.[128,129] Estrogen treatment for 10 or 12 days did, however, increase IMI binding and synaptosomal [³H]5HT uptake in frontal cortex and hypothalamus of OVX rats.[124,128] The response to estrogen may be dose dependent as Wilson and co-workers showed that in both intact and OVX plus EB rats, IMI binding in hypothalamic membranes had a positive correlation with plasma estradiol levels below 40 pg/ml, but a negative correlation at higher estradiol levels.[129] In addition, progesterone effects on [³H]IMI binding were also dependent on estrogen levels, as binding was positively correlated with progesterone levels, but only when estrogen was low (<12 pg/ml).[129] No significant changes were observed throughout the estrous cycle, however, although IMI binding tended to increase during the dark phase at every stage of the cycle.[128,129]

In contrast to its stimulatory effect on 5HTt binding in the hypothalamus, estradiol downregulates uptake sites in hippocampus. One week of estradiol treatment in gonadectomized male and female rats decreased [³H]paroxetine binding in stratum oriens of CA1-CA4 and lacunosum moleculare of CA2 and CA4; no changes in 5HTt binding in parietal cortex were noted in this study, however.[83]

Interestingly, estrogen has been shown to alter binding of [³H]IMI *in vitro* as well as *in vivo*. Estradiol and other estrogen derivatives at relatively high concentrations (1 to 10 μM) inhibit [³H]IMI binding and [³H]5HT uptake *in vitro* in both rat brain and in human platelets,[127-130] although at physiological levels (0.13 nM = 35 pg/ml), estradiol increases [³H]IMI binding.[129] Progesterone effects *in vitro* were modulated by estrogen, with progesterone increasing binding only in the absence or with low levels of estradiol; the *in vitro* effects at physiological concentrations mimic those seen following estrogen and progesterone treatment *in vivo*.[129]

3. Electrophysiology

The effect of estrogen on neuronal responsiveness to 5HT has been examined in a few different regions. Estradiol generally enhances the inhibitory effects of 5HT on these neurons. The resting firing patterns of neurons in the VMN were similar in OVX rats with and without estrogen replacement for at least seven days. Microinjection of 5HT onto these slices was generally inhibitory, and a larger percentage of units from OVX plus EB animals, compared to OVX, responded to 5HT application; responses to norepinephrine and acetylcholine, but not other transmitters, were also enhanced by estrogen treatment.[131] In contrast, using an identical treatment paradigm, the same group found no effect on responsiveness in neurons of the arcuate nucleus, suggesting regional specificity of estrogen effects.[132]

In CA1 neurons of the hippocampus, the decrease in evoked population spike amplitude in response to perfused 5HT and 5-carboxamidotryptamine (5CT) has been shown to be mediated by the $5HT_{1A}$ receptor.[24] Three to six days of estrogen replacement in OVX rats enhanced the ability of 5CT to decrease the CA1 population spike amplitude, shifting the mean EC_{50} for 5CT from 76 nM in vehicle-treated rats to 19 nM with estrogen treatment, without affecting the maximal response.[133] The magnitude of hyperpolarization and the decrease in membrane resistance elicited by perfused 5HT were also enhanced by estrogen replacement, while the resting membrane properties of the pyramidal cells did not differ between OVX and OVX plus EB in the absence of 5HT.[134] These electrophysiological results fit well with the studies on 5HT inhibition of adenylate cyclase activity,[122] suggesting that estrogen enhances responses mediated by 5HT, and specifically $5HT_{1A}$ receptors, in the hippocampus.

4. 5HT-Mediated Behavior

The 5HT system differentially modulates sexual behavior in males and females, presumably through interactions with gonadal steroids. For example, in ovariectomized female rats, the $5HT_{1B/2C}$ agonists TFMPP and mCPP facilitate female sexual behavior (lordosis), but have no effect on OVX rats which have received EB + P; this differs from the effects of these drugs in males (see above). In intact male rats, 5,7-DHT lesion of serotonergic afferents to the hypothalamus facilitates the induction of lordosis after priming with EB with or without P co-administration.[135] However, the effects of 5HT on hormone-mediated sexual behavior in females is controversial, as lesions of hypothalamic 5HT inputs facilitate lordosis induced by EB, but not EB + P,[135,136] while inhibition of 5HT synthesis with PCPA reduces sexual behavior induced by EB + P, but not EB alone[114] (for a detailed review of 5HT and lordosis, see Reference 137).

Head twitches induced by 5HTP, generally believed to be mediated by $5HT_{2A}$ receptors,[34] are also modulated by estrogens and progestins, but not androgens. Pretreatment of female mice with a variety of estrogens reduced the head twitch response to 5HTP administered 4 h later; similar reductions were seen in mice treated with both estrogen and progesterone (Figure 4).[138] In contrast, progesterone alone increased the head twitch response. Another study supports a progesterone-induced increase in $5HT_{2A}$ receptor activity. In OVX rats, sexual behavior in animals primed with EB and P was decreased by PCPA-induced depletion of 5HT; this effect was reversed by 5HTP.[114] Moreover, in rats primed only with high doses of E, neither PCPA nor 5HTP had any effect on lordosis, suggesting that 5HT only played a role in sexual receptivity induced, in part, by progesterone. Since the $5HT_{2A}$ agonist MK 212 also stimulated lordosis, the authors suggest that progesterone may increase sexual receptivity by increasing $5HT_{2A}$ receptor activation.[114]

The effects of testosterone on the sex differences in 8-OH-DPAT-induced hypothermia and CORT secretion are discussed above. In females, OVX diminished the sex differences, making the hypothermic and CORT responses smaller, an effect reversed by administration of estrogen to OVX mice.[84,85] Estrogen also enhanced the CORT response in castrated males.[85] Interestingly, the restoration of both responses is blocked by administration of testosterone to the OVX plus EB mice.[84,85] The hypothermic response to 8-OH-DPAT did not vary across the estrous cycle, although there was cyclicity in the hyperphagic response to 8-OH-DPAT, with the effect most pronounced during diestrus and less evident during proestrus and estrus.[86] Treatment of OVX rats with estradiol also inhibited hyperphagia, while proges-terone had no effect, suggesting that fluctuating estradiol levels during the estrous cycle may be respon-sible for the variations seen in the hyperphagic response to 8-OH-DPAT.[139] Since the hypothermic response is thought to be mediated by postsynaptic $5HT_{1A}$ receptors,[35] while the hyperphagia is attributed to activation of somatodendritic autoreceptors,[36,37] estradiol may differentially regulate these two $5HT_{1A}$ receptor populations throughout the estrous cycle.

D. SUMMARY AND CONCLUSIONS

Overall, androgens tend to inhibit central 5HT activity while estrogens enhance it. Females are generally reported to have higher 5HT and 5HIAA levels and higher 5HT synthesis and turnover rates than males, although not all studies concur on this. Circulating gonadal steroids seem to be important in maintaining the sex differences, as testosterone reverses castration-induced increases in 5HIAA, whereas estrogen and progesterone increase 5HT metabolites in ovariectomized rats. Males also have reduced behavioral responsiveness of the 5HT system compared to females, an effect that is eliminated by castration but restored by endogenous androgens. In male rat groups, dominance associated with elevated testosterone levels and a high incidence of aggression are negatively correlated with 5HT activity, suggesting an inhibitory effect of testosterone on 5HT function.

In contrast, the enhancement of 5HT function by ovarian steroids has been assessed at several different levels. Not only are 5HT and 5HIAA levels generally increased by estrogen and progesterone treatment, electrophysiological responses to 5HT and 5HT analogs are also facilitated by estrogen, as are $5HT_{1A}$-mediated second messenger responses. Estrogens play a role in maintaining natural rhythms in brain 5HT content, as they have been shown to modulate diurnal cyclicity of 5HT in addition to changing 5HT and 5HIAA across the estrous cycle; 5HT input to the hypothalamus and limbic areas, in turn, is required for maintenance of hormonal changes in the estrous cycle.

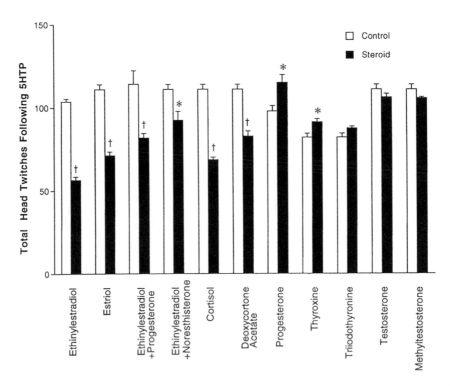

Figure 4 Steroid effects on 5HTP-induced head twitches. Animals were pretreated with steroid or vehicle 4 h prior to administration of 5HTP (200 mg/kg, i.p). Steroid doses are as follows: ethinylestradiol = 0.025 mg/kg; estriol = 0.25 mg/kg; ethinylestradiol + progesterone = 0.25 mg/kg E + 125 mg/kg P; ethinylestradiol + norethisterone = 0.025 mg/kg E + 1.5 mg/kg NE; cortisol = 1 mg/kg; deoxycortone acetate = 1 mg/kg; progesterone = 125 mg/kg; thyroxine = 0.005 mg/kg; triiodothyronine = 0.001 mg/kg; testosterone = 50 mg/kg; methyltestosterone = 50 mg/kg. Head twitches were counted over alternate 2-min intervals for 42 min following 5HTP administration. Total number of head twitches for control and steroid-treated animals were compared by Student t-test; $*p < 0.05$, $†p < 0.005$. (Adapted from Brotherton, C. S. and Doggett, N. S., *Psychopharmacology*, 58, 145, 1978.)

The reported effects of androgens, estrogens, and progestins on 5HT receptors are often contradictory and do not necessarily correspond to changes in receptor-mediated behaviors. Testosterone tends to selectively increase $5HT_{1A}$ binding while reducing that of $5HT_3$ receptors and the 5HT transporter, although the behavioral consequences of these changes are unknown. In fact, the reduced behavioral sensitivity to $5HT_{1A}$ agonists in males compared to females contradicts what would be predicted by reports of comparable levels of $5HT_{1A}$ receptor binding between the sexes. The differences in responsiveness could be accounted for by tighter receptor-effector coupling in females, resulting in an enhanced reactivity in females to an equivalent stimulus. These theories are beginning to be addressed by several groups looking at steroid effects on the electrical properties of neurons and on second messenger systems.

The interactions between estrogen and progesterone appear to be quite complex in some cases and merit further attention as well. For example, although estrogen and progesterone individually increased $5HT_{2A}$ receptor binding, they had no effect when given together. In contrast, $5HT_{2A}$-mediated behaviors are decreased by estrogen alone, or by both steroids together, while progesterone alone enhances these behaviors. The regulation of progesterone receptors by estrogen, and vice versa, may contribute to the complexity of the 5HT responses to these steroids.

IV. ADRENAL HORMONES AND STRESS

A. RECEPTORS AND THEIR DISTRIBUTION

Studies in our laboratory showed that radiolabeled corticosterone was selectively retained by limbic structures in the brain,[140,141] where it was localized to the cell nuclei,[142] indicating an intracellular receptor site. Subsequent work has identified two different types of adrenal steroid receptor: the mineralocorticoid receptor (MR), also known as the type I glucocorticoid receptor, and the glucocorticoid receptor (GR), or type II receptor[143,144] (for reviews see References 145 and 146). Mineralocorticoid receptors have a high affinity for the mineralocorticoid, aldosterone (ALDO), as well as the glucocorticoids, corticosterone (CORT) and cortisol. Glucocorticoid receptors, on the other hand, have a lower affinity for CORT and a much lower affinity for ALDO. It is believed that in the brain MRs are saturated by basal levels of circulating glucocorticoids while GRs become occupied only when CORT levels are very high, such as during the circadian peak and in response to stress.

The two receptor systems have distinct anatomical distributions. Studies measuring mRNA localization and receptor binding have shown that GR binding and message levels are distributed widely throughout the brain, with higher densities in hippocampus, septum, PVN, ARC, and supraoptic nucleus (SON).[147-152] Immunocytochemical studies have also shown GR immunoreactivity in monoaminergic nerve cell bodies, including 5HT neurons in the dorsal raphe and other brainstem cell groups.[153,154] In contrast, the distribution of MRs is more restricted, with binding and mRNA highest in hippocampus, septum, amygdala, olfactory nucleus, layer II of cerebral cortex, and brainstem sensory and motor neurons.[148,151,155] In the hippocampus, both MR and GR mRNA have been shown to be negatively regulated by adrenal steroids during normal circadian rhythms, following surgical and pharmacological manipulations, and following chronic stress.[150,156-158]

B. ADRENAL HORMONES
1. 5HT Levels and Metabolism
a. Adult
Early work by Azmitia and McEwen demonstrated that tryptophan hydroxylase activity in the rat midbrain was inhibited by adrenalectomy (ADX) and augmented by injection of corticosterone.[159-161] The tryptophan hydroxylase protein was also increased in ADX rats given the synthetic glucocorticoid, dexamethasone (DEX).[162] Two hours following ADX, 5HT turnover, measured in several different ways, was decreased in hypothalamus, hippocampus, and brainstem; the effect of ADX was prevented by administration of either CORT or DEX.[163] Interestingly, the levels of 5HT and 5HIAA were largely unaltered despite the decreases in turnover.[163] Similar results were found by De Kloet and colleagues, as 5HT accumulation was decreased in hippocampus 1 h post-ADX,[164] while administration of CORT either before or immediately after ADX restored the 5HT accumulation rate.[164,165] Although in these studies neither DEX nor aldosterone prevented the ADX-induced decrease in 5HT accumulation, pre-treatment with either steroid blocked the restorative effects of CORT.[164,165]

Others have found regionally selective effects of glucocorticoids on 5HT and 5HIAA levels, as 5HT and 5HIAA were decreased in PVN and 5HT was decreased in POA seven days after ADX, while levels of both indoles were increased in SON.[166] Hormone replacement with CORT implants reversed the ADX effects in SON and POA, but not PVN, while decreasing 5HIAA levels in stria terminalis.[166]

b. Developmental Effects
Adrenal steroids may also play a role in the development of the 5HT system. The normal increase in tryptophan hydroxylase activity in neonatal rats has been shown to be blocked by ADX and augmented by CORT injections.[167] In adult mice, however, this group found no effect of either ADX or CORT on tryptophan hydroxylase activity, although ADX did block the increase in enzyme activity induced by reserpine.[167] Another study stimulated CORT secretion in female rat pups via ACTH injections on postnatal days 1 through 7 and found that 5HT uptake in hypothalamus was significantly increased in the ACTH-treated animals both after 7 days of treatment and through adulthood.[168] Although estradiol and progesterone levels were not altered by the ACTH treatment, the treated animals showed delayed sexual maturation and deficits in sexual behavior in adulthood,[168] suggesting that ACTH-induced alterations in serotonergic transmission may have affected the development of reproductive behaviors.

2. 5HT Receptors
Different 5HT receptor subtypes appear to be selectively regulated by adrenal steroids. In subiculum, dentate gyrus, substantia nigra, and dorsal raphe, [^{3}H]5HT binding was increased 1 h after ADX; physiological doses of CORT attenuated the ADX effects and decreased binding in CA1 and presubiculum, while aldosterone had no effect.[169] Subsequent studies have shown that both the $5HT_{1A}$ and $5HT_{1B}$ subtypes are upregulated following ADX: $5HT_{1A}$ binding was increased in CA2, CA3, and CA4 of hippocampus one week after ADX, while both $5HT_{1A}$ and $5HT_{1B}$ binding was increased in dentate gyrus.[170] Administration of physiological levels of CORT decreased $5HT_{1A}$ binding in ADX animals to

control levels while high levels of CORT further decreased binding in dentate gyrus.[170] In ADX rats, $5HT_{1B}$ levels were also decreased in dentate by both low and high doses of CORT.[170] Studies of intact rats have shown that high levels of CORT decrease binding within 48 h in CA4 and dentate with subsequent ADX or termination of CORT administration reversing the CORT effect, resulting in an overcompensatory increase in binding in several regions.[171]

The receptor binding data correlate well with more recent analysis of $5HT_{1A}$ receptor mRNA levels. After either one day or one week ADX, $5HT_{1A}$ mRNA levels were increased throughout the hippocampus and dentate gyrus, with levels increased to a greater extent at the later time point.[172] The increase in mRNA was most pronounced in CA2, CA3, and CA4, intermediate in dentate gyrus, and somewhat lower in CA1. Binding of 8-OH-DPAT to $5HT_{1A}$ receptors also increased in the ADX animals, although the magnitude of the increase was smaller and did not show the regional variability observed in the mRNA levels.[172] Administration of CORT at the time of ADX prevented the increases in both mRNA expression and receptor binding.[172] A similar study found a negative correlation between $5HT_{1A}$ mRNA in dentate gyrus and plasma CORT levels, although the relationship was logarithmic, suggesting a synergistic effect when both MR and GR are occupied by high levels of CORT.[173] In contrast, $5HT_{1A}$ receptor mRNA were found to be decreased in dentate gyrus 8 weeks after ADX;[174] the observation that acute administration of DEX 24 to 72 h prior to sacrifice reversed the ADX effect suggests that the decrease is not due simply to the loss of granule cells seen following long-term ADX.[175]

Levels of $5HT_{1A}$ receptor binding and mRNA have also been shown to correlate with the responsiveness of the hypothalamic-pituitary-adrenal (HPA) axis in three different strains of rat. The Fischer F344/N rat has an exaggerated HPA axis response to stress, whereas the Lewis rat is hyporesponsive, and the Harlan Sprague-Dawley rat lies somewhere in between.[176] Lewis rats had decreased $5HT_{1A}$ binding sites and mRNA levels in hippocampus and frontal cortex when compared to the other two strains, while hippocampal 5HT levels were lower in Lewis and Sprague-Dawley rats compared to the Fischer strain.[176] Adrenalectomy increased $5HT_{1A}$ binding and mRNA in all three strains by approximately the same proportion.[176] The authors of this study suggest that hippocampal $5HT_{1A}$ receptors may play a role in mediating the differences in HPA responsiveness seen in these three strains.

In addition to modulating $5HT_{1A}$ receptors, adrenal steroids have also been shown to affect the $5HT_{2A}$ receptor subtype. Stimulation of CORT secretion by ACTH administration has been shown to increase cortical $5HT_{2A}$ receptor binding sites; similar effects were observed following 10 days of either CORT or DEX administration.[177,178] The effects of ACTH were blocked by ADX, although ADX had no effects on its own.[177] An increase in $5HT_{2A}$ function was also reflected by an enhancement of $5HT_{2A}$-mediated wet-dog shake behavior.[177] Glucocorticoids may potentiate $5HT_{2A}$ receptor function directly, in addition to increasing receptor numbers, as DEX has been shown to increase $5HT_{2A}$-mediated mobilization of Ca^{2+} in C6 glioma cells.[179]

The effects of glucocorticoids on the binding and function of the 5HT transporter have also been examined. One day after ADX, uptake of [^{3}H]5HT was decreased in platelets and in hypothalamic synaptosomes, although uptake in cerebral cortical synaptosomes remained unchanged.[180] Administration of CORT 2 h prior to sacrifice reversed the ADX effect and increased uptake in platelets and hypothalamus in sham-operated rats, while binding of [^{3}H]imipramine was unaffected by any manipulation.[180] In contrast, another group found that ADX increased both the Kd and B_{max} of imipramine binding in platelets but not in brain.[181] Seven days of CORT treatment had the opposite effect, leading to a decrease in platelet Kd and B_{max}, as well as decreasing B_{max} of imipramine binding sites in frontal cortex and hypothalamus.[181] These studies suggest that adrenal steroids are capable of modulating 5HT uptake in both the brain and in the periphery.

3. Electrophysiology

Several studies by Joëls and De Kloet have examined the effects of adrenal steroids on neuronal properties in hippocampal slices from ADX rats. Selective occupation of MRs has been shown to suppress the $5HT_{1A}$-mediated hyperpolarization and resistance decrease in CA1 pyramidal neurons, thus reducing the inhibitory effect of 5HT on neuronal excitability.[182] This effect appears to require protein synthesis, as incubation with cycloheximide blocked the MR modulation of 5HT-induced hyperpolarization, in addition to preventing the GR-mediated increase in afterhyperpolarization following an electrical stimulus.[183] Interestingly, the effects of MR activation could be prevented by concurrent application of the selective GR agonist RU 28362, even though the GR agonist had no effect on its own (Figure 5).[184] Similar results were obtained by perfusing the slice with CORT. At low doses, which would primarily occupy MRs,

Figure 5 Electrophysiological responses of CA1 pyramidal neurons to 5HT. In hippocampal slices from one week ADX rats, application of 30 μM 5HT elicited a response (A), while a 20-min application of aldosterone markedly inhibited these responses (B). Concurrent application of the GR agonist RU 28362 with aldosterone prevented the inhibitory effect seen with aldosterone alone (C). Negative voltage deflections are the response to constant current pulses (0.3 nA, 50 ms, 0.2 Hz). If 5HT induced a large hyperpolarization, membrane potential was temporarily returned to pretreatment level by DC injection to calculate the change in input resistance. (From Joëls, M. and De Kloet, E. R., *Neuroendocrinology*, 55, 334, 1992. With permission.)

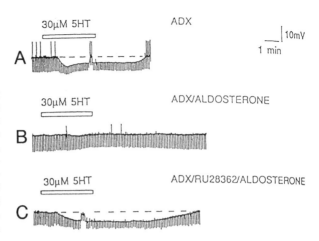

CORT attenuated the 5HT-induced hyperpolarization and resistance decrease, whereas at high doses, sufficient to occupy GRs, the response to 5HT was only transiently decreased.[184] Pyramidal cells in slices from sham-operated rats with relatively high levels of plasma CORT exhibited responses to 5HT similar to those seen in ADX rats, suggesting that GR occupation may antagonize MR effects *in vivo* as well as *in vitro*.[184]

In addition to regulating responses to 5HT, glucocorticoids may also interact with 5HT in modulating input to the hippocampus. Hippocampal theta rhythms can be induced by electrical stimulation of the septum; ADX, CORT administration to intact rats, and destruction of hippocampal 5HT input with 5,7-DHT all change the minimum threshold current for inducing theta rhythm from 7.7 to 6.9 Hz.[185,186] The two systems appear to be intimately connected as CORT can reverse the shift in both ADX and in 5,7-DHT-lesioned animals.[187] In addition, no shift in threshold is seen in ADX rats with 5,7-DHT lesions, although the change to 6.9 Hz can be induced by administration of CORT to these animals.[187] This suggests that 5HT and CORT may act upon common neurons to regulate theta rhythm activity in the hippocampus.

4. 5HT-Mediated Behavior

Glucocorticoid-induced alterations in $5HT_{1A}$ receptor-mediated behavior are consistent with the changes observed in receptor binding. The hypothermic response to 8-OH-DPAT was reduced in mice and rats after 3 days or 10 days of CORT treatment and increased in male rats following ADX, although the 5HT behavioral syndrome induced by 8-OH-DPAT was unaffected.[188,189] Plasma prolactin responses to 8-OH-DPAT and the $5HT_{2C}$ agonist mCPP were also reduced in rats given cortisol for 7 days.[190] $5HT_{1B}$-mediated responses were not changed by CORT treatment.[188]

The effects of adrenal steroids on $5HT_{2A}$-mediated responses are not quite so consistent with the binding data. As mentioned above, glucocorticoids have been shown to increase receptor binding and $5HT_{2A}$-mediated wet-dog shakes.[177,178] Adrenalectomy and corticosterone synthesis inhibition have been shown to block the effects of DOI on glucose metabolism in the hippocampus, suggesting a positive correlation between glucocorticoids and $5HT_{2A}$ function.[191] In contrast, acute treatment with cortisol and deoxycortone reduced 5HTP-induced head twitches in mice (Figure 4),[138] thought to be mediated by $5HT_{2A}$ receptors, while another group found no effect following 3- or 10-day CORT treatment.[188] In guinea pigs, cortisol also inhibited 5HTP-induced myoclonus and abolished diurnal variations in behavioral sensitivity to 5HTP, although after 4 weeks of treatment a subgroup of animals developed a behavioral hypersensitivity to the drug, possibly resulting from changes in postsynaptic receptor systems.[192]

Behavioral responses to nonselective 5HT agonists are also altered by CORT. In animals given CORT for 4 days, responses to the 5HT releaser PCA and the nonselective agonist 5MeODMT were decreased compared to controls, although 5HT and 5HIAA concentrations in hindbrain and striatum were comparable between the two groups.[193] Open field activity was greater than controls 20 h after 1 day of CORT, but was significantly reduced 20 h after 4 days of treatment,[193] suggesting a biphasic effect of CORT on anxiety.

5. Summary of Adrenal Steroid Effects

The many studies of ADX animals suggest that adrenal steroids have a permissive role in modulating 5HT synthesis and metabolism in the brain, such that 5HT activity is diminished in the absence of circulating glucocorticoids. However, function of the $5HT_{1A}$ receptor system, particularly in the hippocampus, is enhanced following ADX and reduced in the presence of adrenal steroids. The glucocorticoid effects on $5HT_{1A}$ receptors have been consistently shown at several different levels. Biochemically, $5HT_{1A}$ receptor binding and mRNA levels are increased in hippocampus following one week ADX, while high levels of CORT have the opposite effect;[170-172] the nonlinear negative correlation between plasma CORT and $5HT_{1A}$ mRNA levels suggests that concomitant activation of MR and GR results in a synergistic effect on receptor expression.[173]

In addition to the direct effects on receptor binding and expression, glucocorticoids reduce the electrophysiological actions of $5HT_{1A}$ receptors as well. The activation of MRs in hippocampal slices from ADX results in an attenuation of the $5HT_{1A}$-mediated hyperpolarization, an effect that is blocked by simultaneous activation of GRs.[182,184] Although binding properties of $5HT_{1A}$ were not changed by the 20-min exposure to CORT or aldosterone in these experiments, it is possible that these hormones may have changed the coupling of the receptors to their second messenger systems.[182] Support for this comes from experiments by Saito and co-workers, who showed that CORT given to either intact or ADX rats decreased the immunoreactivity and mRNA levels of $G_{i\alpha}$, the G protein to which the $5HT_{1A}$ receptor is linked.[194] Harrington and Peroutka, in turn, have suggested that the $5HT_{1A}$ receptor may have the capacity to bind 5HT only when coupled to $G_{i\alpha}$.[195] Thus, by reducing $G_{i\alpha}$, glucocorticoids may reduce $5HT_{1A}$ responses by decreasing the efficacy of both signal transduction and ligand binding.

Behavioral evidence for CORT-induced inhibition of postsynaptic $5HT_{1A}$ responses comes from studies in which the hypothermic response to 8-OH-DPAT was reduced by CORT and enhanced by ADX.[188,189] These results suggest a tonic inhibition of 5HT-mediated responses by basal levels of CORT, with disinhibition occurring when CORT levels are high, such as during the circadian peak or following stress.

In contrast, glucocorticoids have the opposite effect on $5HT_{2A}$ function, as $5HT_{2A}$ receptor binding and function have been shown to be increased by CORT administration,[177-179] although the behavioral effects are variable (see above). Binding to the 5HTt was also increased by CORT,[181] suggesting an increased rate of presynaptic reuptake, possibly in response to increased transmitter release. By differentially regulating both presynaptic and postsynaptic neuronal responsiveness, glucocorticoids appear to play an important role in fine-tuning the function of the 5HT system.

C. STRESS

1. 5HT Levels and Metabolism

The effects of stress on 5HT levels and metabolism have been studied using a variety of different stressors; the results of these studies can be found in Tables 3 and 4. Perhaps the most consistent finding with acute stress is an elevation in 5HIAA levels and the 5HIAA/5HT ratio in the brain, particularly in limbic regions[63,68,69,197-199,201,203,207-215,218,220] although others have found no change[63,202,206,219] or mixed effects in different brain regions;[201] 5HIAA levels have also been shown to increase,[63,91,222,223] decrease,[228,230] or remain unchanged[225,233] following chronic stress. Whether these effects reflect an increase in 5HT synthesis and release is unclear, however, as the stress-induced changes in 5HT levels and synthesis rates are far more variable, with increases,[68,196,197,211,215,219] decreases,[160,200,205,207] and no effects of stress reported,[69,198,202,206,219,220] even when similar stress paradigms were used. Several studies using microdialysis or brain slices to measure 5HT release directly have shown increased 5HT activity in response to stress,[212,215-217,221,229] although chronic isolation and repeated forced swim stress decreased basal 5HT activity.[233]

Sex differences in the 5HT response to stress have been observed, with females generally having a larger increase in 5HT, 5HIAA, and 5HIAA/5HT ratios than males,[68,69] although studies using acute immobilization found the stress-induced increase of 5HIAA was blunted in the frontal cortex of females compared to males.[66] In addition, repeated immobilization caused a decrease in 5HIAA in cortex, hippocampus, and hypothalamus, and an increase in 5HT in hypothalamus in females, while the only change seen in males was an increase in cortical 5HT.[66] Other studies suggest a role for gonadal hormones in modulating stress effects on the 5HT system, as estrogen administration in OVX rats prevented the decrease in 5HT levels in hippocampus, striatum, hypothalamus, and POA caused by 24 h of intermittent footshock.[226]

Table 3 Effects of Acute Stress on 5HT Levels and Metabolism

Condition	Regions examined	Method	Effects	Ref.
Footshock: 6x10 min over 3 h	BST-mesencephalon	Measured 5HT and [³H]5HT in extracts following [³H]Trp injection	↑ Both endogenous 5HT and [³H]5HT formation from [³H]Trp	196
Food withdrawal: 24 h Immobilization: 3 h (+/- ADX)	Whole brain	Organic extraction	↑↑ in brain Trp, 5HIAA; smaller ↑ in 5HT; greater effect with food withdrawal NC in stress effect with ADX	197
Footshock: 3 × 3 s Ether anesthesia: 3x30 s Cold: 4°C for 5, 8, 48 h (+/- ADX)	MBR, forebrain	TPH enzyme assay	↑ TPH activity in MBR blocked by ADX	160
Immobilization: 30–300 min	CTX, STR, HC, BST, HYP-THAL	Organic extraction (+/- probenicid)	↑ 5HIAA in HC, STR, BST, CTX; ↑ 5HT accumulation in CTX; NC in 5HT in any region	198
Immobilization: 3 h (+/- Val +/- Trp, Tyr)	HC	*In vivo* voltammetry	↑ Peak corresponding to released 5-hydroxyindoles effect ↓ by Val, further ↓ by Val + Tyr; restored by Val + Trp	199
Induction of learned helplessness with random footshock	Frontal CTX	Push-pull perfusion and HPLC	Had ↓ 5HT in learned helpless rats compared to nonhelpless rats	200
Ether exposure: 2 × 1.5 min	SCN, PVN, SON, DMN and ARC	HPLC: micropunches	↓ 5HIAA/5HT ratio in PVN, SON and caudal ARC; ↑ 5HIAA/5HT in SCN and rostral ARC	201
Immobilization: 1 h (+/- Lesion of ACe)	PVN, VMN, DMN, AHA, ABl, ACo; AMe	HPLC: micropunches	NC in 5HT, 5HIAA in intact rats; stress reversed ACe lesion-induced ↑ in 5HT, 5HIAA in PVN, AHA and DMN with further ↓ in VMN; NC in lesion-induced ↑ in AMe, ABl ACo	202
Passive avoidance training (footshock)	CTX, HYP, BST	HPLC	↑ 5HIAA/5HT in BST with exposure to apparatus and training; ↑ Trp in BST in learners only	203
Immobilization: 2 h	Whole brain and blood	Measured amino acid concentrations	↑ Free Trp in blood; total Trp, Phe, Val, Leu, Ile ↓ in blood (large neutral amino acids) but ↑ in brain	204
Shock-induced fighting Air puff: 40 min Immobilization: 1 h Footshock: 15 min	DR, MR, STR, HC	HPLC: micropunches	↓ 5HT in all regions with footshock; ↓ 5HT in DR, STR with fighting; ↓ 5HT in MR, HC with air puff; ↓ 5HT, in STR, HC with immobilization	205
Conditioned fear: 3, 12 min Immobilization: 20 min Cold water: 20 min Footshock: 2, 12, 22 min	HYP	HPLC	NC in 5HT or 5HIAA with any stressor	206
Footshock: 15, 30 min	CTX, HYP, BST	HPLC	↓ 5HT in CTX, HYP after 15 but not 30 min; ↑ 5HIAA in CTX at 15 and CTX, HYP at 30 min; ↑ 5HIAA/5HT ratio and Trp in all regions at 15 and 30 min	207
Footshock: 30 min (+/-ADX) CORT (acute)	CTX, STR, Acc; SPT, HYP, BST, HC, AMY	HPLC	↑ 5HIAA/5HT in Acc, SPT; ↑ Trp in all regions ↑ 5HIAA/5HT in STR, AMY and ↓ 5HIAA/5HT in CTX, HYP, BST in stress + ADX mice NC in 5HIAA/5HT with CORT	208
Restraint: 2, 3 h	CTX, AMY, HYP, HC	HPLC (5HT, 5HIAA); 5HTP accumulation after AADC inhibition	↑ 5HIAA in AMY at 2, 3 h; ↑ 5HIAA in HYP at 3 h: ↑ 5HTP accumulation in CTX at 3 h	209

Table 3 (continued) Effects of Acute Stress on 5HT Levels and Metabolism

Condition	Regions examined	Method	Effects	Ref.
Immobilization: 1-6 h	LHA	*In vivo* voltammetry; HPLC of CSF indoles	↑ 5HIAA signal in LHA; ↑ 5HIAA in CSF	210
Immobilization: 2 h	CTX	HPLC	↑ 5HT, 5HIAA, 5HIAA/5HT ratio	211
Tail pinch: 2-5 min	CTX, HC, STR, HYP, OLF	HPLC: all regions microdialysis: STR, ventral HC	↑ 5HIAA/5HT in HC, CTX; ↑ extracellular 5HT and 5HIAA in ventral HC	212
Footshock: 15-60 min Restraint: 30 min Endotoxin injection (+/- Chlorisondamine, an autonomic nervous system blocker)	CTX, HYP, CN, HC, BST	HPLC	↑ 5HIAA/5HT and Trp in CTX, HYP, HC, BST with shock and in CTX, CN, HYP, BST with restraint; ↑ Trp in CTX, HYP, BST and ↑ 5HIAA/5HT in CTX with endotoxin. Almost all ↑ blocked by ANS inhibition	213
Footshock: 30 min (Predictable/unpredictable; M vs. F)	CTX	HPLC	Transient ↑ in 5HIAA, 5HIAA/5HT; lasted longer with unpredictable shock; NC in 5HT; F response > M	69
Footshock: 30, 90 min (controllable/uncontrollable; M vs. F.)	CTX	HPLC	↑ 5HT with uncontrollable shock; ↑ 5HIAA, 5HIAA/5HT after exposure to shock environment with further ↑ after 90 min shock; F response > M	68
Immobilization: 30 min Tail pinch: 30 min	HYP	*In vivo* voltammetry	Have greater ↑ in 5-hydroxyindoles with immobilization than with tail pinch	214
Immobilization: 90 min (+/- Diazepam)	LHA	Microdialysis	↑ 5HT, 5HIAA; blocked by diazepam	215
Elevated plus-maze: 5 min Social interaction test: 2 h (varying conditions)	HC, CTX slices	HPLC: basal and K+-evoked 5HT release; [³H]5HT uptake	↑ Evoked 5HT release and ↓ 5HT uptake in HC with plus-maze; ↑ evoked 5HT release in HC with high-light familiar social interaction; ↑ 5HT uptake in CTX in all social interaction conditions	216
Novel environment: 5 min +/- cat odor Killed immediately or 30 min after stressor	HC, CTX slices	HPLC: basal and K+-evoked 5HT release; [³H]5HT uptake	↑ Basal 5HT release and ↓ 5HT uptake in HC with immediate sac; ↓ basal 5HT release and ↑ uptake in HC and CTX with sac 30 min later; further ↑ uptake in in CTX with cat exposure and delayed sac	217
Immobilization: 10 min Cold: 20 min Forced exercise: 2 h	CTX, DR	*In vivo* voltammetry	↑ 5HIAA by immobilization, cold in CTX, DR; ↑ 5HIAA in CTX with exercise	218
Novel environment: 20 min (young v. old rats)	CTX	HPLC	NC in 5HT, 5HIAA, 5HIAA/5HT with stress and/or age	219
Footshock: 30 min Conditioned fear	CTX, Acc, CN, PVN, AMY, LHA, HC	HPLC: micropunches	↑ Trp in all regions but PVN and ↑ 5HIAA in CTX, Acc, AMY with shock; only ↑ 5HIAA in CTX with fear; NC in 5HT	220
Psychological stress: 20 min (could see shocked rat nearby)	ABl, CTX	Microdialysis	↑ Extracellular 5HT	221

Note: Abbreviations as in Table 1.

Chronic stress appears to induce a sensitization of the 5HT system to subsequent stressors. De Souza and Van Loon found that the increase in 5HIAA in hypothalamus, cerebral cortex, and brainstem induced by restraint stress was further increased by exposure to a second session of restraint 30 or 60, but not 90, minutes later,[222] suggesting a temporal window of hypersensitivity. Similarly, daily sessions of 1-h restraint for 24 days enhanced the 5HIAA response to acute immobilization 20 h after the last restraint session (Figure 6), indicating a prolonged increase in sensitivity following chronic stress.[223] Chronic

Table 4 Effects of Chronic Stress on 5HT Levels and Metabolism

Condition	Regions examined	Method	Effects	Ref.
Chronic + acute stress				
Footshock: 1 h Restraint: 1 h Forced swim: 1 h Social stress: 10-40 days	Whole brain	Fluorescence assay	NC in 5HT with shock; ↑ 5HIAA with shock, restraint, swim, social stress	63
Restraint: 1 or 2 × 2 min	BST, HYP, HC, CTX	Fluorescence assay of 5HT, 5HIAA; 5HTP accumulation after AADC inhibition	↑ 5HT, 5HIAA in HYP, CTX and 5HIAA in BST after 1 session; no further change with second session 90 min later but did get ↑↑ 5HIAA and ↑ 5HTP accumulation in HYP, CTX, BST with second session 30 or 60 min later	222
Immobilization: 1 or 5 × 2 h M vs. F	CTX, HC, HYP	HPLC	↑ 5HIAA in CTX, HC, HYP with acute immobilization; M>F in CTX ↑ 5HT in CTX of M with repeated immobilization; ↓ 5HIAA in CTX, HC, HYP and ↑ 5HT in HYP of F with repeated immobilization	66
Restraint:1 h/day for 24 days +/- Acute immobilization	CTX, MBR, BST, HYP, HC, STR	HPLC	↑ 5HT in all regions and ↑ 5HIAA in CTX, HYP, HC with chronic restraint; NC with acute immobilization only; ↓ 5HT in HC and ↑ 5HIAA in all regions of restraint + immobilization vs. control and ↑ 5HIAA in all regions except STR vs. restraint only	223
Sound: 2 h, acute or 3/wk for 1, 2 or 6 weeks	BST, CTX	TPH assay	↑ TPH activity in CTX, BST immediately but not 1 h after acute stress; ↑ TPH in CTX, BST after chronic stress, persisted 24 h post-stress; further ↑ with chronic + acute	224
Chronic stress				
Immobilization: 2 h/day for 7 days	STR, CTX, HC, HYP, MBR	HPLC	NC in 5HT or 5HIAA	225
Footshock: 24 h (OVX + /–EB in females)	CTX, HC, STR, HYP, AHA-POA	Fluorescence assay	↓ 5HT in all regions: EB prevented ↓ in HC, STR, HYP and AHA-POA	226
Immobilization: 22 h/day for 14 days with 7 day recovery	Thin sections taken through-out brain	Immunocytochemistry for 5HT	↓ 5HT-IR in 5HT cell groups B1, B2, B3, and B7; NC in terminals	227
Exercise and food restriction: 12 days	CTX	HPLC	↓ 5HIAA in CTX	228
Handling: 21 days	CTX, HC	HPLC: basal and K+-evoked 5HT release; [³H]5HT uptake; 5HT content in tissue	↑ 5HT uptake and basal 5HT release in HC, CTX; ↓ 5HT in HC and ↓ 5HIAA in CTX	229
Subordination: 14 days	MBR, BST, POA, CTX, HYP, HC, AMY	HPLC	↑ 5HIAA in subordinates in POA, HC, AMY, entorhinal CTX; ↑ 5HIAA/5HT in MBR and HYP; NC in 5HT	91
Cold: 22 h/day for 5 days	CTX, HYP, THAL, MBR, BST	HPLC	↓ 5HT, 5HIAA in all regions but CTX; NC in 5HIAA/5HT ratios	230
Cold: 24 h	MBR, rest of brain	HPLC (+/- pargyline)	NC in 5HT turnover	231
Forced swim: 15 min + 5 min next day Isolation: 9 months	HC, Acc, CTX slices	HPLC: basal and K+-evoked 5HT release from incubation media and 5HT in tissue	↓ Basal 5HT release in HC, Acc, CTX, ↓ evoked release in HC, ↓ 5HT content in HC, STR with swim and isolation	232
Isolation: 13 wk	HC, CTX, CB	HPLC (+/- AADC inhibition)	NC in 5HT, 5HIAA; ↑ 5HTP accumulation in CTX, CB	233

Note: Abbreviations as in Table 1.

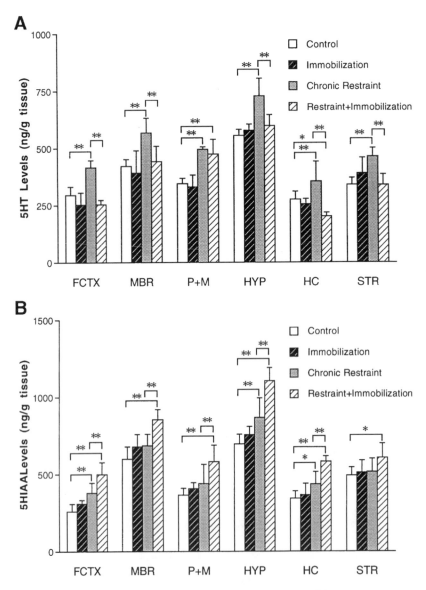

Figure 6 Serotonin levels after acute and chronic stress. Levels of 5HT (A) and 5HIAA (B) were measured by HPLC following 1 h immobilization and/or 1 h per day restraint for 24 days. Values are ng/g tissue ± SEM; n = 6 to 7. Differences between groups tested by one-way ANOVA followed by Duncan's multiple range test; *$p < 0.05$, **$p < 0.01$. FCTX = frontal cortex; MBR = midbrain; P + M = pons + medulla oblongata; HYP = hypothalamus; HC = hippocampus; STR = striatum. (Adapted from Adell et al., *J. Neurochem.*, 50, 1678, 1988.)

sound stress followed by an acute stressor also induced an increase in tryptophan hydroxylase activity in cortex and brainstem above and beyond that induced by either chronic or acute sound stress alone.[224]

The role of glucocorticoids in mediating the 5HT response to stress is unclear. Early work in our lab showed that bilateral ADX abolished the increases in tryptophan hydroxylase activity resulting from footshock, ether anesthesia, and cold exposure.[160] Boadle-Biber and co-workers also found that ADX blocked the increase in tryptophan hydroxylase activity induced by acute or repeated sound stress; however, administration of DEX, but not aldosterone, to ADX rats restored the effect on tryptophan hydroxylase.[234] Interestingly, in the immunocytochemical study by Kitayama and co-workers, chronic immobilization stress resulted in an increase of glucocorticoid receptor immunoreactivity in the cell group B7 (dorsal raphe) where the depletion of 5HT immunoreactivity was most pronounced,[227] suggesting a negative regulation of tryptophan hydroxylase activity by high levels of CORT.

In contrast, others have found no effect of ADX on stress-induced changes in brain tryptophan levels or 5HIAA/5HT ratios;[197,208] in the latter study, ADX actually potentiated the footshock-induced alterations in 5HIAA/5HT ratios in striatum, amygdala, hypothalamus, and brainstem.[208] Further support for a dissociation of 5HT and glucocorticoid responses to stress comes from studies where a variety of stressors were shown to consistently elevate plasma CORT and ACTH concentrations, while having no effect on brain 5HT or 5HIAA content.[202,206,219,233]

Factors other than adrenal steroids seem to be involved in mediating 5HT responses to stress. Activation of the sympathetic nervous system appears to be necessary for inducing changes in brain tryptophan and 5HIAA/5HT ratios, since blockade of the autonomic nervous system prevented, while the β-adrenoceptor antagonist propranolol attenuated, the increases induced by footshock and restraint.[213] Work by Kennett and colleagues has shown that tryptophan availability plays an important role in the 5HT response. Administration of valine, which competes with tryptophan for transport into the brain, blocked the immobilization-induced increase in brain 5HIAA[235] and inhibited the stress-induced release of 5HT in the hippocampus.[199] Valine attenuated the CORT response to stress as well, supporting a role for 5HT in modulating HPA activity.[235] Immobilization also results in increased free plasma tryptophan, brain tryptophan, and tryptophan influx, suggesting that tryptophan transport to the brain is facilitated in response to stress.[204]

2. 5HT Receptors

Stress appears to have variable effects on the $5HT_{1A}$ receptor subtype depending on the duration of the stressor. One 2-h restraint stress led to a significant increase in 8-OH-DPAT binding in CA4 of hippocampus and dentate gyrus.[236] Subsequent autoradiographic studies showed that these stress effects were not seen in ADX rats; in fact, the increase in $5HT_{1A}$ binding seen in CA3, CA4, and dentate following ADX was significantly attenuated by restraint stress, suggesting that acute stress, in the absence of circulation glucocorticoids, may act to reduce $5HT_{1A}$ binding.[170] Stress effects on $5HT_{1A}$ receptors in intact animals do appear to be biphasic, as the increase in binding in CA4 was not apparent after 5 days of daily 2 h restraint sessions[236] and a decrease in $5HT_{1A}$ binding was seen in CA3, CA4, and dentate gyrus following 3 h of immobilization stress for 14 days.[237] This decrease is consistent with what has been observed following CORT treatment (see above), suggesting that prolonged elevations of CORT serve to reduce $5HT_{1A}$ expression or function. Similarly, decreased $5HT_{1A}$ binding throughout the hippocampal formation was observed in stress-responsive subordinates following 14 days of subordination stress (Figure 7), while no changes were seen in hypothalamus, amygdala, or parietal cortex.[238] Reduced binding was also seen in the dominant animals, suggesting that they were somewhat stressed as well. Interestingly, with this stress paradigm, a subgroup of subordinates emerges with little or no CORT response to a subsequent restraint stress; these nonresponsive subordinates do not have a down-regulation of $5HT_{1A}$ binding except in CA3.[238] Because endocrine and physiological data suggest that these animals are the most severely stressed in this paradigm, it is possible that the decrease in binding may be necessary for successful adaptation to chronic stress and the nonresponders no longer have the capacity to cope with further stress.

In contrast to hippocampal $5HT_{1A}$ receptors, cortical $5HT_{2A}$ receptor binding tends to increase in response to stress. Although acute or repeated restraint stress had no effect on $5HT_{2A}$ binding sites in hippocampus or parietal cortex,[236,237] acute immobilization and the combination of activity wheel stress and food restriction did increase the B_{max} for [^{3}H]ketanserin binding to $5HT_{2A}$ receptors in frontal cortex.[211,228] Chronic subordination stress increased binding to $5HT_{2A}$ receptors in layer IV of parietal cortex in both stress-responsive and nonresponsive subordinates (Figure 8).[238] However, 15 min of forced swimming decreased 5HT-stimulated phosphoinositide metabolism in cerebral cortex slices, with no changes in ligand binding characteristics,[239] suggesting that a stress-induced increase in binding sites may be a compensatory response to account for a decrease in functional responsiveness in these receptors.

Few studies have examined the effects of stress on other 5HT receptor subtypes. Work in our laboratory found no changes in $5HT_{2C}$ receptor or 5HTt binding in response to restraint stress, nor were there any changes in $5HT_{1B}$ binding following subordination stress.[236-238] However, Edwards and co-workers have found several changes in the stress-based learned helplessness model of depression. In the hypothalamus of animals that developed learned helplessness, the number of $5HT_{1B}$ and 5HTt binding sites was decreased compared to nonhelpless and naive rats, while $5HT_{1B}$ receptors were increased in cerebral cortex, hippocampus, and septum and 5HTt sites were also increased in cortex.[240,241] In contrast to the studies described above, these investigators found no changes in $5HT_{1A}$ binding;[240] another group also

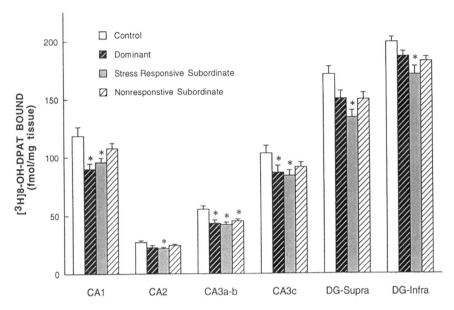

Figure 7 Effects of 14-day subordination stress on 5HT1A receptor binding. Binding of [³H]8-OH-DPAT to $5HT_{1A}$ receptors was measured in CA1, CA2, CA3, and CA4 of hippocampus and the suprapyramidal (DG-supra) and infrapyramidal blades (DG-infra) of dentate gyrus. Nonresponsive subordinates are defined as having less than 10 µg/dL increase in plasma CORT in response to a novel restraint stressor. Data expressed as fmol [³H]8-OH-DPAT bound per milligram tissue; group mean ± SEM. n = 5 for control and dominant; n = 7 for responsive subordinates; and n = 9 for nonresponsive subordinates. Differences between groups were tested by one-way ANOVA followed by Tukey's post-hoc test. *$p < 0.05$ as compared to individually housed controls. (From McKittrick et al., *Biol. Psychiatry,* 37, 383, 1995. With permission.)

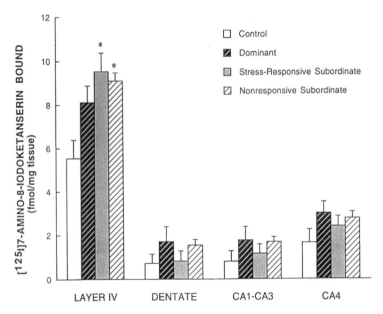

Figure 8 Effects of 14-day subordination stress on $5HT_{2A}$ receptor binding. Binding of [¹²⁵I]7-amino-8-iodo-ketanserin to $5HT_{2A}$ receptors was measured in layer IV of parietal cortex, dentate gyrus, and CA1-CA3 of hippocampus. Data expressed as fmol [¹²⁵I]7-amino-8-iodo-ketanserin bound per milligram tissue; group mean ± SEM. *$p < 0.05$ as compared to individually housed controls. See legend to Figure 7 for definition of nonresponsive subordinates and details on statistical analysis. (From McKittrick et al., *Biol. Psychiatry,* 37, 383, 1995. With permission.)

found no effect of chronic (10-day) footshock stress on $5HT_1$, $5HT_{1A}$, or $5HT_{2A}$ receptors.[242] These differences may be due to the stress paradigms used or the use of tissue homogenates vs. slice autoradiography to measure receptor binding.

3. Electrophysiology

Although several investigators have studied the effects of adrenal steroids on 5HT neuronal activity and responses, few have addressed the consequences of stress. However, in one study examining the activity of 5HT neurons in the cat, neither loud white noise, restraint, nor exposure to a dog altered the firing rates of 5HT neurons in the dorsal raphe, despite physiological evidence that these stimuli were stressful to the animal.[243] Investigation of the effects of stress on postsynaptic responses to 5HT, particularly in the hippocampus, would be useful in determining the physiological relevance of the stress-induced changes in 5HT receptor binding.

4. 5HT-Mediated Behavior

Stress-induced changes in central 5HT systems are suggested by several studies of 5HT-mediated behavior. In a series of experiments by Kennett and colleagues, rats were exposed to repeated immobilization for up to 7 days and their behavioral responses to various serotonergic compounds was measured 24 h later.[225] Seven days, but not 1 or 3 days of immobilization enhanced many of the behaviors induced by the 5HT releaser PCA and the 5HT agonist 5MeODMT, such as forepaw treading and tremor, although hind limb abduction and headweaving were unaffected.[225] Administration of the corticosterone synthesis inhibitor metyrapone before each immobilization session accelerated the behavioral changes so that the enhanced tremor was evident after only 3 days of immobilization and forepaw treading, hind limb abduction, and Straub tail were all increased after 7 days.[244] Deficits in open field behavior, food intake, and growth rate also returned to normal more quickly in animals given metyrapone, suggesting that behavioral adaptation to repeated stress is associated with increased sensitivity of 5HT systems and is accelerated in the presence of metyrapone.[244]

In this paradigm of repeated immobilization, the open field behavior of females adapted less quickly than that of males, although the behavioral deficits in females were initially smaller than in males.[66] In addition, females failed to show the increase in sensitivity to 5MeODMT associated with behavioral adaptation.[66] The CORT response to repeated immobilization was found to be greater in females than in males and reducing the CORT response to the levels of males with metyrapone facilitated adaptation in females and eliminated the sex differences in the behavioral sensitivity of the 5HT system.[245] The effects of stress on 5HT-mediated responses are the opposite of those found following repeated CORT administration, indicating that repeated elevations of glucocorticoids may interfere with adaptive changes associated with increased sensitivity to 5HT.

Similar studies using repeated footshock have also suggested that 5HT hypersensitivity is a characteristic of adaptation to stress. However, since locomotor activity, freezing behavior, and weight gain returned to control levels by day 5, while the behavioral sensitivity to 5MeODMT did not develop until day 10, the authors propose that the 5HT system may be more important in the maintenance, rather than the development, of adaptation.[242] How the changes in behavioral sensitivity relate to receptor levels is unclear, as these authors found no changes in $5HT_{1A}$ or $5HT_{2A}$ receptor binding,[242] although this finding does not preclude alterations in functional responsiveness distal to the receptor, or changes in levels of other 5HT receptor subtypes.

In contrast, to chronic immobilization of footshock, 24 h of exposure to cold reduced the behavioral response to the $5HT_{1A}$ agonist 8-OH-DPAT, although the hypothermic response was unchanged, while the CORT response to 8-OH-DPAT was slightly increased.[231] The behavioral responses to the $5HT_{2C}$ agonist mCPP and the $5HT_{2A}$ antagonist ketanserin remained unaltered. The role of adrenal steroids in mediating the alterations in $5HT_{1A}$ responsiveness in this paradigm is uncertain, however, as there were no differences in plasma CORT levels between cold-exposed and control rats.

5. Summary of Stress Effects

Acute stress has repeatedly been shown to stimulate 5HT release and metabolism; the effects of chronic stress are more inconsistent, with several groups reporting decreases in 5HT activity. It is likely that the 5HT system responds to stress in a biphasic manner, first showing an increase in transmission, followed by a gradual adaptive decrease. A reduction in transmitter levels with chronic stress could then lead to a compensatory upregulation of certain postsynaptic receptor subtypes; this scenario would be consistent with the behavioral hypersensitivity to nonselective 5HT agonists seen by Kennett and colleagues during adaptation to repeated stress.[225] The observation that the $5HT_{2A}$ receptor system has been shown to be upregulated following chronic stress supports this hypothesis.[211,228,238] On the other hand, $5HT_{1A}$ receptors in the hippocampus are reduced following chronic stress. However, $5HT_{1A}$ receptors have not reliably been shown to be upregulated in response to decreased transmitter levels, except in hypothalamus;[78,246] in fact, lesions with 5,7-DHT led to decreases in [^{3}H]8-OH-DPAT binding in raphe and also in the hippocampal regions where binding is decreased in response to stress.[237,238,247]

Gonadal steroids also appear to interact with adrenal steroids in modulating the response to stress. Overall, females appear to be more sensitive to the effects of stress, as stress-induced changes in 5HIAA and 5HT were generally larger in females than in males. The enhanced 5HT response may be due, in part, to the relatively larger CORT response to stress seen in females.[245] There is also a sex difference in behavioral adaptation to repeated stress, in that females were less affected by acute stress, but took longer to adapt to chronic stress;[66] elimination of the sex difference in CORT secretion also eliminated the differences in the behavioral responses, although 5HT metabolism was unchanged.[245]

D. CONCLUSIONS

The most important question regarding the involvement of glucocorticoids in the response of the serotonergic system to stress is whether the adrenal steroids act as mediators of the effects of stress or as a counter-regulatory mechanism that protects the brain from the impact of stressful events. It would appear that adrenal steroids may primarily have a mediating role, acutely facilitating increased 5HT turnover during stress and mediating stress-induced increases in $5HT_{2A}$ receptors during chronic stress. However, a counter-regulatory role is not ruled out. The use of metyrapone to enhance stress induction of serotonin behavioral sensitivity is consistent with such a counter-regulatory role. Moreover, the finding that $5HT_{1A}$ receptors in hippocampus are acutely increased after stress[236] and are downregulated by chronic social stress in stress responsive, but not stress nonresponsive, subordinate rats[238] raises the possibility that there are opposing forces controlling $5HT_{1A}$ receptor expression, in which the suppressive force is the glucocorticoid secretion associated with the stress response. Whether the acute activation of $5HT_{1A}$ receptors by stress is directly dependent on adrenocortical secretion, or is altered by another system that is affected by that secretion, remains to be determined.

V. THYROID HORMONES

A. RECEPTORS AND THEIR DISTRIBUTION

The thyroid hormones thyroxine (T4) and triiodothyronine (T3) have been shown to play a role in the function and development of the brain. Although both hormones are secreted by the thyroid gland, most T4 is monodeiodinated within target tissues to T3, which binds to an intracellular receptor (TR) and mediates the biological functions of the thyroid hormones.[248,249] In the brain, immunoreactive nuclear T3 receptors (IR-TR) can be found in neurons, glia, and ependymal cells.[250] High densities of IR-TRs are located in olfactory bulb, hippocampus, amygdala, and neocortex, whereas hypothalamus has a moderate number of IR-TR and thalamus and striatum have very few.[250] In addition, high levels of IR-TR are found in cerebellum and in the mamillary nuclei.[250] The TR immunoreactivity may not necessarily correspond to functional receptors, however, due to the presence of nonfunctional TR isoforms.

There are different TR isoforms originating from differential splicing of two separate genes (TRα and TRβ), both of which are highly homologous to the viral *erbA* oncogene.[251,252] The two β isoforms (TRβ1 and TRβ2) and the TRα1 isoform are typical of the steroid receptor superfamily in that they have functional DNA-binding and T3-binding domains, can form homo- or heterodimers, bind thyroid hormone response elements (TREs), and activate transcription.[253] The TRα2 isoform, on the other hand, is unable to bind T3 and can inhibit transcriptional activation in the presence of other liganded TRs.[253]

Messenger RNAs coding for the α and β isoforms have distinct regional distributions in the adult brain: TRβ1 mRNA is highest in hippocampus, the parvocellular nucleus of the PVN, and anterior pituitary,[254] while the TRβ2 isoform is not found in brain.[255] High levels of TRα1 mRNA are in olfactory bulb, hippocampus, amygdala, cerebellum, and brainstem, including dorsal raphe, with somewhat lower levels in hypothalamus, neocortex, and anterior pituitary;[254] mRNA for the TRα2 isoform is found in moderate levels in olfactory bulb, hippocampus, dentate gyrus, cerebellum, and brainstem.[254] The relative amounts of the TR mRNAs change during brain development, although changes in expression of TRβ1 and TRα1 mRNAs (the T3-binding isoforms) do not necessarily lead to changes in T3 binding capacity, suggesting possible posttranslational modifications or cofactors needed for functional activation.[255]

B. THYROID HORMONE
1. 5HT Levels and Metabolism
a. Adult
Several investigators have examined the effects of thyroid hormones on the 5HT system using surgical thyroidectomy (TX), thyroid hormone synthesis inhibition with methimazole (MMZ), or propylthiouracil (PTU), or treatment with exogenous T3 or T4. The results have been inconsistent and the effects observed are regionally specific. Most of the early studies in the 1970s suggested that thyroid hormones stimulate 5HT synthesis in brain, as chronic treatment with T4 increased 5HTP accumulation following decarboxylase inhibition in cerebral cortex, striatum, and limbic system; 5HT levels and accumulation following MAO inhibition were also increased.[256-258] However, others found that high levels of T3 decreased 5HT and 5HIAA levels in striatum and cerebellum, without altering tryptophan hydroxylase activity[259] or, conversely, had no effect on 5HT levels, but increased the synthesis rate.[260]

A more recent study looked at the effects of exogenous T4 or thyroid hormone synthesis inhibition with PTU on 5HT metabolism in discrete nuclei of adult rat brain. Savard and co-workers found that high levels of T4 increased 5HT and 5HIAA levels in several regions, including mPOA, infundibulum, medial forebrain bundle, substantia nigra, ventral tegmental area (VTA), and the raphe nuclei; in addition, the 5HIAA/5HT ratio was increased in mPOA and decreased in basolateral amygdala, VTA, and medial habenula.[261] Interestingly, PTU-induced hypothyroidism had similar effects, increasing 5HT and 5HIAA levels and the 5HIAA/5HT ratio in these regions and in several others.[261] Heal and Smith also found that 10-day treatment with T3 increased 5HT and 5HIAA levels in midbrain and hindbrain, and also increased 5HIAA in forebrain, although a single injection of T3 only elevated 5HIAA in midbrain and hindbrain.[262]

Other studies using MMZ or surgical TX also found an increased 5HT turnover in brainstem, which could be reversed by 3.5, but not 1.5 days of treatment with T3; surgical TX also prevented the decreased 5HT turnover seen in streptozotocin-diabetic rats.[263,264] Schwark and Keesey found an increase in 5HT levels in hypothalamus following surgical TX, although no changes were seen in brainstem or basal ganglia.[265] In contrast, others have reported that surgical TX decreased 5HTP accumulation and 5HT levels and accumulation in cerebral cortex, diencephalon, and brainstem.[256,258] Finally, tryptophan uptake into brain following MMZ[266] and 5HT content following surgical TX, TX with T3 replacement, or high T3[267] have been reported to be unaltered in adult rats.

b. Developmental Effects
The developmental effects of thyroid hormone have been widely studied, since fetal hypothyroidism can result in mental retardation, abnormal bone development, and retarded growth. As in adult rats, the effects of thyroid hormone on the developing 5HT system are complex, and may be dependent upon the complement of specific TR isoforms present in a given brain region. Neonatal hypothyroidism induced by MMZ or ^{131}I was shown to result in a decrease in 5HT levels, turnover rate, and tryptophan hydroxylase activity and an increase in MAO activity in one study; these effects were reversed by T3 treatment of neonates, but not adults.[268] Neonatal hyperthyroidism induced by T3 administration had the opposite effect, increasing 5HT turnover of 5HT and tryptophan hydroxylase activity.[259,268]

Conversely, others reported that hypothyroidism induced by PTU increased 5HT levels and tryptophan hydroxylase activity, and selectively decreased MAO activity in basal ganglia, while neonatal hyperthyroidism had no effect on 5HT levels or metabolism.[265,269,270] In addition, this group found an increased synaptosomal uptake of 5HT in hypothyroid rats, which was reversed by T3 treatment from postnatal day 0 to 30, but not from day 25 to 30,[271] suggesting a critical period during which thyroid hormones modulated the 5HT system. A more recent study using micropunches of discrete nuclei also found stimulatory effects with ^{131}I-induced hypothyroidism, with increased 5HT and 5HIAA levels in several brain nuclei, including dorsal raphe, hippocampus, and amygdala; the 5HIAA/5HT was also increased in several regions.[272] Any changes in 5HT synthesis and turnover were reversed by chronic neonatal T4 replacement in ^{131}I-treated rats.[272] Similarly, in rat pups made hypothyroid by suckling from PTU-treated mothers, 5HT and 5HIAA were increased in hypothalamus and cortex, and these increases were potentiated by acute dose of insulin and attenuated by chronic T3 replacement (Figure 9).[273]

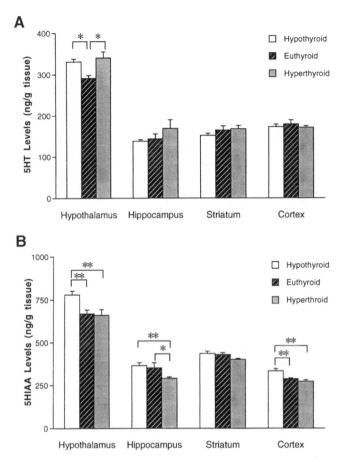

Figure 9 Serotonin levels in hypothyroid, euthyroid, and hyperthyroid rats. Levels of 5HT (A) and 5HIAA (B) in 14-day-old rat pups measured by HPLC after no treatment (euthyroid), propothiouracil (PTU; 160 mg/kg/day, p.o.) given to suckling mothers (hypothyroid), or daily injections of l-triiodothyronine (T3; 10 mg/kg) from birth (hyperthyroid). Values are ng/g tissue ±SEM. Differences between groups were tested by ANOVA followed by the Scheffe F test; $^*p < 0.05$, $^{**}p < 0.01$. (Adapted from Orosco et al., *Psychoneuroendocrinology*, 14, 321, 1989.)

2. 5HT Receptors

a. *Adult*

In vitro, the antithyroid agents MMZ and PTU have been shown to increase [³H]5HT uptake in synaptosomes and [³H]IMI binding to brain membranes, while decreasing $5HT_1$ receptor number and leaving $5HT_2$ receptors unaltered.[274,275] Thirty days of MMZ treatment *in vivo*, however, did not affect [³H]IMI binding in adult rats.[274] Another group did not replicate the *in vitro* effect of PTU on [³H]IMI binding, but found that binding sites were increased in hippocampus following *in vivo* treatment with PTU, whereas binding was decreased in cerebral cortex and hypothalamus.[276] These investigators also observed no effects of T3 either *in vivo* or *in vitro*.[276]

In contrast, a recent autoradiographic study with the more selective ligand [³H]cyanoimipramine did not report any changes in 5HT transporter binding after surgical TX or high-dose T4.[277] $5HT_{1A}$ receptor binding in hippocampus and cerebral cortex was progressively increased by TX, however, with these changes reversed by T4 replacement at physiological doses (Figure 10).[277] Interestingly, supraphysiological doses of T4 also increased [³H]8-OH-DPAT binding to $5HT_{1A}$ receptors in hippocampus and in hypothalamus, as well, while none of the thyroid manipulations alter binding to somatodendritic autoreceptors in dorsal raphe.[277] In addition, in one study, 7 to 10 days of T4 administration have also been shown to increase $5HT_{2A}$ receptor number in cortex, hippocampus, and striatum, while TX had no effect on these receptors.[278] However, others report no change in $5HT_{2A}$ receptor number following a single dose of T3 and a decrease in binding in frontal cortex following 10 days of T3.[262]

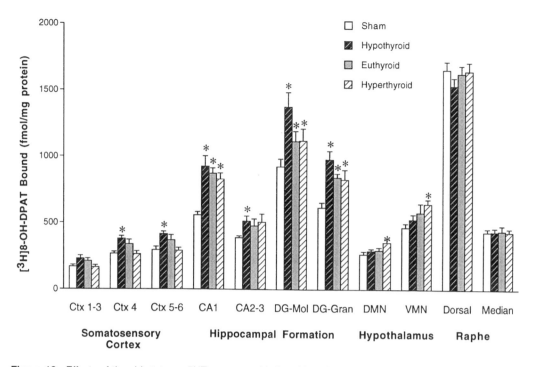

Figure 10 Effects of thyroid state on 5HT$_{1A}$ receptor binding. Male Sprague-Dawley rats were either surgically thyroidectomized or sham-operated (sham). One week after surgery, thyroidectomized rats were given 15 μg/kg thyroxine (T4) in the drinking water (euthyroid), 200 μg/kg T4 (hyperthyroid) or no additional treatment (hypothyroid). Animals were sacrificed 35 days after surgery (28 days after T4 treatment). Values are fmol [3H]8-OH-DPAT bound per milligram protein, mean ± SEM; n = 5 to 7. Data were analyzed by two-way ANOVA followed by Newman-Keuls post-hoc test of effect; *p <0.05 vs. SHAM. Ctx = layers 1 to 3, 4, or 5 to 6 of somatosensory cortex; DG-Mol = molecular layer of dentate gyrus; DG-Gran = granular cell layer of dentate gyrus; VMN = ventromedial hypothalamic nucleus; and DMN = dorsomedial hypothalamic nucleus. (Adapted from Tejani-Butt et al., *Neuroendocrinology*, 57, 1011, 1993.)

b. Developmental Effects

Very few studies have studied the effect of thyroid hormones on the ontogenesis of 5HT receptors and the 5HT transporter. Chronic administration of MMZ to rat pups during the first 30 days of life decreased the number of [3H]IMI binding sites in cerebral cortex, although not in striatum,[274] a result that is inconsistent with the increased synaptosomal 5HT uptake observed by others in neonatal hypothyroid rats.[271] However, since binding to the presynaptic 5HTt is often used as an index of 5HT innervation,[78,246] the reduced number of IMI binding sites may reflect decreased 5HT innervation of the cortex. This hypothesis is supported by the observation that hypothyroid rats have increased 5HT$_1$ and 5HT$_2$ receptor binding in brain, suggesting an adaptive upregulation of postsynaptic receptors in response to reduced availability of 5HT.[279]

3. 5HT-Mediated Behavior
a. Adult

Most of the behavioral studies suggest that thyroid hormones tend to potentiate behavioral responses to 5HT agonists, particularly the head twitch response mediated primarily by 5HT$_{2A}$ receptors. One or three daily injections of T3 or T4 potentiated the head twitch response to 5HTP or 5MeODMT in mice, although there was no change in 5HT$_{2A}$ receptor number (Figure 4).[138,262,280] The observation that norepinephrine uptake blockers, but not 5HT uptake blockers, enhanced the potentiation, while the β-adrenoceptor antagonist penbutolol blocked it, suggests that the enhanced response may be indirectly mediated through changes in noradrenergic transmission.[280] Hyperactivity induced by quipazine, 5MeODMT, and tranylcypromine plus L-tryptophan was also increased by either chronic or acute T3 treatment.[260] However, others found no change in 5HTP-induced head twitches following chronic (10-day) T3, although 5HT$_{2A}$ receptor number in frontal cortex was decreased.[262] In contrast, responses to $5HT_{1A}$ and $5HT_{1B}$ agonists were unaltered after a single dose of T3, but decreased after chronic T3 treatment.[262]

b. Developmental Effects

Although there is evidence suggesting both increased and decreased 5HT function in hypothyroid neonates, few investigators have examined the effects of thyroid hormones on the development of 5HT-mediated behaviors. In one study, rats born to hypothyroid mothers and kept hypothyroid with MMZ had reduced expression of the 5HT syndrome following 5HTP administration,[281] possibly resulting from reduced synthesis and release of 5HT, or postsynaptic subsensitivity, or both, in these animals. However, rat pups treated with either T3 or saline showed comparable developmental patterns of methysergide-induced behavioral disinhibition and similar levels of maximal activity, although hyperthyroid animals were generally more active in the absence of methysergide.[282] These data suggest that neonatal hyperthyroidism does not accelerate the development of 5HT circuits mediating behavioral inhibition, but may increase the activity of other transmitter systems involved in behavioral activation.

C. SUMMARY AND CONCLUSIONS

The effects of thyroid hormones on the 5HT system are difficult to summarize, since thyroid hormones have been shown to increase, decrease, and have no effect on 5HT levels. In addition, some of the most recent studies have shown that both hypo- and hyperthyroid states lead to similar changes in both 5HT levels and metabolism and 5HT receptor binding (c.f. 261 and 277). In the absence of thyroid hormones, HPA axis activity is disregulated;[283] therefore, at least some of the effects of thyroidectomy may be attributable to decreased secretion of adrenal steroids. It is possible that decreased glucocorticoid levels may contribute to the increase in hippocampal $5HT_{1A}$ receptors seen with TX,[277] since it is similar to that seen with ADX,[170] while the increase seen with T4 may be due primarily to the thyroid hormones themselves. The surgical procedures and pharmacological treatments may also be stressful to the animals, as suggested by changes in weight gain and other metabolic parameters,[263] again suggesting that HPA activity may act as a confounding factor in determining the effects of thyroid hormone.

There is also very poor correlation between regulation of 5HT receptor binding and the behaviors these receptors mediate. The study by Heal and Smith is a classic example of this, where the head twitch response to 5MeODMT was altered in the absence of changes in $5HT_{2A}$ receptor binding after 3 days of T3 treatment, but after 10 days $5HT_{2A}$ binding was decreased but the behavioral response was unaffected.[262] Similarly, the increases in $5HT_{1A}$ receptor binding in response to elevated thyroid hormones[277] are not consistent with the decreased behavioral response to 8-OH-DPAT.[262] This lack of correlation suggests that thyroid hormone may be modulating receptor function at a level distal to the receptor, possibly via changes in signal transduction mechanisms. The effects of TH on receptor-mediated responses remains to be addressed at the cellular level using both electrophysiological and biochemical techniques.

VI. CONCLUSIONS AND OUTLOOK

In spite of a vast amount of data showing that gonadal, adrenal, and thyroid hormones affect the serotonergic system, there is very little coherence between data on changes in receptor levels, serotonin turnover, and alterations in behavior. One reason for this is undoubtedly that only a few serotonin receptor subypes have been investigated in detail as regards hormone effects. Another reason for such discrepancies is that serotonin turnover measurements on tissue pieces do not give the most specific assessment of serotonin release from the synaptic terminal or varicosity. Finally, there is very little information that would localize a particular behavioral effect to a specific brain region and, in many cases, there are undoubtedly diffuse or complex circuits involved, most likely involving more than one serotonin receptor subtype. In order to make progress, it is important to concentrate upon more tractable systems, for example, ones in which there is either a high concentration of a particular 5HT receptor subtype or else a highly localized behavioral control system. An example of the former is the $5HT_{1A}$ receptor system concentrated in the hippocampus, and it is hoped that concerted behavioral and neuropharmacological studies of these receptors may begin to provide a more cohesive picture of the effects of hormonal manipulations and stress on 5HT function.

In addition, it is important to bring to bear upon the more tractable systems the power of other techniques, such as microdialysis on awake behaving animals and electrophysiological recording. Microdialysis would help to examine actual synaptic release of 5HT, rather than relying on indirect methods based upon tissue levels of 5HT and 5HIAA. Electrophysiology would help investigators relate hormone-induced changes in receptor binding to the excitatory or inhibitory electrical responses. Moreover, in view of the cross talk between drugs for various serotonin receptor subtypes, it would be useful to explore the use of antisense DNA as a means of selectively ablating particular serotonin receptor subtypes in specific brain regions, as a means of looking more specifically at the physiological roles of the myriad serotonin receptor subtypes.

REFERENCES

1. Dahlström, A. and Fuxe, K., Evidence for the existence of monoamine-containing neurons in the central nervous system. I. Demonstration of monoamines in the cell bodies of brain stem neurons, *Acta Physiol. Scand.,* 62, S232, 5, 1964.
2. Conrad, L. C. A., Leonard, C. M., and Pfaff, D. W., Connections of the median and dorsal raphe nuclei in the rat: an autoradiographic and degeneration study, *J. Comp. Neurol.,* 156, 179, 1974.
3. Kellar, K. J., Brown, P. A., Madrid, J., Bernstein, M., Vernikos-Danellis, J., and Mehler, W. R., Origins of serotonin innervation of forebrain structures, *Exp. Neurol.,* 56, 52, 1977.
4. Azmitia, E. C. and Segal, M., An autoradiographic analysis of the differential ascending projections of the dorsal and median raphe nuclei in the rat, *J. Comp. Neurol.,* 179, 641, 1978.
5. Van de Kar, L. D. and Lorens, S. A., Differential serotonergic innervation of individual hypothalamic nuclei and other forebrain regions by the dorsal and median midbrain raphe nuclei, *Brain Res.,* 162, 45, 1979.
6. Parent, A., Descarries, L., and Beaudet, A., Organization of ascending serotonin systems in the adult rat brain. A radioautographic study after intraventricular administration of [^{3}H]5-hydroxytryptamine, *Neuroscience,* 6, 115, 1981.
7. Steinbusch, H. W. M., Distribution of serotonin-immunoreactivity in the central nervous system of the rat — cell bodies and terminals, *Neuroscience,* 6, 557, 1981.
8. Imai, H., Steindler, D. A., and Kitai, S. T., The organization of divergent axonal projections from the midbrain raphe nuclei in the rat, *J. Comp. Neurol.,* 243, 363, 1986.
9. Peroutka, S. J., Molecular biology of serotonin (5-HT) receptors, *Synapse,* 18, 241, 1994.
10. Chalmers, D. T. and Watson, S. J., Comparative anatomical distribution of 5-HT$_{1A}$ receptor mRNA and 5-HT$_{1A}$ binding in rat brain-a combined *in situ* hybridisation/*in vitro* receptor autoradiographic study, *Brain Res.,* 561, 51, 1991.
11. Pompeiano, M., Palacios, J. M., and Mengod, G., Distribution and cellular localization of mRNA coding for 5-HT$_{1A}$ receptor in the rat brain: correlation with receptor binding, *J. Neurosci.,* 12, 440, 1992.
12. Voight, M. M., Laurie, D. J., Seeburg, P. H., and Bach, A., Molecular cloning and characterization of a rat brain cDNA encoding a 5-hydroxytryptamine$_{1B}$ receptor, *EMBO J.,* 10, 4017, 1991.
13. Bach, A. W., Unger, L., Sprengel, R., Mengod, G., Palacios, J., Seeburg, P. H., and Voight, M. M., Structure, functional expression and spatial distribution of a cloned cDNA encoding a rat 5-HT$_{1D}$-like receptor, *J. Receptor Res.,* 13, 479, 1993.
14. Mengod, G., Pompeiano, M., Martínez-Mir, M. I., and Palacios, J. M., Localization of the mRNA for the 5-HT$_2$ receptor by *in situ* hybridization histochemistry. Correlation with the distribution of receptor sites, *Brain Res.,* 524, 139, 1990.
15. Pompeiano, M., Palacios, J. M., and Mengod, G., Distribution of the serotonin 5-HT$_2$ receptor family mRNAs: comparison between 5-HT$_{2A}$ and 5-HT$_{2C}$ receptors, *Mol. Brain Res.,* 23, 163, 1994.
16. Hoffman, B. J. and Mezey, E., Distribution of serotonin 5-HT$_{1C}$ receptor mRNA in adult rat brain, *FEBS Lett.,* 247, 453, 1989.
17. Kilpatrick, G. J., Jones, B. J., and Tyers, M. B., Identification and distribution of 5-HT$_3$ receptors in rat brain using radioligand binding, *Nature,* 330, 746, 1987.
18. Milburn, C. M. and Peroutka, S. J., Characterization of [^{3}H]quipazine binding to 5-hydroxytryptamine$_3$ receptors in rat brain membranes, *J. Neurochem.,* 52, 1787, 1989.
19. Hoffman, B. J., Mezey, E., and Brownstein, M. J., Cloning of a serotonin transporter affected by antidepressants, *Science,* 254, 578, 1991.
20. Ramamoorthy, R., Bauman, A. L., Moore, K. R., Han, H., Yang-Feng, T., Chang, A. S., Ganapathy, V., and Blakely, R. D., Antidepressant- and cocaine-sensitive human serotonin transporter: molecular cloning, expression and chromosomal localization, *Proc. Natl. Acad. Sci. U.S.A.,* 90, 2542, 1993.
21. De Souza, E. B. and Kuyatt, B. L., Autoradiographic localization of ^{3}H-paroxetine-labeled serotonin uptake sites in rat brain, *Synapse,* 1, 488, 1987.
22. Fujita, M., Shimada, S., Maeno, H., Nishimura, T., and Tohyama, M., Cellular localization of serotonin transporter mRNA in the rat brain, *Neurosci. Lett.,* 162, 59, 1993.
23. Segal, M., Regional differences in neuronal reponses to 5-HT: intracellular studies in hippocampal slices, *J. Physiol., Paris,* 77, 373, 1981.
24. Beck, S. G., Clarke, W. P., and Goldfarb, J., Spiperone differentiates multiple 5-hydroxytryptamine responses in rat hippocampal slices *in vitro, Eur. J. Pharmacol.,* 116, 195, 1985.

25. Andrade, R. and Nicoll, R. A., Pharmacologically distinct actions of serotonin on single pyramidal neurones of the rat hippocampus recorded *in vitro*, *J. Physiol.*, 394, 99, 1987.

26. Ropert, N., Inhibitory action of serotonin in CA1 hippocampal neurons *in vitro*, *Neuroscience*, 26, 69, 1988.

27. Klancnik, J. M., Obenaus, A., Phillips, A. G., and Baimbridge, K. G., The effects of serotonergic compounds on evoked responses in the dentate gyrus and CA1 region of the hippocampal formation of the rat, *Neuropharmacology*, 30, 1201, 1991.

28. Colino, A. and Halliwell, J. V., Differential modulation of three separate K-conductances in hippocampal CA1 neurons by serotonin, *Nature*, 328, 73, 1987.

29. Baskys, A., Niesen, C. E., Davies, M. F., and Carlen, P. L., Modulatory actions of serotonin on ionic conductances of hippocampal dentate granule cells, *Neuroscience*, 29, 443, 1989.

30. Beck, S. G., 5-hydroxytryptamine increases excitability of CA1 hippocampal pyramidal cells, *Synapse*, 10, 334, 1992.

31. Newberry, N. R., 5-HT$_{1A}$ receptors activate a potassium conductance in rat ventromedial hypothalamic neurones, *Eur. J. Pharmacol.*, 210, 209, 1992.

32. Saphier, D. and Zhang, K., Inhibition by the serotonin$_{1A}$ agonist, 8-hydroxy-2-(di-*n*-propylamino)tetralin, of antidromically identified paraventricular nucleus neurons in the rat, *Brain Res.*, 615, 7, 1993.

33. Kawano, S., Osaka, T., Kannan, H., and Yamashita, H., Excitation of hypothalamic paraventricular neurons by stimulation of the raphe nuclei, *Brain Res. Bull.*, 28, 573, 1992.

34. Green, A. R., 5-HT-mediated behavior, *Neuropharmacology*, 23, 1521, 1984.

35. Hutson, P. H., Donohoe, T. P., and Curzon, G., Hypothermia induced by the putative 5-HT$_{1A}$ agonists LY165163 and 8-OH-DPAT is not prevented by 5-HT depletion, *Eur. J. Pharmacol.*, 143, 221, 1987.

36. Hutson, P. H., Dourish, C. T., and Curzon, G., Neurochemical and behavioral evidence for mediation of the hyperphagic action of 8-OH-DPAT by 5-HT cell body autoreceptors, *Eur. J. Pharmacol.*, 129, 347, 1986.

37. Dourish, C. T., Hutson, P. H., and Curzon, G., Para-chlorophenylalanine prevents feeding induced by the serotonin agonist 8-hydroxy-2-(di-*n*-propylamino)tetralin (8-OH-DPAT), *Psychopharmacology*, 89, 467, 1986.

38. Heaton, J. C. P., Njung'e, K., and Handley, S. L., Behavioural profile of 1-(2,5-dimethoxy-4-iodophenyl)-2-aminopropane (DOI), a selective 5-HT$_2$ agonist, *Br. J. Pharmacol.*, 94, 388P, 1988.

39. Pranzatelli, M. R., Evidence for involvement of 5-HT$_2$ and 5-HT$_{1C}$ receptors in the behavioral effects of the 5-HT agonist 1-(2,5-dimethoxy-4-iodophenyl aminopropane)-2 (DOI), *Neurosci. Lett.*, 115, 74, 1990.

40. Lucki, I., Ward, H. R., and Frazer, A., Effect of 1-(m-chlorophenyl)piperazine and 1-(m-trifluoromethylphenyl)piperazine on locomotor activity, *J. Pharmacol. Exp. Ther.*, 249, 155, 1989.

41. Eison, A. S. and Yocca, F. D., Reduction in cortical 5-HT$_2$ receptor sensitivity after continuous gepirone treatment, *Eur. J. Pharmacol.*, 111, 389, 1985.

42. Nash, J. F., Jr., Meltzer, H. Y., and Gudelsky, G. A., Selective cross-tolerance to 5-HT$_{1A}$ and 5-HT$_2$ receptor-mediated temperature and corticosterone responses, *Pharmacol. Biochem. Behav.*, 33, 781, 1989.

43. Backus, L. I., Sharp, T., and Grahame-Smith, D. G., Behavioural evidence for a functional interaction between central 5-HT$_2$ and 5-HT$_{1A}$ receptors, *Br. J. Pharmacol.*, 100, 793, 1990.

44. Berendsen, H. H. G. and Broekkamp, C. L. E., Behavioural evidence for functional interactions between 5-HT-receptor subtypes in rats and mice, *Br. J. Pharmacol.*, 101, 667, 1990.

45. Darmani, N. A., Martin, B. R., Pandey, U., and Glennon, R. A., Do functional relationships exist betweeen 5-HT$_{1A}$ and 5-HT$_2$ receptors? *Pharmacol. Biochem. Behav.*, 36, 901, 1990.

46. Lund, A. and Mjellem, N., Chronic, combined treatment with desipramine and mianserin: enhanced 5-HT$_{1A}$ receptor function and altered 5-HT$_{1A}$/5-HT$_2$ receptor interaction in rats, *Pharmacol. Biochem. Behav.*, 45, 777, 1993.

47. Van de Kar, L. D., Neuroendocrine pharmacology of serotonergic (5-HT) neurons, *Annu. Rev. Pharmacol. Toxicol.*, 31, 289, 1991.

48. Whitaker-Azmitia, P. W. and Peroutka, S. J., Eds.,*The Neuropharmacology of Serotonin* (Annals NYAS, Vol. 600), New York Academy of Sciences, New York, 1990.

49. Bevan, R., Cools, A. R., and Archer, T., Eds., *Behavioural Pharmacology of 5-HT*, Lawrence Erlbaum, Hillsdale, NJ, 1989.

50. Pfaff, D. and Keiner, M., Atlas of estradiol-concentrating cells in the central nervous system of the female rat, *J. Comp. Neurol.*, 151, 121, 1973.

51. Stumpf, W. E., Binding of estrogens, progestagens, androgens and corticosteroids in the central nervous system, in *Interaction between the Nervous and the Endocrine Systems*, Verhofstad, A. A. J. and van Kemenade, J. A. M., Eds., University of Nijmegen Press, Sweden, 1979, 103.

52. Simerly, R. B., Chang, C., Muramatsu, M., and Swanson, L. W., Distribution of androgen and estrogen receptor mRNA-containing cells in the rat brain: an *in situ* hybridization study, *J. Comp. Neurol.*, 294, 76, 1990.

53. Rainbow, T. C., Parsons, B., and McEwen, B. S., Sex differences in rat brain estrogen and progestin receptors, *Nature*, 300, 648, 1982.

54. Munn, A. R., Sar, M., and Stumpf, W. E., Topographic distribution of progestin target cells in hamster brain and pituitary after injection of [³H]R5020, *Brain Res.*, 274, 1, 1983.

55. MacLusky, N. J. and McEwen, B. S., Progestin receptors in rat brain: distribution and properties of cytoplasmic progestin-binding sites, *Endocrinology*, 106, 192, 1980.

56. Parsons, B., Rainbow, T. C., MacLusky, N. J., and McEwen, B. S., Progestin receptor levels in rat hypothalamic and limbic nuclei, *J. Neurosci.,* 10, 1446, 1982.

57. Bethea, C. L., Fahrenbach, W. H., Sprangers, S. A., and Freesh, F., Immunocytochemical localization of progestin receptors in monkey hypothalamus: effect of estrogen and progestin, *Endocrinology*, 130, 895, 1992.

58. Bethea, C. L., Colocalization of progestin receptors with serotonin in raphe neurons of macaque, *Neuroendocrinology*, 57, 1, 1993.

59. Bitar, M. S., Ota, M., Linnoila, M., and Shapiro, B. H., Modification of gonadectomy-induced increases in brain monoamine metabolism by steroid hormones in male and female rats, *Psychoneuroendocrinology*, 16, 547, 1991.

60. Gabriel, S. M., Clark, J. T., Kalra, P. S., Kalra, S. P., and Simpkins, J. W., Chronic morphine and testosterone treatment: effects on norepinephrine and serotonin metabolism and gonadotrophin secretion in male rats, *Brain Res.,* 447, 200, 1988.

61. Bonson, K. R., Johnson, R. G., Fiorella, D., Rabin, R. A., and Winter, J. C., Serotonin control of androgen-induced dominance, *Pharmacol. Biochem. Behav.,* 49, 313, 1994.

62. Goudsmit, E., Feenstra, M. G. P., and Swaab, D. F., Central monoamine metabolism in the Brown-Norway rat in relation to aging and testosterone, *Brain Res. Bull.,* 25, 755, 1990.

63. Bliss, E. L., Thatcher, W., and Ailion, J., Relationship of stress to brain serotonin and 5-hydroxyindoleacetic acid, *J. Psychiatr. Res.,* 9, 71, 1972.

64. Carlsson, M., Svennson, K., Erikkson, E., and Carlsson, A., Rat brain serotonin: biochemical and functional evidence for a sex difference, *J. Neural Transm.,* 63, 297, 1985.

65. Dickinson, S. L. and Curzon, G., 5-hydroxytryptamine-mediated behavior in male and female rats, *Neuropharmacology*, 25, 771, 1986.

66. Kennett, G. A., Chaouloff, F., Marcou, M., and Curzon, G., Female rats are more vulnerable than males in an animal model of depression: the possible role of serotonin, *Brain Res.,* 382, 416, 1986.

67. Haleem, D. J., Kennett, G. A., and Curzon, G., Hippocampal 5-hydroxytryptamine synthesis is greater in female rats than in males and more decreased by the 5-HT$_{1A}$ agonist 8-OH-DPAT, *J. Neural Transm.,* 79, 93, 1990.

68. Heinsbroek, R. P. W., Van Haaren, F., Feenstra, M. G. P., Boon, P., and Van de Poll, N. E., Controllable and uncontrollable footshock and monoaminergic activity in the frontal cortex of male and female rats, *Brain Res.,* 551, 247, 1991.

69. Heinsbroek, R. P. W., Van Haaren, F., Feenstra, M. G. P., Endert, E., and Van de Poll, N. E., Sex- and time-dependent changes in neurochemical and hormonal variables induced by predictable and unpredictable footshock, *Physiol. Behav.,* 49, 1251, 1991.

70. Carlsson, M. and Carlsson, A., *In vivo* evidence for a greater brain tryptophan hydroxylase capacity in female than in male rats, *Naunyn-Schmiedeberg Arch. Pharmacol.,* 338, 345, 1988.

71. González, M. I. and Leret, M. L., Extrahypothalamic serotonergic modification after masculinization induced by neonatal gonadal hormones, *Pharmacol. Biochem. Behav.,* 41, 329, 1992.

72. Simerly, R. B., Swanson, L. W., and Gorski, R. A., Demonstration of a sexual dimorphism in the distribution of serotonin-immunoreactive fibers in the medial preoptic nucleus of the rat, *J. Comp. Neurol.,* 225, 151, 1984.

73. Simerly, R. B., Swanson, L. W., and Gorski, R. A., Reversal of the sexual dimorphic distribution of serotonin-immunoreactive fibers in the medial preoptic nucleus by treatment with perinatal androgen, *Brain Res.,* 340, 91, 1985.

74. Hardin, C. M., Sex differences in serotonin synthesis from 5-hydroxytryptophan in neonatal rat brain, *Brain Res.,* 59, 437, 1973.

75. Hardin, C. M., Sex differences and the effects of testosterone injections on biogenic amine levels of neonatal rat brain, *Brain Res.,* 62, 286, 1973.

76. Watts, A. G. and Stanley, H. F., Indoleamines in the hypothalamus and area of the midbrain raphe nuclei of male and female rats throughout postnatal development, *Neuroendocrinology*, 38, 461, 1984.

77. Fischette, C. T., Biegon, A., and McEwen, B. S., Sex differences in serotonin 1 binding in rat brain, *Science*, 222, 333, 1983.

78. Frankfurt, M., McKittrick, C. R., Mendelson, S. D., and McEwen, B. S., Effect of 5,7-dihydroxytryptamine, ovariectomy and gonadal steroids on serotonin receptor binding in rat brain, *Neuroendocrinology*, 59, 245, 1994.

79. Mendelson, S. D. and McEwen, B. S., Testosterone increases the concentration of [³H]8-hydroxy-2-(di-*n*-propylamino)tetralin binding at 5-HT$_{1A}$ receptors in the medial preoptic nucleus of the castrated male rat, *Eur. J. Pharmacol.,* 181, 329, 1990.

80. Saudou, F., Aït Amara, D., Dierich, A., Lemeur, M., Ramboz, S., Segu, L., Buhot, M. C., and Hen, R., Enhanced aggressive behavior in mice lacking 5-HT$_{1B}$ receptor, *Science*, 265, 1175, 1994.

81. Mendelson, S. D. and McEwen, B. S., Chronic testosterone propionate treatment decreases the concentration of [³H]quipazine binding at 5-HT$_3$ receptor in the amygdala of the castrated male rat, *Brain Res.,* 528, 339, 1990.

82. Sandrini, M., Vergoni, A. V., and Bertolini, A., [³H]Imipramine binding in discrete brain areas is affected by castration in male rats, *Brain Res.,* 496, 29, 1989.

83. Mendelson, S. D., McKittrick, C. R., and McEwen, B. S., Autoradiographic analyses of the effects of estradiol benzoate on [³H]paroxetine binding in the cerebral cortex and dorsal hippocampus of gonadectomized male and female rats, *Brain Res.,* 601, 299, 1993.

84. Matsuda, T., Nakano, Y., Kanda, T., Iwata, H., and Baba, A., Gonadal hormones affect the hypothermia induced by serotonin$_1$ (5-HT$_{1A}$) receptor activation, *Life Sci.*, 48, 1627, 1991.

85. Matsuda, T., Nakano, Y., Kanda, T., Iwata, H., and Baba, A., Gonadectomy changes the pituitary-adrenocortical response in mice to 5-HT$_{1A}$ receptor agonists, *Eur. J. Pharmacol.*, 200, 299, 1991.

86. Uphouse, L., Salamanca, S., and Caldarola-Pastuszka, M., Gender and estrous cycle differences in the response to the 5-HT$_{1A}$ agonist 8-OH-DPAT, *Pharmacol. Biochem. Behav.*, 40, 901, 1991.

87. Haleem, D. J., Kennett, G. A., Whitton, P. S., and Curzon, G., 8-OH-DPAT increases corticosterone but not other 5-HT$_{1A}$ receptor-dependent responses more in females, *Eur. J. Pharmacol.*, 164, 435, 1989.

88. Fischette, C. T., Biegon, A., and McEwen, B. S., Sex steroid modulation of the serotonin behavioral syndrome, *Life Sci.*, 35, 1197, 1984.

89. Mendelson, S. D. and Gorzalka, B. B., Sex differences in the effects of 1-(m-trifluoromethylphenyl)piperazine and 1-(m-chlorophenyl)piperazine on copulatory behavior in the rat, *Neuropharmacology*, 29, 783, 1990.

90. Gonzalez, M. I., Farabollini, F., Albonetti, E., and Wilson, C. A., Interactions between 5-hydroxytryptamine (5-HT) and testosterone in the control of sexual and nonsexual behavior in male and female rats, *Pharmacol. Biochem. Behav.*, 47, 591, 1994.

91. Blanchard, D. C., Cholvanich, P., Blanchard, R. J., Clow, D. W., Hammer, R. P., Jr., Rowlett, J. K., and Bardo, M. T., Serotonin, but not dopamine, metabolites are increased in selected brain regions of subordinate male rats in a colony environment, *Brain Res.*, 568, 61, 1991.

92. Higley, J. D., Mehlman, P. T., Taub, D. M., Higley, S. B., Suomi, S. J., Vickers, J. H., and Linnoila, M., Cerebrospinal fluid monoamine and adrenal correlates of aggression in free-ranging rhesus monkeys, *Arch. Gen. Psychiatry*, 49, 436, 1992.

93. Raleigh, M. J., McGuire, M. T., Brammer, G. L., and Yuwiler, A., Social and environmental influences on blood serotonin concentrations in monkeys, *Arch. Gen. Psychiatry*, 41, 405, 1984.

94. Raleigh, M. J., Brammer, G. L., McGuire, M. T., and Yuwiler, A., Dominant social status facilitates the behavioral effects of serotonergic agonists, *Brain Res.*, 348, 274, 1985.

95. Raleigh, M. J., McGuire, M. T., Brammer, G. L., Pollack, D. B., and Yuwiler, A., Serotonergic mechanisms promote dominance acquisition in adult male vervet monkeys, *Brain Res.*, 559, 181, 1991.

96. Ladisich, W., Effect of progesterone on regional 5-hydroxytryptamine metabolism in the rat brain, *Neuropharmacology*, 13, 877, 1974.

97. Wirz-Justice, A., Hackmann, E., and Lichtsteiner, M., The effect of oestradiol dipropionate and progesterone on monoamine uptake in rat brain, *J. Neurochem.*, 22, 187, 1974.

98. Cone, R. I., Davis, G. A., and Goy, R. W., Effects of ovarian steroids on serotonin metabolism within grossly dissected and microdissected brain regions of the ovariectomized rat, *Brain Res. Bull.*, 7, 639, 1981.

99. Héry, M., Faudon, M., Dusticier, G., and Héry, F., Daily variations in serotonin metabolism in the suprachiasmatic nucleus of the rat: influence of oestradiol impregnation, *J. Endocrinol.*, 94, 157, 1982.

100. Biegon, A., Reches, A., Snyder, L., and McEwen, B. S., Serotonin and noradrenergic receptors in the rat brain: modulation by chronic exposure to ovarian hormones, *Life Sci.*, 32, 2015, 1983.

101. Di Paolo, T., Daigle, M., Picard, V., and Barden, N., Effect of acute and chronic 17β-estradiol treatment on serotonin and 5-hydroxyindole acetic acid content of discrete brain nuclei of ovariectomized rat, *Exp. Brain Res.*, 51, 73, 1983.

102. Johnson, M. D. and Crowley, W. R., Acute effects of estradiol on circulating luteinizing hormone and prolactin concentrations and on serotonin turnover in individual brain nuclei, *Endocrinology*, 113, 1935, 1983.

103. Di Paolo, T., Lévesque, D., and Daigle, M., A physiological dose of progesterone affects rat striatum biogenic amine metabolism, *Eur. J. Pharmacol.*, 125, 11, 1986.

104. King, T. S., Steger, R. W., and Morgan, W. W., Effect of ovarian steroids to stimulate region-specific hypothalamic 5-hydroxytryptamine synthesis in ovariectomized rats, *Neuroendocrinology*, 42, 344, 1986.

105. Renner, K. J., Allen, D. L., and Luine, V. N., Monoamine levels and turnover in brain: relationship to priming actions of estrogen, *Brain Res. Bull.*, 16, 469, 1986.

106. Renner, K. and Luine, V., Analysis of temporal and dose-dependent effects of estrogen on monoamines in brain nuclei, *Brain Res.*, 366, 64, 1986.

107. Renner, K. J., Krey, L. C., and Luine, V. N., Effect of progesterone on monamine turnover in the brain of the estrogen-primed rat, *Brain Res. Bull.*, 19, 195, 1987.

108. Cohen, I. R. and Wise, P. M., Effects of estradiol on the diurnal rhythm of serotonin activity in microdissected brain areas of ovariectomized rats, *Endocrinology*, 122, 2619, 1988.

109. James, M. D., Hole, D. R., and Wilson, C. A., Differential involvement of 5-hydroxytryptamine (5HT) in specific hypothalamic areas in the mediation of steroid-induced changes in gonadotrophin release and sexual behavior in female rats, *Neuroendocrinology*, 49, 561, 1989.

110. Morrissette, M., Lévesque, D., Bélanger, A., and Di Paolo, T., A physiological dose of estradiol with progesterone affects striatum biogenic amines, *Can. J. Physiol. Pharmacol.*, 68, 1520, 1990.

111. Gereau, R. W., IV, Kedzie, K. A., and Renner, K. J., Effect of progesterone on serotonin turnover in rats primed with estrogen implants into the ventromedial hypothalamus, *Brain Res. Bull.*, 32, 293, 1993.

112. Desan, P. H., Woodmansee, W. W., Ryan, S. M., Smock, T. K., and Maier, S. F., Monoamine neurotransmitters and metabolites during the estrous cycle, pregnancy and the postpartum period, *Pharmacol. Biochem. Behav.*, 30, 563, 1988.

113. Luine, V. N., Serotonin, catecholamines and metabolites in discrete brain areas in relation to lordotic responding on proestrus, *Neuroendocrinology*, 57, 946, 1993.
114. Wilson, C. A. and Hunter, A. J., Progesterone stimulates sexual behavior in female rats by increasing 5-HT activity on 5-HT$_2$ receptors, *Brain Res.*, 333, 223, 1985.
115. Meyer, D. C., Singh, J., and Jiminez, A. E., Uptake of serotonin and norepinephrine in hypothalamic and limbic brain regions during the estrous cycle and the effect of neurotoxic lesions on estrous cyclicity, *Brain Res. Bull.*, 10, 639, 1983.
116. Tomogane, H., Mizoguchi, K., and Yokoyama, A., Effects of progesterone on concentrations on monoamines in hypothalamic areas and plasma prolactin levels in rats, *Proc. Soc. Exp. Biol. Med.*, 195, 208, 1990.
117. Moguilevsky, J. A., Arias, P., Szwarcfarb, B., Carbone, S., and Rondina, D., Sexual maturation modifies the catecholaminergic control of gonadotrophin secretion and the effect of ovarian hormones on hypothalamic neurotransmitters in female rats, *Neuroendocrinology*, 52, 393, 1990.
118. Biegon, A. and McEwen, B. S., Modulation by estradiol of serotonin$_1$ receptors in brain, *J. Neurosci.*, 2, 199, 1982.
119. Biegon, A., Fischette, C. T., Rainbow, T. C., and McEwen, B. S., Serotonin receptor modulation by estrogen in discrete brain nuclei, *Neuroendocrinology*, 35, 287, 1982.
120. Williams, J. and Uphouse, L., Serotonin binding sites during proestrus and following estradiol treatment, *Pharmacol. Biochem. Behav.*, 33, 615, 1989.
121. Halpern, R. E., Black, E. W., and Lakoski, J. M., Aging but not estrogen alters regional [^{3}H]5-HT binding in the CNS of Fischer 344 rats, *Neurosci. Lett.*, 101, 293, 1989.
122. Clarke, W. P. and Maayani, S., Estrogen effects on 5-HT$_{1A}$ receptors in hippocampal membranes from ovariectomized rats: functional and binding studies, *Brain Res.*, 518, 287, 1990.
123. Goetz, C., Bourgoin, S., Cesselin, F., Brandi, A., Bression, D., Martinet, M., Peillon, F., and Hamon, M., Alterations in central neurotransmitter receptor binding sites following estradiol implantation in female rats, *Neurochem. Int.*, 5, 375, 1983.
124. Ravizza, L., Nicoletti, F., Pozzi, O., and Barbaccia, M. L., Repeated daily treatments with estradiol benzoate increase the [^{3}H]imipramine binding in male rat frontal cortex, *Eur. J. Pharmacol.*, 107, 395, 1985.
125. Biegon, A., Bercovitz, H., and Samuel, D., Serotonin receptor concentration during the estrous cycle of the rat, *Brain Res.*, 187, 221, 1980.
126. Uphouse, L., Williams, J., Eckols, K., and Sierra, V., Variations in binding of [^{3}H]5-HT to cortical membranes during the female rat estrous cycle, *Brain Res.*, 381, 376, 1986.
127. Stockert, M. and De Robertis, E., Effect of ovariectomy and estrogen on [^{3}H]imipramine binding to different regions of rat brain, *Eur. J. Pharmacol.*, 119, 255, 1985.
128. Rehavi, M., Sepcuti, H., and Weizman, A., Upregulation of imipramine binding and serotonin uptake by estradiol in female rat brain, *Brain Res.*, 410, 135, 1987.
129. Wilson, M. A., Dwyer, K. D., and Roy, E. J., Direct effects of ovarian hormones on antidepressant binding sites, *Brain Res. Bull.*, 22, 181, 1989.
130. Michel, M. C., Rother, A., Hiemke, C., and Ghraf, R., Inhibition of synaptosomal high-affinity uptake of dopamine and serotonin by estrogen agonists and antagonists, *Biochem. Pharmacol.*, 36, 3175, 1987.
131. Kow, L. M. and Pfaff, D. W., Estrogen effects on neuronal responsiveness to electrical and neurotransmitter stimulation: an *in vitro* study on the ventromedial nucleus of the hypothalamus, *Brain Res.*, 347, 1, 1985.
132. Pan, J. T., Kow, L. M., and Pfaff, D. W., Single-unit activity of hypothalamic arcuate neurons in brain tissue slices, *Neuroendocrinology*, 43, 189, 1986.
133. Clarke, W. P. and Goldfarb, J., Estrogen enhances a 5-HT$_{1A}$ response in hippocampal slices from female rats, *Eur. J. Pharmacol.*, 160, 195, 1989.
134. Beck, S. G., Clarke, W. P., and Goldfarb, J., Chronic estrogen effects on 5-hydroxytryptamine-mediated responses in hippocampal pyramidal cells of female rats, *Neurosci. Lett.*, 106, 181, 1989.
135. Moreines, J., Kelton, M., Luine, V. N., Pfaff, D. W., and McEwen, B. S., Hypothalamic serotonin lesions unmask hormone responsiveness of lordosis behavior in adult male rats, *Neuroendocrinology*, 47, 453, 1988.
136. Luine, V. N., Frankfurt, M., Rainbow, T. C., Biegon, A., and Azmitia, E., Intrahypothalamic 5,7-dihydroxytryptamine facilitates feminine sexual behavior and decreases [^{3}H]imipramine binding and 5-HT uptake, *Brain Res.*, 264, 344, 1983.
137. Mendelson, S. D., A review and reevaluation of the role of serotonin in the modulation of lordosis behavior in the female rat, *Neurosci. Biobehav. Rev.*, 16, 309, 1992.
138. Brotherton, C. S. and Doggett, N. S., Modification of the 5-hydroxytryptophan-induced head-twitch response by exogenous endocrine agents, *Psychopharmacology*, 58, 145, 1978.
139. Salamanca, S. and Uphouse, L., Estradiol modulation of the hyperphagia induced by the 5-HT$_{1A}$ agonist, 8-OH-DPAT, *Pharmacol. Biochem. Behav.*, 43, 953, 1992.
140. McEwen, B. S., Weiss, J. M., and Schwartz, L. S., Selective retention of corticosterone by limbic structures in rat brain, *Nature*, 220, 911, 1968.
141. Gerlach, J. L. and McEwen, B. S., Rat brain binds adrenal steroid hormone: radioautography of hippocampus with corticosterone, *Science*, 175, 1133, 1972.
142. McEwen, B. S., Weiss, J. M., and Schwartz, L. S., Retention of corticosterone by cell nuclei from brain regions of adrenalectomized rats, *Brain Res.*, 17, 471, 1970.

143. De Kloet, E. R., Wallach, G., and McEwen, B. S., Differences in corticosterone and dexamethasone binding to rat brain and pituitary, *Endocrinology*, 96, 598, 1975.

144. Reul, J. M. H. M. and De Kloet, E. R., Two receptor systems for corticosterone in rat brain: microdistribution and differential occupation, *Endocrinology*, 117, 2505, 1985.

145. McEwen, B. S., De Kloet, E. R., and Rostene, W., Adrenal steroid receptors and actions in the nervous system, *Physiol. Rev.*, 66, 1121, 1986.

146. De Kloet, E. R., Brain corticosteroid receptor balance and homeostatic control, *Front. Neuroendocrinol.*, 12, 95, 1991.

147. Aronsson, M., Fuxe, K., Dong, Y., Agnati, L. F., Okret, S., and Gustafsson, J. A., Localization of glucocorticoid receptor mRNA in the male rat brain by *in situ* hybridization, *Proc. Natl. Acad. Sci. U.S.A.*, 85, 9331, 1988.

148. Chao, H. M., Choo, P. H., and McEwen, B. S., Glucocorticoid and mineralocorticoid receptor mRNA expression in rat brain, *Neuroendocrinology*, 50, 365, 1989.

149. Yang, G., Matocha, M. F., and Rapoport, S. I., Localization of glucocorticoid receptor messenger ribonucleic acid in hippocampus of rat brain using *in situ* hybridization, *Mol. Endocrinol.*, 2, 682, 1988.

150. Herman, J. P., Watson, S. J., Chao, H. M., Coirini, H., and McEwen, B. S., Diurnal regulation of glucocorticoid receptor and mineralocorticoid receptor mRNAs in rat hippocampus, *Mol. Cell. Neurosci.*, 4, 181, 1993.

151. Van Eekelen, J. A. M., Jiang, W., De Kloet, E. R., and Bohn, M. C., Distribution of the mineralocorticoid and the glucocorticoid receptor mRNAs in the rat hippocampus, *J. Neurosci. Res.*, 21, 88, 1988.

152. Kiss, J. Z., Van Eekelen, J. A. M., Reul, J. M. H. M., Westphal, H. M., and De Kloet, E. R., Glucocorticoid receptor in magnocellular neurosecretory cells, *Endocrinology*, 122, 444, 1988.

153. Fuxe, K., Härfstrand, A., Agnati, L. F., Yu, Z. Y., Cintra, A., Wikström, A. C., Okret, S., Cantoni, E., and Gustafsson, J. A., Immunocytochemical studies on the localization of glucocorticoid receptor immunoreactive nerve cells in the lower brain stem and spinal cord of the male rat using a monoclonal antibody against rat liver glucocorticoid receptor, *Neurosci. Lett.*, 60, 1, 1985.

154. Härfstrand, A., Fuxe, K., Cintra, A., Agnati, L. F., Zni, I., Wikström, A. C., Okret, S., Yu, Z. Y., Goldstein, M., Steinbusch, H., Verhofstad, A., and Gustafsson, J. A., Glucocorticoid receptor immunoreactivity in monoaminergic neurons of rat brain, *Proc. Natl. Acad. Sci. U.S.A.*, 83, 9779, 1986.

155. Arriza, J. L., Simerly, R. B., Swanson, L. W., and Evans, R. M., The neuronal mineralocorticoid receptor as a mediator of glucocorticoid response, *Neuron*, 1, 887, 1988.

156. Herman, J. P., Patel, P. D., Akil, H., and Watson, S. J., Localization and regulation of glucocorticoid and mineralo-corticoid receptor messenger RNAs in the hippocampal formation of the rat, *Mol. Endocrinol.*, 3, 1886, 1989.

157. Reul, J. M. H. M., Pearce, P. T., Funder, J. W., and Krozowski, Z. S., Type I and type II corticosteroid receptor gene expression in the rat: effect of adrenalectomy and dexamethasone administration, *Mol. Endocrinol.*, 3, 1674, 1989.

158. Chao, H. M., Blanchard, D. C., Blanchard, R. J., McEwen, B. S., and Sakai, R. R., The effect of social stress on hippocampal gene expression, *Mol. Cell. Neurosci.*, 4, 543, 1993.

159. Azmitia, E. C. and McEwen, B. S., Corticosterone regulation of tryptophan hydroxylase in midbrain of the rat, *Science*, 166, 1274, 1969.

160. Azmitia, E. C. and McEwen, B. S., Adrenalcortical influence on rat brain tryptophan hydroxylase activity, *Brain Res.*, 78, 291, 1974.

161. Azmitia, E. C. and McEwen, B. S., Early response of rat brain tryptophan hydroxylase activity to cycloheximide, puromycin and corticosterone, *J. Neurochem.*, 27, 773, 1976.

162. Azmitia, E. C., Liao, B., and Chen, Y., Increase of tryptophan hydroxylase enzyme protein by dexamethasone in adrenalectomized rat midbrain, *J. Neurosci.*, 13, 5041, 1993.

163. Van Loon, G. R., Shum, M., and Sole, M. J., Decreased brain serotonin turnover after short term (two-hour) adrenalectomy in rats: a comparison of four turnover methods, *Endocrinology*, 108, 1392, 1981.

164. De Kloet, E. R., Kovacs, G. L., Szabó, G., Telegdy, G., Bohus, B., and Versteeg, D. H. G., Decreased serotonin turnover in the dorsal hippocampus of rat brain shortly after adrenalectomy: selective normalization after cortico-sterone substitution, *Brain Res.*, 239, 659, 1982.

165. De Kloet, E. R., Versteeg, D. H. G., and Kovacs, G. L., Aldosterone blocks the response to corticosterone in the raphe-hippocampal serotonin system, *Brain Res.*, 264, 323, 1983.

166. Jhanwar-Uniyal, M., Renner, K., Bailo, M., Luine, V. N., and Leibowitz, S. F., Serotonin and 5-hydroxyindoleacetic acid levels in discrete hypothalamic areas of the rat brain: relation to circulating corticosterone, *Neurosci. Lett.*, 79, 145, 1987.

167. Sze, P. Y., Neckers, L., and Towle, A. C., Glucocorticoids act as regulatory factor for brain tryptophan hydroxylase, *J. Neurochem.*, 26, 169, 1976.

168. Alves, S. E., Akbari, H. M., Azmitia, E. C., and Strand, F. L., Neonatal ACTH and corticosterone alter hypothalamic monoamine innervation and reproductive parameters in the female rat, *Peptides*, 14, 379, 1993.

169. De Kloet, E. R., Sybesma, H., and Reul, H. M. H. M., Selective control by corticosterone of serotonin$_1$ receptor capacity in raphe-hippocampal system, *Neuroendocrinology*, 42, 513, 1986.

170. Mendelson, S. D. and McEwen, B. S., Autoradiographic analyses of the effects of adrenalectomy and corticosterone of 5-HT$_{1A}$ and 5-HT$_{1B}$ receptors in the dorsal hippocampus and cortex of the rat, *Neuroendocrinology*, 55, 444, 1992.

171. Mendelson, S. D. and McEwen, B. S., Quantitative autoradiographic analyses of the time course and reversibility of corticosterone-induced decreases in binding at 5-HT$_{1A}$ receptors in rat forebrain, *Neuroendocrinology*, 56, 881, 1992.

172. Chalmers, D. T., Kwak, S. P., Mansour, A., Akil, H., and Watson, S. J., Corticosteroids regulate brain hippocampal 5-HT$_{1A}$ receptor mRNA expression, *J. Neurosci.*, 13, 914, 1993.

173. Meijer, O. C. and De Kloet, E. R., Corticosterone suppresses the expression of 5-HT$_{1A}$ receptor mRNA in rat dentate gyrus, *Eur. J. Pharmacol.*, 266, 255, 1994.

174. Liao, B., Miesak, B., and Azmitia, E. C., Loss of 5-HT$_{1A}$ receptor mRNA in the dentate gyrus of the long-term adrenalectomized rats and rapid reversal by dexamethasone, *Mol. Brain Res.*, 19, 328, 1993.

175. Sloviter, R. S., Valiquette, G., Abrams, G. M., Ronk, E. C., Sollas, A. L., Paul, L. A., and Neubort, S., Selective loss of hippocampal granule cells in the mature rat brain after adrenalectomy, *Science*, 243, 535, 1989.

176. Burnet, P. W. J., Mefford, I. N., Smith, C. C., Gold, P. W., and Sternberg, E. M., Hippocampal 8-[^{3}H]hydroxy-2-(di-*n*-propylamino)tetralin binding site densities, serotonin receptor (5-HT$_{1A}$) messenger ribonucleic acid abundance, and serotonin levels parallel the activity of the hypothalamopituitary-adrenal axis in rat, *J. Neurochem.*, 59, 1062, 1992.

177. Kuroda, Y., Mikuni, M., Ogawa, T., and Takahashi, K., Effect of ACTH, adrenalectomy and the combination treatment of the density of 5-HT$_2$ receptor binding sites in neocortex of rat forebrain and 5-HT$_2$ receptor-mediated wet-dog shake behaviors, *Psychopharmacology*, 108, 27, 1992.

178. Kuroda, Y., Mikuni, M., Nomura, N., and Takahashi, K., Differential effect of subchronic dexamethasone treatment on serotonin-2 and β-adrenergic receptors in the rat cerebral cortex and hippocampus, *Neurosci. Lett.*, 155, 195, 1993.

179. Muraoka, S. I., Mikuni, M., Kagaya, A., Saitoh, K., and Takahashi, K., Dexamethasone potentiates serotonin-2 receptor mediated intracellular Ca^{2+} mobilization in C6 glioma cells, *Neuroendocrinology*, 57, 322, 1993.

180. Lee, P. H. K. and Chan, M. Y., Changes induced by corticosterone and adrenalectomy in synaptsosomal and platelet uptake and binding of 5-HT: the relationship to [^{3}H]imipramine binding, *Neuropharmacology*, 24, 1043, 1985.

181. Arora, R. C. and Meltzer, H. Y., Effect of adrenalectomy and corticosterone on [^{3}H]imipramine binding in rat blood platelets and brain, *Eur. J. Pharmacol.*, 123, 415, 1986.

182. Joëls, M., Hesen, W., and De Kloet, E. R., Mineralocorticoid hormones suppress serotonin-induced hyperpolarization of rat hippocampal CA1 neurons, *J. Neurosci.*, 11, 2288, 1991.

183. Karst, H. and Joëls, M., The induction of corticosteroid actions on membrane properties of hippocampal CA1 neurons requires protein synthesis, *Neurosci. Lett.*, 130, 27, 1991.

184. Joëls, M. and De Kloet, E. R., Coordinative mineralocorticoid and glucocorticoid receptor-mediated control of responses to serotonin in rat hippocampus, *Neuroendocrinology*, 55, 344, 1992.

185. Valero, I., Stewart, J., McNaughton, N., and Gray, J. A., Septal driving of the hippocampal theta rhythm as a function of frequency in the male rat: effects of adreno-pituitary hormones, *Neuroscience*, 2, 1029, 1977.

186. McNaughton, N., Azmitia, E. C., Williams, J. H., Buchan, A., and Gray, J. A., Septal elicitation of hippocampal theta rhythm after localized de-afferentation of serotoninergic fibers, *Brain Res.*, 200, 259, 1980.

187. Azmitia, E. C., McNaughton, N., Tsaltas, L., Fillenz, M., and Gray, J. A., Interactions between hippocampal serotonin and the pituitary-adrenal axis in the septal driving of hippocampal theta rhythm, *Neuroendocrinology*, 39, 471, 1984.

188. Young, A. H., MacDonald, L. M., St. John, H., Dick, H., and Goodwin, G. M., The effects of corticosterone on 5-HT receptor function in rodents, *Neuropharmacology*, 31, 433, 1992.

189. Young, A. H., Dow, R. C., Goodwin, G. M., and Fink, G., The effects of adrenalectomy and ovariectomy on the behavioral and hypothermic responses of rats to 8-hydroxy-2(di-*n*-propylamino)tetralin (8-OH-DPAT), *Neuropharmacology*, 32, 653, 1993.

190. Bagdy, G., Calogero, A. E., Aulakh, C. S., Szemeredi, K., and Murphy, D. L., Long-term cortisol treatment impairs behavioral and neuroendocrine responses to 5-HT$_1$ agonists in rat, *Neuroendocrinology*, 50, 241, 1989.

191. Freo, U., Holloway, H. W., Kalogeras, K., Rapoport, S. I., and Soncrant, T. T., Adrenalectomy or metyrapone-pretreatment abolishes cerebral metabolic responses to the serotonin agonist 1-(2,5-dimethoxy-4-iodophenyl)-2-aminopropane (DOI) in the hippocampus, *Brain Res.*, 586, 256, 1992.

192. Nausieda, P. A., Carvey, P. M., and Weiner, W. J., Modification of central serotonergic and dopaminergic behaviors in the course of chronic corticosteroid administration, *Eur. J. Pharmacol.*, 78, 335, 1982.

193. Dickinson, S. L., Kennett, G. A., and Curzon, G., Reduced 5-hydroxytryptamine-dependent behavior in rats following chronic corticosterone treatment, *Brain Res.*, 345, 10, 1985.

194. Saito, N., Guitart, X., Hayward, M., Tallman, J. F., Duman, R. S., and Nestler, E. J., Corticosterone differentially regulates the expression of G$_{s\alpha}$ and G$_{i\alpha}$ messenger RNA and protein in rat cerebral cortex, *Proc. Natl. Acad. Sci. U.S.A.*, 86, 3906, 1989.

195. Harrington, M. A. and Peroutka, S. J., Modulation of 5-hydroxytryptamine$_{1A}$ receptor density by nonhydrolyzable GTP analogs, *J. Neurochem.*, 54, 294, 1990.

196. Thierry, A. M., Fekete, M., and Glowinski, J., Effects of stress on the metabolism of noradrenaline, dopamine and serotonin (5HT) in the central nervous system of the rat. II. Modifications of serotonin metabolism, *Eur. J. Pharmacol.*, 4, 384, 1968.

197. Curzon, G., Joseph, M. H., and Knott, P. J., Effects of immobilization and food deprivation on rat brain tryptophan metabolism, *J. Neurochem.*, 19, 1967, 1972.

198. Morgan, W. W., Rudeen, P. K., and Pfeil, K. A., Effect of immobilization stress on serotonin content and turnover in regions of the rat brain, *Life Sci.*, 17, 143, 1975.

199. Joseph, M. H. and Kennett, G. A., Stress-induced release of 5-HT in the hippocampus and its dependence on increased tryptophan availability: an *in vivo* electrochemical study, *Brain Res.*, 270, 251, 1983.

200. Petty, F. and Sherman, A. D., Learned helplessness induction decreases *in vivo* cortical serotonin release, *Pharmacol. Biochem. Behav.*, 18, 649, 1983.

201. Johnston, C. A., Spinedi, E. J., and Negro-Vilar, A., Effect of acute ether stress on monoamine metabolism in median eminence and discrete hypothalamic nuclei of the rat brain and on anterior pituitary hormone secretion, *Neuroendocrinology*, 41, 83, 1985.

202. Beaulieu, S., Di Paolo, T., and Barden, N., Control of ACTH secretion by the central nucleus of the amygdala implication of the serotoninergic system and its relevance to the glucocorticoid delayed negative feedback mechanism, *Neuroendocrinology*, 44, 247, 1986.

203. Dunn, A. J., Elfvin, K. L., and Berridge, C. W., Changes in plasma corticosterone and cerebral biogenic amines and their catabolites during training and testing of mice in passive avoidance behavior, *Behav. Neural Biol.*, 46, 410, 1986.

204. Kennett, G. A., Curzon, G., Hunt, A., and Patel, A. J., Immobilization decreases amino acid concentrations in plasma but maintains or increases them in brain, *J. Neurochem.*, 46, 208, 1986.

205. Lee, E. H. Y., Lin, H. H., and Yin, H. M., Differential influences of different stressors upon midbrain raphe neurons in rats, *Neurosci. Lett.*, 80, 115, 1987.

206. Paris, J. M., Lorens, S. A., Van de Kar, L. D., Urban, J. H., Richardson-Morton, K. D., and Bethea, C. L., A comparison of acute stress paradigms: hormonal responses and hypothalamic serotonin, *Physiol. Behav.*, 39, 33, 1987.

207. Dunn, A. J., Changes in plasma and brain tryptophan and brain serotonin and 5-hydroxyindoleacetic acid after footshock stress, *Life Sci.*, 42, 1847, 1988.

208. Dunn, A. J., Stress-related changes in cerebral catecholamine and indoleamine metabolism: lack of effect of adrenalectomy and corticosterone, *J. Neurochem.*, 51, 406, 1988.

209. Mitchell, S. N. and Thomas, P. J., Effect of restraint stress and anxiolytics on 5-HT turnover in rat brain, *Pharmacology*, 37, 105, 1988.

210. Shimizu, N., Oomura, Y., and Aoyagi, K., Electrochemical analysis of hypothalamic serotonin metabolism accompanied by immobilization stress in rats, *Physiol. Behav.*, 46, 829, 1989.

211. Torda, T., Murgas, K., Cechova, E., Kiss, A., and Saavedra, J. M., Adrenergic regulation of [³H]ketanserin binding sites during immobilization stress in the rat frontal cortex, *Brain Res.*, 527, 198, 1990.

212. Pei, Q., Zetterström, T., and Fillenz, M., Tail pinch-induced changes in the turnover and release of dopamine and 5-hydroxytryptamine in different brain regions of the rat, *Neuroscience*, 35, 133, 1990.

213. Dunn, A. J. and Welch, J., Stress- and endotoxin-induced increases in brain tryptophan and serotonin metabolism depend on sympathetic nervous system activity, *J. Neurochem.*, 57, 1615, 1991.

214. Houdouin, F., Cespuglio, R., Gharib, A., Sarda, N., and Jouvet, M., Detection of the release of 5-hydroxyindole compounds in the hypothalamus and the n. raphe dorsalis throughout the sleep-waking cycle and during stressful situations in the rat: a polygraphic and voltammetric approach, *Exp. Brain Res.*, 85, 153, 1991.

215. Shimizu, N., Take, S., Hori, T., and Oomura, Y., *In vivo* measurement of hypothalamic serotonin release by intracerebroventricular microdialysis: significant enhancement by immobilization stress in rats, *Brain Res. Bull.*, 28, 727, 1992.

216. File, S. E., Zangrossi, H., and Andrews, N., Social interaction and elevated plus-maze tests: changes in release and uptake of 5-HT and GABA, *Neuropharmacology*, 32, 217, 1993.

217. File, S. E., Zangrossi, H., and Andrews, N., Novel environment and cat odor change GABA and 5-HT release and uptake in the rat, *Pharmacol. Biochem. Behav.*, 45, 931, 1993.

218. Clement, H. W., Schäfer, F., Ruwe, C., Gemsa, D., and Wesemann, W., Stress-induced changes of extracellular 5-hydroxyindoleacetic acid concentrations followed in the nucleus raphe dorsalis and the frontal cortex of the rat, *Brain Res.*, 614, 117, 1993.

219. Handa, R. J., Cross, M. K., George, M., Gordon, B. H., Burgess, L. H., Cabrera, T. M., Hata, N., Campbell, D. B., and Lorens, S. A., Neuroendocrine and neurochemical responses to novelty stress in young and old male F344 rats: effects of *d*-fenfluramine treatment, *Pharmacol. Biochem. Behav.*, 46, 101, 1993.

220. Inoue, T., Koyama, T., and Yamashita, I., Effect of conditioned fear stress on serotonin metabolism in the rat brain, *Pharmacol. Biochem. Behav.*, 44, 371, 1993.

221. Kawahara, H., Yoshida, M., Yokoo, H., Nishi, M., and Tanaka, M., Psychological stress increases serotonin release in the rat amygdala and prefrontal cortex assessed by *in vivo* microdialysis, *Neurosci. Lett.*, 162, 81, 1993.

222. De Souza, E. B. and Van Loon, G. R., Brain serotonin and catecholamine responses to repeated stress in rats, *Brain Res.*, 367, 77, 1986.

223. Adell, A., Garcia-Marquez, C., Armario, A., and Gelpi, E., Chronic stress increases serotonin and noradrenaline in rat brain and sensitizes their responses to a further acute stress, *J. Neurochem.*, 50, 1678, 1988.

224. Boadle-Biber, M. C., Corley, K. C., Graves, L., Phan, T. H., and Rosecrans, J., Increase in the activity of tryptophan hydroxylase from cortex and midbrain of male Fischer 344 rats in response to acute or repeated sound stress, *Brain Res.*, 482, 306, 1989.

225. Kennett, G. A., Dickinson, S. L., and Curzon, G., Enhancement of some 5-HT-dependent behavioral responses following repeated immobilization in rats, *Brain Res.*, 330, 253, 1985.

226. Chomicka, L. K., Effect of oestradiol on the responses of regional brain serotonin to stress in the ovariectomized rat, *J. Neural Transm.*, 67, 267, 1986.

227. Kitayama, I., Cintra, A., Janson, A. M., Fuxe, K., Agnati, L. F., Eneroth, P., Aronsson, M., Härfstrand, A., Steinbusch, H. W. M., Visser, T. J., Goldstein, M., Vale, W., and Gustafsson, J. A., Chronic immobilization stress: evidence for decreases of 5-hydroxy-tryptamine immunoreactivity and for increases of glucocorticoid receptor immunoreactivity in various brain regions of the male rat, *J. Neural Transm.*, 77, 93, 1989.

228. Mayeda, A. R., Simon, J. R., Hingtgen, J. N., Hofstetter, J. R., and Aprison, M. H., Activity-wheel stress and serotonergic hypersensitivity in rats, *Pharmacol. Biochem. Behav.*, 33, 349, 1989.

229. File, S. E., Andrews, N., and Zharkovsky, A., Handling habituation and chlordiazepoxide have different effects on GABA and 5-HT function in the frontal cortex and hippocampus, *Eur. J. Pharmacol.*, 190, 229, 1990.

230. Hata, T., Itoh, E., and Kawabata, A., Changes in CNS levels of serotonin and its metabolite in SART-stressed (repeatedly cold-stressed) rats, *Jpn. J. Pharmacol.*, 56, 101, 1991.

231. Zamfir, O., Broqua, P., Baudrie, V., and Chaouloff, F., Effects of cold stress on some 5-HT$_{1A}$, 5-HT$_{1C}$ and 5-HT$_2$ receptor-mediated responses, *Eur. J. Pharmacol.*, 219, 261, 1992.

232. Jaffe, E. H., De Frias, V., and Ibarra, C., Changes in basal and stimulated release of endogenous serotonin from different nuclei of rats subjected to two models of depression, *Neurosci. Lett.*, 162, 157, 1993.

233. Miachon, S., Rochet, T., Mathian, B., Barbagli, B., and Claustrat, B., Long-term isolation of Wistar rats alters brain monoamine turnover, blood corticosterone, and ACTH, *Brain Res. Bull.*, 32, 611, 1993.

234. Singh, V. B., Corley, K. C., Phan, T. H., and Boadle-Biber, M. C., Increases in the activity of tryptophan hydroxylase from rat cortex and midbrain in response to acute or repeated sound stress are blocked by adrenalectomy and restored by dexamethasone treatment, *Brain Res.*, 516, 66, 1990.

235. Kennett, G. A. and Joseph, M. H., The functional importance of increased brain tryptophan in the serotonergic response to restraint stress, *Neuropharmacology*, 20, 39, 1981.

236. Mendelson, S. D. and McEwen, B. S., Autoradiographic analyses of the effects of restraint-induced stress on 5-HT$_{1A}$, 5-HT$_{1B}$, and 5-HT$_2$ receptors in the dorsal hippocampus of male and female rats, *Neuroendocrinology*, 54, 454, 1991.

237. Watanabe, Y., Sakai, R. R., McEwen, B. S., and Mendelson, S. D., Stress and antidepressant effects on hippocampal and cortical 5-HT$_{1A}$ and 5-HT$_2$ receptors and transport sites for serotonin, *Brain Res.*, 615, 87, 1993.

238. McKittrick, C. R., Blanchard, D. C., Blanchard, R. J., McEwen, B. S., and Sakai, R. R., Serotonin receptor binding in a colony model of chronic social stress, *Biol. Psychiatry*, 37, 383, 1995.

239. Kawanami, T., Morinobu, S., Totsuka, S., and Endoh, M., Influence of stress and antidepressant treatment on 5-HT-stimulated phosphoinositide hydrolysis in rat brain, *Eur. J. Pharmacol.*, 26, 385, 1992.

240. Edwards, E., Harkins, K., Wright, G., and Henn, F. A., 5-HT$_{1B}$ receptors in an animal model of depression, *Neuropharmacology*, 30, 101, 1991.

241. Edwards, E., Harkins, K., Wright, G., and Henn, F., Modulation of [^{3}H]paroxetine binding to the 5-hydroxytryptamine uptake site in an animal model of depression, *J. Neurochem.*, 56, 1581, 1991.

242. Ohi, K., Mikuni, M., and Takahashi, K., Stress adaptation and hypersensitivity in 5-HT neuronal systems after repeated footshock, *Pharmacol. Biochem. Behav.*, 34, 603, 1989.

243. Wilkinson, L. O. and Jacobs, B. L., Lack of response of serotonergic neurons in the dorsal raphe nuclei of freely moving rats to stressful stimuli, *Exp. Neurol.*, 101, 445, 1988.

244. Kennett, G. A., Dickinson, S. L., and Curzon, G., Central serotonergic responses and behavioral adaptation to repeated immobilisation: the effect of the corticosterone inhibitor metyrapone, *Eur. J. Pharmacol.*, 119, 143, 1985.

245. Haleem, D. J., Kennett, G., and Curzon, G., Adaptation of female rats to stress: shift to male pattern by inhibition of corticosterone synthesis, *Brain Res.*, 458, 339, 1988.

246. Frankfurt, M., Mendelson, S. D., McKittrick, C. R., and McEwen, B. S., Alterations of serotonin receptor binding in the hypothalamus following acute denervation, *Brain Res.*, 601, 349, 1993.

247. Frazer, A. and Hensler, J. G., 5-HT$_{1A}$ receptors and 5-HT$_{1A}$-mediated responses: effect of treatments that modify serotonergic neurotransmission, *Ann. N.Y. Acad. Sci.*, 600, 460, 1990.

248. Oppenheimer, J. H., Schwartz, H. L., Mariash, C. N., Kinlaw, W. B., Wong, N. C. W., and Freake, H. C., Advances in our understanding of thyroid hormone action at the cellular levels, *Endocrine Rev.*, 8, 288, 1987.

249. Samuels, H. H., Forman, B. M., Horowitz, Z. D., and Ye, Z. S., Regulation of gene expression by thyroid hormone, *J. Clin. Invest.*, 81, 957, 1988.

250. Puymirat, J., Miehe, M., Marchand, R., Sarlieve, L., and Dussault, J. H., Immunocytochemical localization of thyroid hormone receptors in the adult rat brain, *Thyroid*, 1, 173, 1991.

251. Weinberger, C., Thompson, C. C., Ong, E. S., Lebo, R., Gruol, D. J., and Evans, R. M., The c-*erb*-A gene encodes a thyroid hormone receptor, *Nature*, 324, 641, 1986.

252. Sap, J., Muñoz, A., Damm, K., Goldberg, Y., Ghysdael, J., Leutz, A., Beug, H., and Vennström, B., The c-*erb*-A protein is a high-affinity receptor for thyroid hormone, *Nature*, 324, 635, 1986.

253. Lazar, M. A., Thyroid hormone receptors: multiple forms, multiple possibilities, *Endocrine Rev.*, 14, 184, 1993.

254. Bradley, D. J., Young, W. S., and Weinberger, C., Differential expression of α and β thyroid hormone receptor genes in rat brain and pituitary, *Proc. Natl. Acad. Sci. U.S.A.*, 86, 7250, 1989.

255. Puymirat, J., Thyroid receptors in the rat brain, *Prog. Neurobiol.*, 39, 281, 1992.

256. Jacoby, J. H., Mueller, G., and Wurtman, R. J., Thyroid state and brain monoamine metabolism, *Endocrinology*, 97, 1332, 1975.

257. Strömbom, U., Svensson, T. H., Jackson, D. M., and Engström, G., Hyperthyroidism: specifically increased response to central NA-(α)receptor stimulation and generally increased monoamine turnover, *J. Neural Transm.*, 41, 73, 1977.

258. Ito, J. M., Valcana, T., and Timiras, P. S., Effect of hypo- and hyperthyroidism on regional metabolism in the adult rat brain, *Neuroendocrinology*, 24, 55, 1977.

259. Rastogi, R. B. and Singhal, R. L., Influence of neonatal and adult hyperthyroidism on behavior and biosynthetic capacity for norepinephrine, dopamine and 5-hydroxytryptamine in rat brain, *J. Pharmacol. Exp. Ther.*, 198, 609, 1976.

260. Atterwill, C. K., Effect of acute and chronic tri-iodothyronine (T_3) administration to rats on central 5-HT and dopamine-mediated behavioral responses and related brain biochemistry, *Neuropharmacology*, 20, 131, 1981.

261. Savard, P., Mérand, Y., Di Paolo, T., and Dupont, A., Effects of thyroid state on serotonin, 5-hydroxyindoleacetic acid and substance P contents in discrete brain nuclei of adult rats, *Neuroscience*, 10, 1399, 1983.

262. Heal, D. J. and Smith, S. L., The effects of acute and repeated administration of T_3 to mice on 5-HT$_1$ and 5-HT$_2$ function in the brain and its influence on the actions of repeated electroconvulsive shock, *Neuropharmacology*, 27, 1239, 1988.

263. Henley, W. N., Chen, X., Klettner, C., Bellush, L. L., and Notestine, M. A., Hypothyroidism increases serotonin turnover and sympathetic activity in the adult rat, *Can. J. Physiol. Pharmacol.*, 69, 205, 1991.

264. Henley, W. N. and Bellush, L. L., Streptozotocin-induced decreases in serotonin turnover are prevented by thyroidectomy, *Neuroendocrinology*, 56, 354, 1992.

265. Schwark, W. S. and Keesey, R. R., Thyroid hormone control of serotonin in developing rat brain, *Res. Commun. Chem. Pathol. Pharmacol.*, 10, 37, 1975.

266. Mooradian, A. D., Metabolic fuel and amino acid transport into the brain in experimental hypothyroidism, *Acta Endocrinol.*, 122, 156, 1990.

267. Azam, M. and Baquer, N. Z., Modulation of insulin receptors and catecholamines in rat brain in hyperthyroidism and hypothyroidism, *Biochem. Int.*, 20, 1141, 1990.

268. Singhal, R. L., Rastogi, R. B., and Hrdina, P. D., Brain biogenic amines and altered thyroid function, *Life Sci.*, 17, 1617, 1975.

269. Schwark, W. S. and Keesey, R. R., Influence of thyroid hormone on norepinephrine metabolism in rat brain during maturation, *Res. Commun. Chem. Pathol. Pharmacol.*, 3, 673, 1976.

270. Schwark, W. S. and Keesey, R. R., Cretinism: influence on rate-limiting enzymes of amine synthesis in rat brain, *Life Sci.*, 19, 1699, 1976.

271. Schwark, W. S. and Keesey, R. R., Altered thyroid function and synaptosomal uptake of serotonin in developing rat brain, *J. Neurochem.*, 30, 1583, 1978.

272. Savard, P., Mérand, Y., Di Paolo, T., and Dupont, A., Effect of neonatal hypothyroidism on the serotonin system of the rat brain, *Brain Res.*, 292, 99, 1984.

273. Orosco, M., Rouch, C., Jacquot, C., Gripois, D., Valens, M., and Roffi, J., Effects of insulin on brain serotonin in the young rat: influence of thyroid status, *Psychoneuroendocrinology*, 14, 321, 1989.

274. Vaccari, A., Effects of neonatal antithyroid treatment on brain [^{3}H]imipramine binding sites, *Br. J. Pharmacol.*, 84, 773, 1985.

275. Biassoni, R. and Vaccari, A., Selective effects of thiol reagents on the binding sites for imipramine and neurotransmitter amines in the rat brain, *Br. J. Pharmacol.*, 85, 447, 1985.

276. Sandrini, M., Marrama, D., Vergoni, A. V., and Bertolini, A., Effects of thyroid status on the characteristics of alpha$_1$-, alpha$_2$-, beta, imipramine and GABA receptors in the rat brain, *Life Sci.*, 48, 659, 1991.

277. Tejani-Butt, S. M., Yang, J., and Kaviani, A., Time course of altered thyroid states on 5-HT$_{1A}$ receptors and 5-HT uptake sites in rat brain: an autoradiographic analysis, *Neuroendocrinology*, 57, 1011, 1993.

278. Mason, G. A., Bondy, S. C., Nemeroff, C. B., Walker, C. H., and Prange, A. J., Jr., The effects of thyroid state on beta-adrenergic and serotonergic receptors in rat brain, *Psychoneuroendocrinology*, 12, 261, 1987.

279. Vaccari, A., Biassoni, R., and Timiras, P. S., Effects of neonatal dysthyroidism on serotonin type 1 and type 2 receptors, *Eur. J. Pharmacol.*, 95, 53, 1983.

280. Brochet, D., Martin, P., Soubrie, P., and Simon, P., Effects of triiodothyronine on the 5-hydroxytryptophan-induced head twitch and its potentiation by antidepressants in mice, *Eur. J. Pharmacol.*, 112, 411, 1985.

281. Vaccari, A., Decreased central serotonin function in hypothyroidism, *Eur. J. Pharmacol.*, 82, 93, 1982.

282. Nagy, Z. M. and Forster, M. J., Development of serotonin-mediated behavioral inhibtion in the hyperthyroid mouse, *Pharmacol. Biochem. Behav.*, 16, 203, 1982.

283. Ottenweller, J. E. and Hedge, G. A., Thryoid hormones are required for daily rhythms of plasma corticosterone and prolactin concentration, *Life Sci.*, 28, 1033, 1981.

Chapter 4

Modulation of Ion Channel Function by Steroids

Henk Karst and Marian Joëls

CONTENTS

I. INTRODUCTION

Cholesterol is the precursor of all steroids in mammals. The steroid hormones are synthesized in specialized organs such as adrenal glands (mineralo- and glucocorticoids), gonads, and placenta (sex hormones). Because of their lipophilicity these hormones easily penetrate cells and organs, including the central nervous system (CNS). Steroid actions in the CNS produce diverse, rapid and delayed effects.[1-4]

In addition to the peripheral glands, steroids and their metabolites can also be synthesized in the CNS. The steroids of central origin have been termed "neurosteroids."[5] The role of the neurosteroids in the CNS is not fully understood, but it becomes more and more evident that they, like the hormones of peripheral origin, play an important role as neuromodulators.

The classical mechanism of steroid hormone action involves the activation of intracellular receptors which then modulate gene expression and the synthesis of new proteins in the brain. Alternatively, steroids activate membrane receptors for the hormone or bind to recognition sites, such as on GABA receptors. These events, in contrast to the genomic effects, are fast. An example of such a fast event is the anesthetic role of progesterone.[6]

During the past few years several laboratories have studied the genomic and nongenomic mode actions of steroids on ion conductances. This review addresses some of the recent observations in this field.

II. GENOMIC ACTION OF STEROIDS

In the classical genomic mechanism, steroids exert long-term effects lasting hours to days, by activating intracellular receptors that regulate transcription and protein synthesis. In the nervous system, the effects are usually restricted to particular groups of cells that contain intracellular steroid receptors. Examples of this well-established genomic mode of steroid action are the induction of the enzyme choline acetyltransferase by estradiol in the basal forebrain[7] and the induction of receptors for progesterone[8] and oxytocin[9] by estradiol in the ventromedial hypothalamus.

Similarly, slow gene-mediated effects of corticosteroids have been described with respect to regulation of cellular proteins, transmitter systems, and animal behavior.[10]

Potentially, the steroid-induced changes in gene products comprise changes in channel function. The latter represents a key feature of neuronal cells, since ion channel function is directly involved in the transmission of signals, and thus information, in the brain. Steroid-evoked changes in ion channels, resulting from alterations in cellular protein content, may partly underlie the behavioral effects of the hormones.

0-8493-7633-5/96/$0.00+$.50
© 1996 by CRC Press, Inc.

Recent studies have supplied ample evidence that steroids may alter channel function. It should be noted, though, that the evidence is as yet at the level of ion conductances: only in one case have steroid actions been demonstrated at the single channel level. Clearly, extension of these single channel data remains one of the targets for future research.

A. EFFECT OF CORTICOSTERONE ON EXCITABILITY

The rat adrenal cortex hormone corticosterone can cross the blood-brain barrier and bind to two intracellular receptor types in the brain: the mineralocorticoid (MR) and glucocorticoid receptors (GR).[11,12] It was shown that the affinity of MRs for both corticosterone and aldosterone is high (Kd ~ 0.2 nM) and that the affinity of GRs for corticosterone is about one order of magnitude lower (Kd ~ 3 nM).[10,12,13] The MRs are located mainly in structures of the limbic system: the hippocampus and septum. The GR is a common receptor, widely distributed in the CNS.[14] In the hippocampus, especially the CA1 pyramidal neurones, both receptors are co-localized.

The plasma corticosterone level varies throughout the circadian cycle and during periods of stress. Given the differential affinity of the MRs and GRs for corticosterone, the occupation of steroid receptors also varies during the day and after stress. In the morning, when the steroid level in rats is low, mainly MRs will be occupied, while during periods of stress and at the peak of the circadian cycle both MRs and GRs will be occupied.

After binding of corticosterone to the intracellular receptors, the steroid-receptor complex moves to the nucleus where it affects gene transcription; via changes in the protein synthesis several membrane properties can be affected. These membrane properties comprise receptor-mediated neurotransmitter responses (for review see Joëls and De Kloet[15,16]) and also ion channels. We will focus on the effects that have been described for corticosterone on ionic conductances.

1. Corticosteroid Action on Potassium Currents

An important feature of corticosteroid hormone action is that the steroids have no effects on passive membrane properties, such as resting membrane potential or input resistance as studied with the current-clamp technique.[17-19] Similarly, in voltage-clamp studies it was shown that the steroids did not affect the capacity or leak conductance of the membrane.[20,21] Only when the cells are shifted toward a depolarized or hyperpolarized voltage level do steroid-induced effects become apparent.

In a recent study we showed a long-lasting modulation by corticosterone of a voltage-dependent potassium current. It appeared that the inwardly rectifying K current (I_Q) in hippocampal CA1 neurons is affected by steroid treatment.[20] This current, which is carried mainly by K^+ and to a lesser degree by Na^+, is activated at potentials negative to –80 mV (Figure 1). The I_Q amplitude was found to be small when recorded 1 to 3 h after activation of MRs and increased when GRs are additionally occupied.[20,21] Changes in the amplitude were not accompanied by changes in the voltage dependency or kinetic properties of this current. Cycloheximide, a protein synthesis inhibiting agent, prevents these effects on the Q current,[20,22] suggesting a genomic mode of action.

Interestingly, selective activation of GRs only with RU28362 did not result on a large I_Q amplitude, showing that simultaneous occupation of both MR and GR is required to evoke a large I_Q conductance (Figure 1).

Other voltage-dependent K conductances did not depend on MR/GR occupation. The transient outward current I_A and the delayed rectifier (I_K) in hippocampal neurons are similar in tissue from adrenalectomized (ADX) or sham-operated controls, or in tissue from ADX rats in which only MRs, GRs, or both receptor types are occupied.[20] The M current showed a tendency toward steroid sensitivity, but the differences between the various experimental groups were relatively small.[21]

Potassium currents depending on intracellular calcium rather than voltage are probably also subject to steroid regulation. Thus far this was only indirectly demonstrated in current-clamp studies. Thus during a long depolarization (500 ms, 0.5 nA) CA1 hippocampal neurons tend to slow down the firing of action potentials. This phenomenon, accommodation, is due to the activation of a slow Ca-dependent K conductance.[23-26] Due to slow deactivation of this K conductance at the end of the depolarization the membrane will remain hyperpolarized for several seconds, a phenomenon known as the after-hyperpolarizing potential (AHP). Both the accommodation and the AHP will eventually suppress the transmission of excitatory signals in the CA1 region. It appeared that neurons in slices from ADX rats display a significantly smaller AHP amplitude and duration than neurons from the sham-operated controls.[17,18] Selective occupation of MRs and GRs provided insight into putative variation of AHP/accommodation

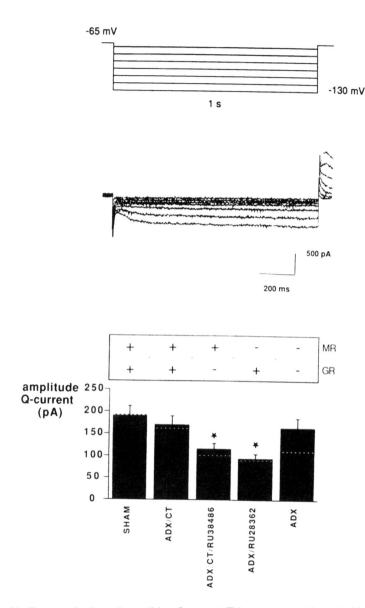

Figure 1 Steroid effects on the inwardly rectifying Q current. This current, mainly carried by potassium ions, is activated with hyperpolarizing command potentials (voltage protocol at top). Current traces (middle) show that the inward current is slowly activated and noninactivating. Selective occupation of the steroid receptors reveals a decrease of the current when MR is activated, while additional occupation of the GR restores the current to the level of sham-operated animals. Selective GR activation (i.e., no MRs activated) also induces small current amplitudes. Averaged amplitudes ± S.E.M. are shown at the bottom. MR activation: 500 nM RU38486 (GR antagonist) for 20 min. followed by incubation of both RU38486 (500 nM) and corticosterone (CT, 30 nM) for an additional 20 min. GR activation: 30 nM RU28362 (GR agonist) during 20 min. MR plus GR activation: incubation of the slices with 30 nM CT for 20 min. All recordings were made 1 to 3 h after steroid treatment. (Adapted from Karst et al., *Brain Res.*, 612, 172, 1993. With permission.)

throughout the day, as a function of MR and GR occupation. Selective activation of MRs is associated with a small AHP, poor accommodation,[17] and short AHP duration[19] when compared to the untreated ADX tissue. This MR-mediated effect peaks around 1 h after steroid application.[27] Additional GR occupation overrides and eventually (after 2 h) reverses this action and results in an increase of the AHP/accommodation.[17,18,27] The latter effect can also be achieved with GR occupation only[17] and depends on *de novo* protein synthesis.[22]

2. Corticosteroid Action on Calcium Currents

Knowing that the slow Ca-dependent K conductance is probably modulated by corticosteroids, the question arises whether this conductance is affected by changes in the Ca homeostasis or by direct modulation of the K channels. This led to studies focusing on steroid regulation of Ca conductances.

Current-clamp investigations revealed that the duration of high threshold Ca spike in CA1 pyramidal neurons is enhanced in neurons from rats with an intact HPA axis as opposed to neurons from ADX rats;[18] similarly, high doses of the GR agonist RU28362 (7 μM) administered *in vitro* increased the slow phase amplitude and duration of the Ca spike.[28] In the same study it was shown that RU28362 also increases the amplitude of the high threshold N- and L-type Ca conductances in hippocampal CA1 neurons (Figure 2) through a protein synthesis requiring process; voltage dependency and steady-state inactivation of the Ca conductances remained unaltered by the steroids.[28] In cultured pituitary CH3 cells, which only express GR receptors, incubation with glucocorticoids (synthetic hydrocorticosterone or dexamethasone) was also found to increase Ca currents, both the low- and high-threshold Ca currents.[29]

We recently performed a study using the whole cell voltage clamp technique in hippocampal slices from ADX rats (± 150 g) in which particularly the role of MR activation on Ca currents was studied. With this technique, we observed, in addition to the high-threshold Ca current, also a low-threshold, presumably T-type Ca current. In the adult hippocampal CA1 neurons this T-type current is probably located on the dendrites.[30,31] One to three hours after a brief application of 30 nM corticosterone in the presence of the GR antagonist RU38486, thus occupying mainly the MRs, both low- and high-threshold Ca conductances were small in amplitude (Figure 2).[32] Voltage dependency and kinetic properties of the Ca conductances were not affected by the steroids; neither a shift in the steady-state inactivation nor changes in activation were observed. Additional GR activation restored the Ca conductances to the level of the sham-operated controls. Therefore, the data that are available thus far indicate that Ca conductances are small with predominant MR activation and increase when GRs are additionally activated.

We also studied the effect of selective GR activation with 30 nM of the agonist. GR activation with this concentration was not sufficient to yield large Ca currents (Figure 2), pointing to a mechanism of MR and GR cooperatively, also seen for the inward rectifier. Higher doses of RU28362, however, do result in large Ca currents.[28]

Remarkably, we found that adrenalectomy itself causes an elevation of the Ca conductances. From three days after adrenalectomy a clear increase is seen in low- and high-threshold Ca currents in the slice preparation;[32] the increase of the high-threshold currents is also observed in acutely dissociated hippocampal CA1 neurons.[70]

B. SEX STEROIDS AFFECT ION CONDUCTANCES

Like the adrenal corticosteroids, the female sex steroid hormones estradiol and progesterone act via intracellular receptors. The receptors are located in the rat hypothalamus, preoptic area, and some limbic structures. Many electrophysiological studies were performed in the classical target areas for sex steroids, i.e., the hypothalamus, preoptic area, and amygdala. It has become clear that sex steroids have multiple effects on neuronal responses to transmitters.[33] With the current-clamp technique it was further shown that intrinsic membrane properties of neurons in these areas are also subject to regulation by sex steroids.[34,35] Although these studies do not allow definitive conclusions about involvement of specific ion conductances, indirect evidence supports steroid modulation of at least K conductances.[34] Recent studies using voltage clamp recording with microelectrodes or patch clamp recording have supplied more evidence for modulation of specific ion conductances by sex steroids.

1. Progesterone Affects Ca Currents after Estrogen Priming

McEwen and colleagues showed that treatment of ovariectomized (OVX) rats with estradiol and progesterone increased the density of dendritic spines on distal dendrites of CA1 pyramidal neurons when compared to the neurons in tissue from untreated OVX-rats.[36] Maximal differences were observed after 2 to 3 days of estradiol treatment followed by 2 to 6 h of progesterone application (Figure 3).[37] Differences were also observed during various stages of the estrous cycle.[38,39] The steroid dependency of dendritic spine density may be linked to changes in intrinsic membrane properties and synaptic responses of CA1 neurons. In accordance with the latter, Wong and Moss[40] reported that 2 days after treatment with estradiol the duration of the excitatory postsynaptic potential and repetitive firing induced by synaptic stimulation were enhanced compared to neurons in untreated OVX tissue. Intrinsic membrane properties, however, also appeared to be affected by treatment with sex steroids. While passive membrane

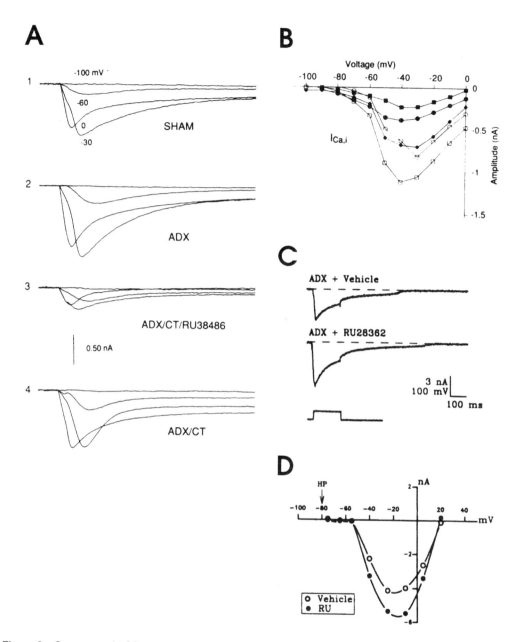

Figure 2 Ca currents in CA1 neurons of rat hippocampal slices. A hyperpolarizing prepulse of −130 mV for 3 s removes the steady-state inactivation. Depolarizing command potentials from −100 to 0 mV (200 ms) activates inward Ca currents. (A) These typical examples show that, compared to the sham-operated rats, the Ca current in CA1 neurons of ADX rats is increased. MR activation *in vitro* (500 nM RU38486 and 30 nM CT) drastically decreases the calcium currents. Additional occupation of the GR restores the current to sham amplitudes. (B) The results are summarized in the I-V plot for the $I_{Ca,i}$, a rapidly inactivating calcium current, and shows that the reduction of the Ca currents by MR activation is not voltage dependent. The I-V plot shows that GR activation alone with 30 nM RU28362 also results in a reduction of the Ca currents (○) Sham, (□) ADX, (■) ADX + CT + RU38486, (●) ADX + RU28362, (♦) ADX + CT. (C) Ca currents in hippocampal CA1 neurons in slices of 3- to 4-month-old male Fischer 344 rats. Compared to vehicle treatment RU38486 in a high dose (7 μM) causes an elevation in current amplitude. In (D) the I-V curve of ADX neurons with vehicle and RU28362 treatment are shown. Only high-threshold Ca currents were elicited with depolarizing command potentials from −75 to −20 mV from a holding potential of −80 mV. In contrast to the results in B, RU28362 treatment enhances the Ca currents. (A and B adapted from Karst et al., *Brain Res.,* 649, 234, 1994; and C and D adapted from Kerr et al., *Proc. Natl. Acad. Sci. U.S.A.,* 89, 8527, 1992. With permission.)

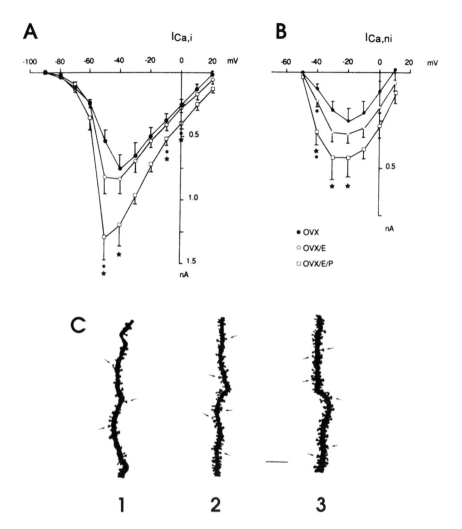

Figure 3 Effect of the sex steroids progesterone and estrogen on the Ca conductance. After estrogen priming for 2 days with estradiol benzoate (10 μg/0.1 ml peanut oil) in OVX female rats, progesterone (500 μg/0.1 ml peanut oil) treatment 2 to 6 h before decapitation results in an increase of Ca currents with (A, $I_{Ca,i}$) and without inactivating properties (B, $I_{Ca,ni}$). Progesterone treatment without priming had no effect (not shown here, see Joëls and Karst[41]). * indicate significant differences ($p < 0.05$) with OVX control and (●) with the estradiol-treated group. (C) Parallel to the effects on Ca currents, the spine density in female rat hippocampal CA1 neurones is increased by estradiol treatment. Further treatment with progesterone augments the effect of estradiol. Camera lucida drawings of representative dendritic segments from ovariectomized rats treated with oil vehicle (1), free estradiol (2), and estradiol plus progesterone (3). Some dendritic spines are indicated by arrows. Scale bar = 10 μm. (A and B adapted from Joëls, M. and Karst, H., *J. Neurosci.,* 15, 4289, 1995; and C is adapted from Woolley, C. S. and McEwen, B. S., *J. Comp. Neurol.,* 336, 293, 1993. Reprinted by permission of Wiley-Liss, Inc., a subsidiary of John Wiley & Sons, Inc.)

properties were not altered after estradiol treatment, voltage-gated Ca conductances did display steroid sensitivity.[41] Treatment with estradiol for two days, starting three days after OVX, caused small changes in the conductance of low- and high-threshold Ca currents in the rat hippocampal CA1 neurons compared to the nontreated animals (Figure 3). However, progesterone treatment 4 h before the experiment had a remarkable effect on the Ca currents in estradiol-primed OVX rats. Both the low- and high-threshold currents were increased in amplitude. Estradiol priming induces the synthesis of progesterone receptors. Therefore these results suggest that the effects of progesterone on the Ca channels is caused by the newly synthesized progesterone receptors. Indeed, without estrogen priming, progesterone did not enhance the Ca currents.

Besides the effect on the Ca conductances and the effect of these female sex hormones on two voltage-dependent potassium currents, the transient I_A and the delayed rectifier (I_K) were also studied. It appeared that these K conductances were not affected by estradiol/progesterone treatment.

III. FAST NONGENOMIC EFFECTS

In addition to the slow, gene-mediated effects by steroids, there are also reports of fast physiological responses caused by direct corticosterone application. Direct actions of cortisol and corticosterone were seen in the paraventricular nucleus. The majority of the paraventricular neurons were, within seconds, inhibited, both *in vivo*[42,43] and *in vitro*.[43,44] Some of these inhibitory effects were blocked by the GR antagonist RU38486, suggesting that the membrane receptor has chemical similarity to the cytosolic receptor. The synthetic glucocorticoid dexamethasone excited rather than inhibited the neural activity.[43] Also neurons in the brainstem, the raphe nuclei,[45] and pontis region,[46] were predominantly excited by corticosterone.

In isolated guinea pig coeliac ganglia, cortisol induced a rapid membrane hyperpolarization, which became prominent with concentrations exceeding 100 nM.[47] The input resistance was decreased and the hyperpolarization persisted in the presence of low Ca/high Mg.[43] These data indicate that corticosteroid hormones evoke a rapid postsynaptic increase of K conductances. They were blocked by RU38486.[47]

Rapid corticosteroid actions on GABA-mediated responses have also been described. GABA responses of primary afferent neurons in isolated bullfrog spinal ganglia were diminished in the presence of high doses (5 µM to 1 mM) of natural or synthetic glucocorticoids.[48] This depression was not due to diminished GABA uptake or facilitated desensitization but rather to noncompetitive antagonism of the GABA-induced Cl conductance.

At present, it is not quite clear what receptors mediate these fast steroid actions. The effective dose suggests that there may be a low-affinity membrane receptor for steroids in the brain. A recognition site for corticosterone on membranes in amphibian brain,[49] which is coupled to G proteins,[50] was already described. However, thus far, attempts to demonstrate a similar membrane receptor in mammalian brain have been unsuccessful.

A. NEUROSTEROIDS

The neurosteroids that were originally discovered and characterized are pregnenolone (PE) and dehydroepiandrosterone, their sulfate derivatives (pregnenolone sulfate (PES) and dehydroepiandrosterone sulfate), and their fatty acid esters.[5,51,52] These steroids were found in the brain at concentrations much higher than those in the plasma, suggesting that they play a functional role in the CNS. Originally only PE, dehydroepiandrosterone, and their sulfon and lipid derivatives were regarded as neurosteroids, but now it seems appropriate to include into this term all the steroids formed in the CNS. Thus not only the steroids synthesized in the CNS itself, but also the metabolites of progesterone, deoxycorticosterone, and testosterone.

The brain contains enzymes which are required for further metabolism of pregnenolone and dehydroepiandrosterone. Pregnenolone can be converted in the brain to progesterone by 3β-hydroxy-steroid oxireductase, 5-ene-isomerase.[53] The enzymes 5α-reductase and 3α-oxidoreductase are located mainly in glial cells.[54,55] These enzymes reduce progesterone to 5α-pregnane-3α-ol-20-one (tetrahydroprogesterone). 5α-pregnane-3α,21-diol-20-one (THDOC) is derived from desoxycorticosterone. These steroids are the so-called A ring reduced (or pregnane) steroids, which are in general the steroids that most effectively modulate GABA-induced Cl flux.

1. Modulation of the GABA Responses

Although the action of steroids on ligand-gated ion channels is not the main issue of this chapter, we will briefly address the effect of the neurosteroids on these ion channels.

The A ring reduced steroids modulate GABA-induced Cl flux in a manner that resembles that of the barbiturates (for review see Majewska[56]). GABA enhancing effects can be observed with nanomolar concentrations of active steroids (10 to 30 nM). The GABA$_a$ receptor active steroids prolong the response to GABA due to increased burst duration of channel currents elicited by GABA.[57-59] In contrast to tetrahydroprogesterone, THDOC, or androsterone which act as allosteric agonists of the GABA$_a$ receptor, some neurosteroids behave as noncompetitive antagonists of this receptor.[60-62] For example, pregnenolone sulfate bimodally alters binding of the GABA agonist [³H]-muscimol to the GABA$_a$ receptor in synaptosomal membranes, increasing binding at nanomolar concentrations and decreasing it at micromolar concentrations.[63] It slightly potentiates benzodiazepine binding but also inhibits barbiturate-induced enhancement of benzodiazepine binding.[60] Recently it was found that dehydroepiandrosterone sulfate also inhibits GABA-induced currents in a noncompetitive manner (IC_{50} about 10 μM).[64,65] There are, however, differences between these steroids in their modes and presumably, sites of action. Dehydroepiandrosterone sulfate, unlike PES, does not enhance [^{3}H]-benzodiazepine binding, but rather reduces it at micromolar concentrations.[65]

2. Direct Actions of Neurosteroids on Ca Currents

ffrench-Mullen and colleagues[67,68] provided convincing evidence for the role of neurosteroids such as PE as modulators of Ca currents in guinea pig hippocampal CA1 neurons. Evidence suggests that the reduction of Ca currents is through surface receptors coupled to the Ca channel via a G protein-dependent mechanism that may be direct or via intracellular mediators. The neurosteroids allotetrahydrocorticosterone, PE, and PES inhibited whole cell Ca currents during rapid superfusion in a concentration-dependent manner in the nanomolar to micromolar range (Figure 4). The inhibition of the Ca currents by THCC and PES was voltage dependent; the percentage inhibition occurred at the more negative test potentials (Figure 4).

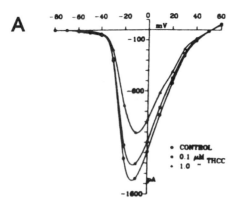

Figure 4 (A) Neurosteroids inhibit Ca currents. In this figure the action of allotetrahydrocorticosterone (THCC) is taken as an example of this mode of action. The inhibition of Ca currents in guinea pig hippocampal dissociated CA1 neurons by THCC is voltage dependent. The inhibition is more intense at negative potentials. (B) An example of a current trace (200 ms depolarization from −80 to −10 mV) showing the inhibition by THCC. CgTX, an N-type Ca channel blocker, did not further inhibit the Ca current, indicating that THCC primarily affects N-type Ca currents. (C) Nifidipine, an L-type Ca channel blocker, further inhibited the Ca current in the presence of THCC, suggesting that THCC does not block the L-type Ca current. (Adapted from ffrench-Mullen et al., *J. Neurosci.*, 14, 1963, 1994. With permission.)

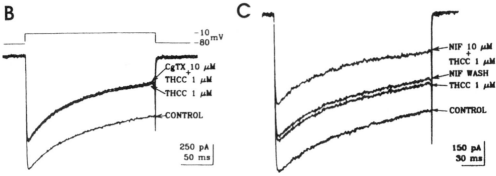

To test which Ca channel type in adult guinea pig hippocampal cells, i.e., the N- or L-type, was affected by the neurosteroids, ffrench-Mullen et al.[67,68] studied the neurosteroids in the presence of ω-conotoxin (N-type antagonist) and nifidipine (L-type antagonist). THCC in the presence of ω-conotoxin did not further inhibit the Ca currents (Figure 4). However, after addition of THCC nifidipine reversibly inhibited the Ca current additionally by an extra 21%. Thus THCC appeared to inhibit primarily the N- but not the L-type current. Allopregnenolone also blocked the N-type current. PE, however, nonselectively blocked the N- and L-type currents.

In an attempt to distinguish between an external membrane receptor protein and/or a channel protein site vs. an intracellular site of action, 0.1 mM PES was added to the intracellular solutions. There was no difference in effect on peak Ca currents between normal and PES dialyzed neurons following extracellular application of PES. These data suggest that PES interacts with a binding site on the external surface.

Pertussis toxin significantly diminished the effect of PES, PE, and THCC suggesting that G proteins are involved in the inhibitory mode of action (Figure 4). It was also shown that protein phosphorylation by PKC during the inhibition of the Ca currents was needed.

These data suggest that the neurosteroids modulate Ca currents in hippocampal cells through a G protein-coupled membrane receptor. Modulation of the gating kinetics could occur due to G protein and/or intracellular mediators.

IV. SUMMARY

In this chapter evidence was presented showing that both peripheral and neurosteroids are able to modulate ion channel function via genomic and nongenomic mechanisms.

It was previously hypothesized that genomic and nongenomic mechanisms of action of steroids might operate within the same neuron and might even interact through the same molecular target.[66,69] Indeed, Wong and Moss[40] presented evidence that a genomic and a direct membrane mechanism are indeed activated in CA1 neurons by estrogen. While the intrinsic membrane properties of the hippocampal CA1 cells were not altered, estrogen priming appeared to increase the duration of the excited postsynaptic potentials (EPSP) and induce repetitive firing in response to Schaffer colateral stimulation. Yet, superfusion of 17β-estradiol but not 17α-estradiol caused a rapid and reversible increase in the amplitude of the Schaffer collateral-activated EPSP (non-NMDA sensitive). So, neuronal excitability in hippocampus is subject to both long- and short-term excitatory modulation by estrogen, emphasizing that one compound can affect information over a considerable range of time. Quite in contrast to this divergence in time, the data presented here also demonstrate that there is a remarkable convergence of steroid actions on specific ion conductances.

The most intensively studied target of these steroids in ion channel modulation is the Ca conductance. The neurosteroids, the short-term, direct modulating agents, all proved to inhibit Ca currents. Cortico-sterone either inhibits Ca currents at low concentrations (MR occupation only), or enhances it at concentrations high enough to activate both the MRs and GRs. Progesterone injected in estrogen-primed female rats, proved to increase the Ca currents. No effect by progesterone was observed in rats without estrogen priming. Because the metabolites of progesterone, e.g., PE and THCC, are formed in the brain, one might expect that during the proestrous modulation of neuronal excitability via both a direct or long-lasting pathway should be possible.

In conclusion, recent technical advances have allowed an in-depth investigation of steroid actions on ion conductances in the brain. It has become evident that corticosteroids, sex steroids, and neurosteroids can exert multiple effects on ion conductances, over a time span of seconds to hours. These effects on intrinsic membrane properties of central neurons potentially regulate neuronal excitability for consider-able periods of time. This may have important consequences for information processing in those circuits which are targets for steroid hormone action and, consequently, for the functional processes in which these circuits play a role.

REFERENCES

1. McEwen, B. S., Davis, P. G., Parsons, B., and Pfaff, D. W., The brain as a target for steroid hormone action, *Annu. Rev. Neurosci.,* 2, 65, 1979.
2. McEwen, B. S., Glucocorticoids and hippocampus: receptors in search of function, in *Adrenal Actions on Brain,* Ganten, D. and Pfaff, D., Eds., Springer, Berlin, 1982, 1–22.
3. Riker, W. F., Jr., Baker, T., and Sastre, A., Electrophysiologic and clinical aspects of glucocorticoids on certain neural systems, in *Adrenal Actions on Brain,* Ganten, D. and Pfaff, D., Eds., Springer, Berlin, 1982, 69–106.
4. Parsons, B. and Pfaff, D., Progesterone receptors in CNS correlated with reproductive behavior, in *Actions of Progesterone on the Brain,* Ganten, D. and Pfaff, D., Springer, Berlin, 1985, 103–140.
5. Baulieu, E. E., Robel, P., Vatier, O., Haug, A., Le Coascogne, C., and Bourreau, E., Neurosteroids: pregnenolone and dehydroepiandrosterone in the rat brain, in *Receptor-Receptor Interaction, A New Intramembrane Intergrative Mechanism,* Fuxe, K. and Agnati, L. F., Eds., MacMillan, Basingstoke, 1987, 89–104.
6. Holzbauer, M., Physiological aspects of steroids with anesthetic properties, *Med. Biol.,* 54, 227, 1976.
7. Luine, V. N. and McEwen, B. S., Sex differences in cholinergic enzymes of diagonal band nuclei in the rat preoptic area, *Neuroendocrinology,* 36, 475, 1983.
8. Parsons, B., Macksky, N. J., Krey, L., Pfaff, D. W., and McEwen, B. S., The temporal relationship between estrogen-inducible progestin receptors in the female rat brain and the time course of estrogen activation of mating behavior, *Endocrinology,* 107, 774, 1980.

9. De Kloet, E. R., Voorhuis, T. A. M., and Elands, J., Estradiol induces oxytocin binding sites in rat hypothalamic ventromedial nucleus, *J. Eur. J. Pharmacol.*, 118, 185, 1985.

10. De Kloet, E. R., Brain corticosteroid receptor balance and homeostatic control, *Front. Neuroendocrinol.*, 12, 95, 1991.

11. McEwen, B. S., Weiss, J. M., and Schwartz, L. S., Selective retention of corticosterone by limbic structures in rat brain, *Nature (London)*, 220, 911, 1968.

12. McEwen, B. S., De Kloet, E. R., and Rostene, W., Adrenal steroid receptors and actions in the nervous system, *Physiol. Rev.*, 66, 1121, 1986.

13. Reul, J. M. H. M. and De Kloet, E. R., Two receptor systems for corticosterone in rat brain; microdistribution and differential occupation, *Endocrinology*, 117, 2505, 1985.

14. Van Eekelen, J. A. M., Jiang, W., De Kloet, E. R., and Bohn, M. C., Distribution of the mineralocorticoid and glucocorticoid receptor mRNAs in the rat hippocampus, *J. Neurosci. Res.*, 21, 88, 1988.

15. Joëls, M. and De Kloet, E. R., Control of neuronal excitability by corticosteroid hormones, *Trends Neurosci.*, 15, 25, 1992.

16. Joëls, M. and De Kloet, E. R., Mineralocorticoid and glucocorticoid receptors in the brain. Implications for ion permeability and transmitter systems, *Prog. Neurobiol.*, 43, 1, 1994.

17. Joëls, M. and De Kloet, E. R., Effects of glucocorticoids and norepinephrine on excitability in hippocampus, *Science*, 245, 1502–1505.

18. Kerr, D. S., Campbell, L. W., Hao, S.-Y., and Landfield, P. W., Corticosteroid modulation of hippocampal potentials: increased effect with aging, *Science*, 245, 1505, 1989.

19. Beck, S. G., List, T., and Choi, K. C., Long- and short-term administration of corticosterone alters CA1 hippocampal neuronal properties, *Neuroendocrinology*, 60, 261, 1994.

20. Karst, H., Wadman, W. J., and Joëls, M., Long-term control by corticosteroids of the inward rectifier in rat CA1 pyramidal neurons, in vitro, *Brain Res.*, 612, 172, 1993.

21. Hesen, W. and Joëls, M., Corticosteroid-mediated modulation of carbachol responsiveness in CA1 pyramidal neurons: a voltage clamp analysis, *Synapse*, 20, 299, 1995.

22. Karst, H. and Joëls, M., The induction of corticosteroid actions on membrane properties of hippocampal CA1 neurons requires protein synthesis, *Neurosci. Lett.*, 130, 27, 1991.

23. Hotson, J. R. and Prince, D. A., A calcium-activated hyperpolarization follows repetitive firing in hippocampal neurons, *J. Neurophysiol.*, 43, 409, 1980.

24. Gustaffson, B. and Wigstrom, H., Evidence for two types of afterhyperpolarization in CA1 pyramidal cells in hippocampus, *Brain Res.*, 206, 462, 1981.

25. Madison, D. V. and Nicoll, R. A., Control of the repititive discharge of rat CA1 pyramidal neurones in vitro, *J. Physiol.*, 354, 319, 1984.

26. Lancaster, B. and Adams, P. R., Calcium-dependent current generating the afterhyperpolarization of hippocampal neurons, *J. Neurophysiol.*, 55, 1268, 1986.

27. Joëls, M. and De Kloet, E. R., Mineralocorticoid receptor mediated changes in membrane properties of rat CA1 pyramidal neurons in vitro, *Proc. Natl. Acad. Sci. U.S.A.*, 87, 4495, 1990.

28. Kerr, D. S., Campbell, L. W., Thibault, O., and Landfield, P. W., Hippocampal glucocorticoid receptor activation enhances voltage-dependent Ca-conductances: relevance to brain aging, *Proc. Natl. Acad. Sci. U.S.A.*, 89, 8527, 1992.

29. Fomina, A. F., Kostyuk, P. G., and Sedova, M. B., Glucocorticoids modulation of calcium currents in growth hormone 3 cells, *Neuroscience*, 55, 721, 1993.

30. Karst, H., Joëls, M., and Wadman, W. J., Low-threshold calcium current in dendrites of the adult rat hippocampus, *Neurosci. Lett.*, 164, 154, 1993.

31. Eliot, L. S., Christie, B. R., Ito, K., Miyakawa, H., and Johnston, D., Contribution of different channel types to spike-induced calcium influx in hippocampal CA1 pyramidal neurons, *Neurosci. Abstr.*, 279.5, 1994.

32. Karst, H., Wadman, W. J., and Joëls, M., Corticosteroid receptor dependent modulation of calcium currents in rat hippocampal CA1 neurons, *Brain Res.*, 649, 234, 1994.

33. Kow, L. and Pfaff, D. W., Estrogen effects on neuronal responsiveness to electrical and neurotransmitter stimulation: an in vitro study on the ventromedial nucleus of the hypothalamus, *Brain Res.*, 347, 1, 1985.

34. Minami, T., Oomura, Y., Nabekura, J., and Fukuda, A., 17-β-Estradiol depolarization of hypothalamic neurons is mediated by cyclic AMP, *Brain Res.*, 519, 301, 1990.

35. Schiess, M. C., Joëls, M., and Shinnick-Gallagher, P., Estrogen priming affects active membrane properties of medial amygdala neurons, *Brain Res.*, 440, 380, 1988.

36. Gould, E., Woolley, C. S., Frankfurt, M., and McEwen, B. S., Gonadal steroids regulate dendritic spine density in hippocampal pyramidal cells in adulthood, *J. Neurosci.*, 10, 256, 1990.

37. Woolley, C. S. and McEwen, B. S., Roles of estradiol and progesterone in regulation of hippocampal dendritic spine density during estrous cycle in the rat, *J. Comp. Neurol.*, 336, 293, 1993.

38. Woolley, C. S. and McEwen, B. S., Estradiol mediates fluctuation in hippocampal synapse density during the estrous cycle in the adult rat, *J. Neurosci.*, 12, 2549, 1992.

39. Woolley, C. S., Gould, E., Frankfurt, M., and McEwen, B. S., Naturally occurring fluctuation in dendritic spine density on adult hippocampal pyramidal neurons, *J. Neurosci.*, 10, 4035, 1990.

40. Wong, M. and Moss, R. L., Long-term and short-term electrophysiological effects of estrogen on the synaptic properties of hippocampal CA1 neurons, *J. Neurosci.,* 12, 3217, 1992.

41. Joëls, M. and Karst, H., Effects of estradiol and progesterone on voltage-gated calcium and potassium conductances in rat CA1 hippocampal neurons, *J. Neurosci.,* 15, 4289, 1995.

42. Saphier, D. and Feldman, S., Iontophoretic application of glucocorticoids inhibits identified neurones in the rat paraventricular nucleus, *Brain Res.,* 453, 183, 1988.

43. Chen, Y.-Z., Hua, S.-Y., Wang, C. A., Wu, L. G., Gu, Q., and Xing, B. R., An electrophysiological study on the membrane receptor-mediated action of glucocorticoids on mammalian neurons, *Neuroendocrinology,* 53, S25, 1991.

44. Kasai, M. and Yamashita, H., Inhibition by cortisol of neurons in the paraventricular nucleus of the hypothalamus in adrenalectomized rats: an *in vitro* study, *Neurosci. Lett.,* 91, 59, 1988.

45. Avanzino, G. L., Ermirio, R., Ruggeri, P., and Cogo, C. E., Effect of microelectrophoretically applied corticosterone on raphe neurones in the rat, *Neurosci. Lett.,* 50, 307, 1984.

46. Dubrovsky, B., Williams, D., and Kraulis, I., Effects of corticosterone and 5α-dihydrocorticosterone on brain excitability in rat, *J. Neurosci. Res.,* 14, 118, 1985.

47. Hua, S.-Y. and Chen, Y.-Z., Membrane receptor-mediated electrophysiological effects of glucocorticoid on mammalian neurons, *Endocrinology,* 124, 687, 1989.

48. Ariyoshi, M. and Akasu, T., Voltage-clamp studies of the inhibition of γ-aminobutyric acid response by glucocorticoids in bullfrog primary afferent neurons, *Brain Res.,* 435, 241, 1987.

49. Orchinik, M., Murray, T. F., and Moore, F. L. A., Corticosteroid receptor in neuronal membranes, *Science,* 252, 1848, 1991.

50. Orchinik, M., Murray, T. F., Franklin, P. H., and Moore, F. L., Guanyl nucleotids modulate binding to steroid receptors in neuronal membranes, *Proc. Natl. Acad. Sci. U.S.A.,* 89, 3830, 1992.

51. Corpechot, C., Robel, P., Axelson, M., Sjovall, J., and Baulieu, E. E., Characterization and measurement of dehydroepiandrosterone sulfate in rat brain, *Proc. Natl. Acad. Sci. U.S.A.,* 78, 4704, 1981.

52. Corpechot, C., Synguelakis, M., Talha, S., Axelson, M., Sjovall, J., Vihko, R., Baulieu, E. E., and Robel, P., Pregnenolone and its sulfate ester in the rat brain, *Brain Res.,* 270, 119, 1983.

53. Weidenfeld, J., Siegel, R. A., and Chowers, I., *In vitro* conversion of pregnenolone to progesterone by discrete brain areas of the male rat, *J. Steroid Biochem.,* 13, 961, 1980.

54. Canick, J. A., Vaccaro, D. E., Livingston, E. M., Leeman, S. E., Ryan, K. J., and Fox, T. O., Localization of aromatase and 5α-reductase to neuronal and non-neuronal cells in the fetal rat hypothalamus, *Brain Res.,* 372, 277, 1986.

55. Krieger, N. R. and Scott, R. G., Nonneuronal localization for steroid converting enzyme: 3α-hydroxysteroid oxireductase in olfactory tubercle of rat brain, *J. Neurochem.,* 52, 1866, 1989.

56. Majewska, M. D., Neurosteroids: endogenous bimodal modulators of the GABAa receptor. Mechanism of action and physiological significance, *Prog. Neurobiol.,* 38, 379, 1992.

57. Twyman, R. E. and MacDonald, R. L., Neurosteroid regulation of GABAa receptor single-channel kinetic properties of mouse spinal cord neurons in culture, *J. Physiol.,* 456, 215, 1992.

58. Harrison, N. L., Majewska, M. D., Harrington, J. W., and Barker, J. L., Structure-activity relationships for steroid interaction with the γ-aminobutyric acid$_a$ receptor complex, *J. Pharmacol. Exp. Ther.,* 241, 346, 1987.

59. Barker, J. L., Harrison, N. L., Lange, G. D., Majewska, M. D., and Owen, D. G., Voltage-clamp studies of the potentiation of the GABA-activated chloride conductance by alphaxalone and reduced metabolite of progesterone, *J. Physiol.,* London, 377, 83P, 1986.

60. Majewska, M. D. and Schwartz, R. D., Pregnenolone-sulfate: an endogenous antagonist of the γ-aminobutyric acid receptor complex in brain?, *Brain Res.,* 404, 355, 1987.

61. Majewska, M. D., Mienville, J. M., and Vicini, S., Neurosteroid pregnenolone sulfate antagonizes electrophysiological responses to GABA in neurons, *Neurosci. Lett.,* 90, 279, 1988.

62. Mienville, J. M. and Vicini, S., Pregnenolone sulfate antagonizes GABA$_a$ receptor-mediated currents via a reduction of channel opening frequency, *Brain Res.,* 489, 190, 1989.

63. Majewska, M. D., Bisserbe, J. C., and Eskay, R. E., Glucocorticoids are modulators of GABAa receptors in brain, *Brain Res.,* 339, 178, 1985.

64. Majewska, M. D., Demirgoren, S., and London, E. D., Binding of pregnenolone sulfate to rat brain membranes suggest multiple sites of steroid action of the GABA receptor, *Eur. J. Pharmacol. Mol. Pharmacol. Sect.,* 189, 307, 1990.

65. Demirgoren, S., Majewska, M. D., Spivak, Ch. E., and London, E. D., Receptor binding and electrophysiological effects of dehydroepiandrosterone sulfate, an antagonist of the GABA$_A$ receptor, *Neuroscience,* 45, 127, 1991.

66. Schumacher, M., Rapid membrane effects of steroid hormones: an emerging concept in neuroendocrinology, *Trends Neurosci.,* 13, 359, 1990.

67. ffrench-Mullen, M. H. and Spence, K. T., Neurosteroids block CA^{2+} channel current in freshly isolated hippocampal CA1 neurons, *Eur. J. Pharmacol.,* 202, 269, 1991.

68. ffrench-Mullen, J. M. H., Danks, P., and Spence, K. T., Neurosteroids modulate calcium currents in hippocampal CA1 neurons via a pertussis toxin-sensitive G-protein-coupled mechanism, *J. Neurosci.,* 141, 1963, 1994.

69. McEwen, B. S., Non-genomic and genomic effects of steroids on neural activity, *Trends Pharmacol. Sci.,* 12, 141, 1991.

70. Werkman, T. C., Van Der Linden, S., Struik, M., and Joëls, M., Corticosteroid actions on ionic conductances in acutely dissociated rat hippocampal neurons, *Neurosci. Abstr.,* 619.11, 1995.

Chapter 5

Electrophysiological Studies of Steroid Action in the CNS

B. Dubrovsky, A. Yoo, and J. Harris

CONTENTS

I. INTRODUCTION

Steroid hormones affect central nervous system (CNS)[1] functions through different mechanisms and throughout the life cycle.[2,3] Organizational and activational effects[2,3] have been classically described. More recently, trophic as well as injury-producing effects on CNSs have also been identified.[4-6]

Electrophysiological methods allow the investigation of some of these effects, in particular, those related to control of neuronal excitability. Diverse phenomena in the CNS can be explored with these methods, e.g., hormonal effects on plasticity. These are changes in synaptic efficacy, either increases or decreases as a result of previous use or disuse.[7-9] Thus, ocular dominance shift, a phenomenon which occurs in visual cortical regions following monocular deprivation in the experimental subject, is inhibited, although not entirely, by cortisol.[10]

Another important manifestation of neuronal plasticity amenable to study with electrophysiological methods is long-term potentiation (LTP).[11] It too can be modulated by steroid hormones.[8-13] We will address these issues in this chapter. Prior to this, some general considerations that bear on the interpretation of results on the electrophysiological effects of steroid hormones on nervous systems are in order.

Discussions about the electrophysiological effects of steroid hormones on nervous systems are often marred by three points of confusion which stem from characterizing as identical very different measuring processes.[14] These processes are quantitation (numerical quantification) and measurement, magnitude (quantity) and scale, and objectifier (index) with operational definition.[14]

According to the indices selected, different, even contradictory, conclusions on the effects of steroids on the CNS can be reached. Woodbury,[15] measuring the amount of current necessary to produce electroshock convulsions in rats, reported that glucocorticoids increased excitability. They lowered the amount of current necessary to induce convulsions. In contrast, mineralocorticoids like deoxycorticosterone (DOC), increased the threshold.[15] Hall[16] also found increased excitability with glucocorticoid chronic treatment. He used the area of polysynaptic responses recorded at a lumbrosacral ventral root in response to single supramaximal peripheral nerve stimulation (e.g., triceps surae) as an index of the effect. Yet when studying other parameters, accommodation and afterhyperpolarization (AHP), exposure to high corticosterone levels decreased excitability in hippocampal neurons.[17,18] On the other hand, Ehlers et al.[19] treated rats with daily oral doses of corticosterone for 2 to 3 months. They could not detect gross electrophysiological changes in EEG (cortical or hippocampal), auditory evoked potentials, or open field and locomotor behavior with the treatment. The results questioned reports of excitability changes with glucocorticoid treatment. While doses and sites of action are paramount in determining steroid hormonal effects, the electrophysiological indices employed, e.g., evoked potentials, threshold for seizure induction, accommodation, etc. should also be carefully defined and evaluated. Only indices that can be identified as representing identical underlying processes should be compared.

Like other biologically active substances, e.g., lidocaine, the effects of a large number of steroids on CNS excitability have been shown to be dose dependent.[20,21]

Studies on chemical evolution of corticosteroid hormones indicate that chemical steroid structures were selected and stabilized at a very early stage of vertebrate evolution.[22,23] This suggests that diversification of actions and metabolism of corticosteroid hormones during the evolutionary process were probably met by adaptive evolution at peripheral target organs, rather than by evolution of their own molecular structure. An example of this is the nasal glands, an acquisition of marine birds that allows them to excrete body fluids with higher salt content than either the blood plasma or the urine. This is under the regulation of existing steroids. Furthermore, specificity at target sites may be enzymatically and not receptor mediated.[24] Thus, in mammalian cells, the enzyme 11β-hydroxy-steroid dehydrogenase (11β-OHDS) in concert with 11 oxoreductase acts to catalyze the reversible conversion of corticosterone/cortisol to their 11 oxo metabolites which show low affinity for the mineralocorticoid receptor (MR).[18,25] This fact, and the presence of intracellular corticosteroid-binding globulin (CBG or transcortin), which competes with MR for binding of corticosterone, are part of the set of factors that will determine the effects of a particular dose of GC hormones.[18,26] Enzymes that mediate metabolic conversion of steroids at A ring sites are fundamental components of this set. 5α-reductase, significantly concentrated in the white matter,[27] is a rate-limiting enzyme for hormonal steroids. Jointly with 3 ol dehydrogenase, 5α-reductase can produce from progesterone (P) and DOC, two reduced pregnanes, tetrahydro P and tetrahydro DOC,[28] both powerful $GABA_A$ agonists.[29,30] It was believed that these 3α-hydroxy A ring reduced steroids did not interact with intracellular receptors. However, recent evidence, from Rupprecht et al.,[31] indicated that after oxidation these A ring reduced pregnanes can regulate gene expression via the progesterone receptor.

This new molecular framework on the mechanism(s) of action of corticosteroids on neurons provided a new perspective from which to reevaluate some conflicting data.

II. ELECTROPHYSIOLOGICAL EFFECTS OF ADRENAL STEROIDS

Early studies (reviewed in Reference 32) revealed the significant and complex effects of adrenocortical hormones on the CNS. They can affect neuronal activity, basal, and evoked,[33-39] as well as sensory thresholds and discrimination.[40] These effects are, among other factors (e.g., dose, endocrine priming), subjected to regional characteristics. In the hippocampus, rostral regions of the brainstem reticular formation, and in the paraventricular nucleus (PVN), for example, corticosterone exerted predominantly inhibitory actions, i.e., decreased firing rates of tonically active neurons.[33,41] In contrast, in the caudal pontine regions of the reticular formation, corticosterone increased firing rates.[33,42]

This consensus, however, is not unanimous. In anesthetized rats, with iontophoretic release of dexamethasone[43] or cortisol and corticosterone, no changes in hippocampal neuronal activity were detected with extracellular recordings. The steroids in these experiments were used in their salt-conjugated form. It is conceivable that small configurational changes in the hormone structure could give rise, in these rigid molecules, to noneffective compounds.

Alternative explanations emerged when other indices of neuronal activity, i.e., characteristics of responses, were considered.[45] Mandelbrodt et al.[45] showed that apparently insensitive neurons in the hypothalamus, i.e., those whose tonic firing was not affected by cortisol, could change their responses to afferent stimuli with the hormone. The changes could even lead to a reversal of the response, i.e., neurons were inhibited rather than excited by afferent stimuli. Further, these effects were more marked with hippocampal input to the hypothalamus than with sensory modality afferents.

These studies established that, although not directly activated or inhibited by steroids, neurons can significantly change not only the intensity but also the quality of responses to afferent stimuli when influenced by these hormones. They are also in keeping with the notion that mechanisms related to response and spike generation in neurons can be selectively affected by steroid hormones. Hall[16] showed that the synthetic steroid methylprednisolone increased the excitability of the axon hillock but decreased excitability in somadendritic regions.

A well-established action of corticosterone and other GCs is the inhibition both of ongoing activity and of responses to incoming stimuli by neurons in the PVN. The parvocellular region of this nucleus is a main component of the negative feedback effect of glucocorticoids. Feldman's group[46] interpreted the electrophysiological correlates of PVN discharge inhibition with corticosterone, as corresponding with the decrease of corticotrophin releasing factor secretion by the parvocellular neurons of the nucleus.

With the advent of the isolated slice technique, investigators have taken advantage of this method to redefine cellular effects of steroids in nervous systems. Thus, Reiheld et al.[37] measured evoked potentials following afferent stimulation with 4 to 7 nM corticosterone, concentrations corresponding to morning and evening resting state and 15 nM (stress levels) in the bathing solution. They found that all of these treatments increased excitability from CA1 neurons in hippocampal slices from adrenalectomized (ADX) rats. The responses to the hormone were essentially the same 20 and 60 min after exposure. In contrast, Vidal et al.[39] found that a nonphysiological corticosterone concentration in the bathing solution (1 μM) produced inhibitory effects in hippocampal slices. These results were independent from the GABA inhibitory system. In these experiments, preparations were made from intact, non-ADX animals. That means that MRs were most likely saturated with circulating steroids. Besides, the high concentration of corticosterone used (1 μM) would have surely activated glucocorticoid receptors (GRs). And as will be shown later,[17,18] with a similar experimental paradigm, GR activation elicited inhibitory related phenomena, hence the difference in results between Reinheld et al.[37] and Vidal et al.[39]

Taking advantage of the high affinity of corticosterone for the MR (Kd about 0.5 nM) in contrast to the GR (which has a 10-fold lower affinity for the hormone), and of the availability of synthetic specific agonists and antagonists for these receptors, various research groups set out to investigate the cellular electrophysiological behavior resulting from their selective activation. Rey et al. studied the effects of corticosterone on hippocampal slices of both intact and ADX Balb/c mice.[47,48] Evoked potentials were recorded from CA1 regions. Low concentrations of the hormone, 5 to 200 pM in slices from intact mice, and 5 pM in those from ADX mice, produced an increase in the evoked population spike responses. Shorter latencies and facilitation of paired pulses were also noted. In contrast, moderate, 2000 pM, and even more with higher doses, 5000 to 10,000 pM in slices from intact animals produced a decrease in the population spike amplitude. In slices from ADX mice, decreases were observed with concentrations of 40 to 100 pM of corticosterone and increases with 5 pM. Rey et al.[47,48] interpreted their results in terms of the different receptor types activated with the different concentrations of the hormone: type I with low and type II with high corticosterone levels. In work from the same laboratories, Talmi et al.[49] showed that the inhibitory actions mediated by high concentrations of corticosterone were significantly enhanced by increased extracellular Ca^{2+} levels. Aldosterone (5 nM), a pure type I agonist, produced an increase in the population spikes, decreased latencies, and facilitation of paired pulses.[50]

Joëls and de Kloet used intracellular techniques in their studies. Addition of corticosterone to the medium in isolated hippocampal slices produced neither changes in the resting membrane potential nor changes in the membrane resistance from CA1 hippocampal neurons (Table 1).[17,18,36]

Table 1 Summary of Experimental Effects of Corticosteroids on Resting Membrane Properties

Experiment	RMP	Input resistance	Ref.
Rat hippocampus *in vitro;* corticosterone 1 μM, aldosterone 1 nM; 30–90 min	No change	No change	18
Rat hippocampus *in vitro;* corticosterone 7.5 mg/kg body weight, 14 and 1.5 h prior to study	No change	No change	52
Guinea pig celiac ganglia *in vitro;* cortisol 10^{-8}–10^{-6} M; 1–2 min	26% of cells hyperpolarized; effect persistent in low Ca^{2+}/high Mg^{2+}	Increase in input resistance	58
Rat hippocampus and neocortex *in vitro;* corticosterone 10^{-7}–10^{-5} M, 10–30 min	No change	No change	51

RMP: resting membrane potential.

However, accommodation, the phenomenon by which a temporarily depolarized neuron (by the passage of current via an intracellular electrode) decreases and finally silences its firing rate, was significantly affected by adrenal steroid hormones. A low concentration (1 nM) of corticosterone as well as of the mineralocorticoid aldosterone, applied for at least 20 min to slices of ADX rats, diminished accommodation, probably mediated by the MC receptor. However, in the case of corticosterone, reduction of accommodation gradually reversed until a total inversion occurred, i.e., accommodation was increased.[18] In contrast to this electrophysiological behavior with low corticosterone concentration, exposure of hippocampal slices to 1 mM of corticosterone concentration or the selective GC agonist RU28362 enhanced accommodation.

The phenomenon of accommodation corresponds, at least in part, with the activation of a slow Ca^{2+}-dependent K conductance. At the end of the depolarization period, the K^+ conductance is slowly inactivated, giving rise to a protracted AHP referred to as the slow AHP.

This phenomenon, AHP, was also highly sensitive to the effects of corticosteroids. Low concentrations of corticosterone (1 nM) and aldosterone (1 nM) had similar effects; reduction of the AHP. In contrast, high levels of corticosterone, i.e., μM, significantly enhanced both amplitude and duration of AHP. Although not unanimous,[51] there is agreement that the steroids affect the slow component of AHP. Increased amplitude and prolongation of the AHP with high corticosterone concentration have also been consistently observed by Kerr et al. in a similar preparation.[52]

Changes leading to enhancement of accommodation and increases in the amplitude and duration of AHP are considered as "an intrinsic mechanism by which neurons dampen excessive excitation."[18] Taken together, these results led Joëls and de Kloet to propose that aldosterone and low levels of corticosterone, through mechanisms mediated via the MR, (type I), enhance neuronal excitability. In turn, high corticosterone levels acting through the GR, (type II), decrease neuronal excitability.

The Ca^{2+}-dependent slow AHP is greater in aged rats, and as the Ca^{2+} action potential is decreased in ADX animals, it has been hypothesized that a primary effect of corticosteroids may be on Ca^{2+} conductance.[54] The similarity of effects of aging and corticosterone treatment on electrophysiological properties of hippocampal neurons has also led to the hypothesis that a part of aged-induced functional disturbances is mediated by glucocorticoids.[55,56]

Zeise et al.[51] reported results that differ from those described above. These workers did not observe an increase in slow AHP or in neuronal accommodation with corticosterone in the hippocampal slice bath.

However, Zeise et al.[51] reported, both in neocortical and hippocampal CA1 neurons in slices from adult rats, that corticosterone (1 μM) reduced both the early and late component of the orthodromically evoked inhibitory postsynaptic potentials (IPSP). In addition to a reduction in synaptic inhibition, Zeise et al.[51] found a decrease of electrical excitability in the hippocampus, but not in neocortical neurons (Table 2).

Table 2 Summary of Experimental Effects of Corticosteroids on Synaptic Activity

Experiment	Synaptic changes	Ref.
Rat hippocampus CA1 *in vitro;* corticosterone 30 nM	Decreased EPSP; decreased slow IPSP; decreased probability of synaptically drive action potentials; increased AHP	36
Rat hippocampus CA1 *in vitro;* aldosterone 3 nM	Decreased AHP	36
Rat hippocampus *in vitro;* corticosterone 7.5 mg/kg body weight, 14 and 1.5 h prior to study	Increased AHP	52
Rat hippocampus and neocortex *in vitro;* corticosterone 10^{-7}–10^{-5} M, 10–30 min	No change in EPSP or AHP; decreased IPSP	51

In agreement with Joëls and de Kloet[17,18] and with Kerr et al.,[52] Zeise et al.[51] did not observe changes in resting membrane properties (membrane potential and input resistance) with corticosterone.

The described processes regulating neuronal excitability are believed to be genomically mediated and/or at least require protein synthesis. In the presence of cycloheximide,[57] a protein synthesis inhibitor, the described changes in membrane properties failed to occur.[51,57] However, more than one mechanism appeared to be in place for the different membrane phenomena.

Steroids modulate both inhibitory (GABA-mediated) and excitatory (glutamate-mediated) effects in CA1 hippocampal regions.[36] While corticosterone (30 nM), a concentration which occupies GRs as well as MRs, reduced the amplitude of the excitatory postsynaptic potentials (EPSP), the slow IPSP, and the firing probability of synaptically driven action potentials within 20 min of exposure to the hormone, the GR-mediated increase in AHP developed only after a 1 to 4 h time lag. The different time course between synaptic potentials and slow AHP effects may thus represent different processes by which glucocorticoids can affect excitability in the CA1 region and which require protein synthesis.[36,51]

While the studies reviewed thus far have all been performed in neurons of the somatic nervous system, Chen et al. studied the actions of glucocorticoids on cell membranes of neurons from the superfused celiac ganglia of the guinea pig.[58] In this preparation, with intracellular recording, these authors detected steroid-induced electrophysiological changes in nonsynaptic cellular phenomena, i.e., resting membrane potential and input resistance. They found that cortisol succinate hyperpolarized the membrane potential (range 2 to 25 mV increase) of 47 of 79 neurons studied. In most cases, this phenomenon was accompanied by an increase in input resistance (IR) of the membrane. These changes took place within 1 to 2 min after superfusion of the ganglion with the steroid. The changes in membrane potential (MP) and IR persisted when the steroid was conjugated to bovine serum albumin (BSA), a condition that significantly impairs the passage of the steroid through membranes.[59] These effects had a latency of 1 to 2 min and disappeared soon after washout of the steroid. A dose response relationship of the hyperpolarization in the concentrations of the cortisol BSA was also observed. Taken together, all these changes are consistent with a membrane effect of the steroid. Moreover, the glucocorticoid actions were abolished by RU38486, a competitive antagonist of the glucocorticoid cytosolic receptor. As RU38486 has no agonist effects, the results suggest a possible relationship between membrane and cytosolic mechanisms for some aspects of glucocorticoid activity (see tamoxifen effects below); Table 1.[58]

Csaba suggested that if hormones are considered as signals "given to the cell by its environment, it seems self-evident that primordial receptor development took place on the surface of contact."[60] In turn, Csaba added, "cytosol receptors [could] arise by internalization of membrane receptors."

The existence of membrane mechanisms for steroid effects[61] in no way precludes the well-established two step process: receptor binding (intracellularly) and transcriptional changes. Although chronologically the first to be described,[23,60] genomic effects are not necessarily the first to appear phylogenetically. In the *Xenopus laevis* oocyte system, for instance, progesterone can act at the cell membrane level,[62] promoting the reinitiation of meiosis. Testosterone and deoxycorticosterone worked as well as progesterone in this phenomenon, but estradiol did not.[62] Corticosterone, but not dexamethasone, suppressed rapidly and specifically courtship clasping in an amphibian, an effect likely mediated by neural membranes in the caudal neuraxis.[63] Rapid facilitatory effects on lordosis behavior have also been interpreted as membrane mediated.[61] Specific binding sites[62,62a,64,65] and putative membrane receptors for various steroids have been reported. However, time overlaps between ascribed membrane and/or genomic effects (about 15 min) is one of the factors that hampers reaching a consensus in answering the question, "How do steroid hormones affect electrophysiological properties of neurons?"

In many instances, glucocorticoids (e.g., corticosterone) and mineralocorticoids (e.g., aldosterone) can antagonize each other's effects in what relates to nervous system excitability.[15,18,66] Differential central effects of minerolo- and glucocorticoids, the former increasing and the latter lowering blood pressure, have also been reported.[66] These data have been interpreted as supporting Selye's stress theory.[67]

In Selye's view, "stress is the nonspecific response of the body to any demand." One of the essential components of the theory in this phenomenon is the activation of the hypophyseal adrenal axis. As identified by Selye, the triphasic evolution in time of stress responses is (1) the alarm, (2) the stage of resistance, and (3) the stage of exhaustion. The final outcome results, in Selye's view, from the predominance of one or the other group of specific corticoadrenal hormones, gluco- or mineralocorticoids. There has been a tendency to extrapolate results from *in vitro* effects of high corticosterone in solution on neurons and to identify them with the effects of stress in the living, behaving animal. This is not only dangerous, but questionable for a variety of reasons.

For one, gluco- and mineralocorticoids do not always behave as counteracting hormones. Thus, e.g., cortisol, the principal endogenous glucocorticoid in humans, decreases rapid eye movement sleep (REM) and increases slow wave sleep.[68] On the other hand, the mineralocorticoid DOC did not substantially affect the sleep EEG patterns.[69] This was expected to be the case because the major metabolite of DOC, tetrahydro DOC, is one of the strongest $GABA_A$ agonists.[28-30] The stress response is a more complex and specific phenomenon than that envisaged by Selye.

Further, while exogenous administration of corticosterone produced, for example, a decrease in hippocampal glutamate binding, certain types of stress produced an increase in glutamate binding in intact[12] or ADX rats.[70] Thus, a variety of neuromodulatory substances released during various types of stress can override, or act independently of, corticosterone. This is consistent with modern views of stress responses expressed principally through the studies and writings of Mason's school.[71] Mason expanded the scope of hormonal substances examined during responses to stress, and a very different picture than that originally proposed by Selye emerged.

The study of multihormonal patterns indicates that rather than a "nonspecific response of the body to any demand," these patterns appear to be organized in a rather specific or selective manner. That is, depending upon the particular stimulus under study, and in relation to the complex interdependencies in hormonal actions at the metabolic level, different and selective stressors will elicit a specific multihormonal pattern response from the body. Further, the mode or time of demand, e.g., sudden, acute heat exposure from 20 to 30°C as opposed to a gradual, 24 to 36 h shift in temperature of the same magnitude, produced contrasting responses in the hypothalamic hypophiso adrenal axis. The selectivity of response extends even to the nature of the ACTH secretagog released, e.g., CRH, arginine, vasopressin, oxytocin, or catecholamines. That is while ACTH release is multifactorial, different secretagogs are involved for different stressors (e.g., hypoglycemia, hypotension, hemorrhage).[72] It is, therefore, misleading to equate stress responses solely with an increase in glucocorticoids because (1) stress responses are selective and multihormonal and (2) in most instances steroid effects are state dependent. The impact of hormones on behavior has multiple determinants, only one of them being their circulating levels. Thus, for example, progesterone potentiation of GABA-mediated inhibition is significantly higher in the presence of high, longstanding background levels of estradiol.[73,74] A possible mechanism for this effect could be an increase in GABA production as well as changes in $GABA_A$ receptor binding induced by estradiol treatment.[74,75] Under acute conditions, however, what appeared to be more relevant, was the relative estrogen/progesterone ratio. This is revealed by the fact that administration of high doses of estradiol acutely reverses the suppressant effect of progesterone to a facilitatory effect.[74]

Background steroid milieu is, therefore, an important factor in determining the degree of steroid modulation of nervous systems.

III. ELECTROPHYSIOLOGICAL EFFECTS OF GONADAL STEROIDS

The existence of cyclical changes in CNS excitability with the estrous and menstrual cycle were recognized early.[76] Catamenial seizures, which are aggravated during the first half of the menstrual cycle, are a case in point.[76-78] Although the causal relationship between mood disturbances and hormonal cyclicity is still controversial,[79] the bulk of data supports such a relationship.[74,77] Progesterone and its related A ring reduced metabolites exert mostly depressant actions at extrahypothalamic sites.[73,74] The anesthetic effects of large doses of progesterone were also recognized early.[73,74] More powerful anesthetic steroids, such as the P derivate, tetrahydro P, have been shown to be produced *in situ* in the CNS, as well as secreted directly by the adrenocortical gland.[29,80-82]

Within physiological dose range, progestins also have anticonvulsant properties. Landgren et al.[83] studied the effects of P and its A ring reduced metabolites on penicillin-induced epileptic foci. They injected the steroid, dissolved in Nutralipid®, via the lingual artery. This procedure allows for lateralization of injected substances to one hemisphere making it possible to use the other hemisphere as control within the same experiment.[84] A ring reduced pregnanes from P were significantly more powerful in reducing or arresting ictal discharges than the parent compound P.

Smith showed that physiological levels of circulating P enhance GABA-evoked inhibition of cerebellar Purkinje cells. The effects were observed 3 to 7 min after i.v. or i.p. injection and remained for 20 to 50 min poststeroid injection.[73,74]

The suggestion has been made that the active compound producing the depressant effects of P is the tetrahydro-reduced metabolite of the hormone 3α-5α-THP, not P itself.[74] Evidence supporting this contention comes from different sources. For one, 3α-5α-THP and not the parent compound P, produces immediate (40 to 80 s) effects on neurons. P produced significant enhancement of GABA-mediated inhibition only after 9 min of exposure to the hormone (local application). Further, 4M, a drug that blocks 5α-reductase, entirely prevented the potentiation effect of P on GABA inhibition.

Configurational changes in steroid hormones are associated with significant functional changes. These become evident in the type of electrophysiological changes induced by the gonadal steroid 7β-estradiol in contrast to the adrenal steroids, e.g., corticosterone and aldosterone. While an important measure of control of neuronal excitability by adrenal steroids is modulated through changes in the slow AHP and accommodation phenomena, β-estradiol did not affect either slow AHP or accommodation.[85] However, as revealed by clinical as well as experimental electrophysiological indices, estrogens play a significant role in modulating CNS excitability. Therefore they must use alternative paths to exercise this capacity.

Estradiol actions are characterized by a high degree of stereospecificity. 17β-estradiol appears as the most potent steroid in increasing CNS excitability. 17β-estradiol is ineffective except at very high doses.[86]

Smith et al.,[74,87,88] using extracellular recording from cerebellar Purkinje neurons, examined the effects of estradiol (100 to 300 ng/kg) i.v. and i.p. She evaluated the effect of the hormone on QUIS (quisqualate)/AMPA (α-amino-3 hydroxy-5-methyl-isotazole-4-propionate) and glutamate responses.

The results showed that responses increased an average of 100% with QUIS/AMPA and 67 to 96% with glutamate. No effects on kainate responses were observed. The effects appeared 15 to 35 min after injection. Pressure injection of 17β-estradiol produced similar, but faster, onset effects (3 to 5 min). In both systemic and local applications, the increased responsiveness persisted for at least 6 h post administration.

Tamoxifen, an antiestrogen agent which blocks many genomic-mediated effects of estrogens can, although not consistently, block early electrophysiological changes in neuronal responses, presumably membrane mediated. Evidence has also been advanced as to the fact that cyclic AMP[89-90] can also mediate estrogen effects in CNS.

While corticosteroids did not affect resting membrane properties in hippocampal slices, neurons from hippocampal slices exposed to 17β-estradiol developed a reversible depolarization and increased IR.[85,91] These effects had an onset latency of less than 1 min, therefore, implicating membrane mechanisms.

Neither pharmacological nor electrophysiological studies indicated, overall, a significant role for estradiol inhibitory effects in acute experiments.[92,93] However, estradiol treatment can suppress hyperpolarization induced by μ-opioids and GABA$_B$ receptor stimulation in hypothalamic neurons.[94] Occasionally, in the hippocampus, hyperpolarization responses with estradiol were observed, but this was certainly the exception.[85] Studies at neuronal levels then substantiate earlier work with field potentials in hippocampal slices[38,86] showing that estrogens can rapidly facilitate glutaminergic neurotransmission,[95] and, further, that this effect is not believed to be similar for both male and female rats.[74]

In contrast to priming with adrenocortical steroids, which significantly affect intrinsic membrane properties of neurons in hippocampal slices,[18,56] estrogen priming (10 μg of estrogen benzoate 2 to 3 days previous to experiment) did not alter any of these properties.[85] Thus no changes were detected in resting MP, IR, or on active membrane properties such as amplitude and half duration of action potentials, characteristics of AHP and accommodation. Dihydrotestosterone, the reduced metabolite of testosterone, has been shown to reduce threshold stimulation in a neuronal population subset of ventromedial hypothalamus.[96]

IV. ELECTROPHYSIOLOGICAL EFFECTS OF NEUROSTEROIDS

It has been known for some time now that nervous systems can synthesize steroids.[97] These steroids (biosynthesized in the glia) are endowed with significant power to act on neurons, particularly at the GABA complex receptor sites.[75,81,82,98] Originally only pregnenolone, dehydroepiandrosterone (DHEAS), and their sulfates and lipid derivatives were considered neurosteroids. More recently Majewska[82] proposed that not only steroids synthesized *de novo* in the CNS, but also tetrahydro metabolites of progesterone and DOC as well as androsterone. These metabolites are both synthesized by the endocrine glands and enzymatically produced *in situ* in the CNS.[80,81,97-99] They, too, exercise profound effects on the GABA receptor complex. As is the case with the glucocorticoid corticosterone, neurosteroid effects on nervous systems are concentration dependent. The most potent of the allosteric GABA agonist neurosteroids are tetrahydroprogesterone, tetrahydrodeoxycorticosterone, and androsterone.[82,98] *In vitro* studies (concentrations of 10 to 30 nM) of these steroids enhanced GABA$_A$-operated Cl channels. At higher concentrations they directly open the Cl channel in neurons.[28,87] Pregnenolone sulfate (PS) exerts a bimodal effect on the GABA$_A$ receptor complex, agonist-like effects at low (nanomolar) and antagonistic effects at high (micromolar) concentrations. When injected intracerebroventricularly (8 μg/10 μl), PS reduced pentobarbital-induced sleep time in rats, while at lower doses it prolonged sleep time.[100] DHEAS inhibits GABA-induced currents in neurons in a noncompetitive manner. When injected in doses from 100 to 150 mg/kg, it produced tonic-clonic seizures in animals (20), however, in lower doses it elicited sedation.[82] In the hippocampal slice DHEAS did increase excitability of CA1 neurons in response to Schaffer collateral stimulation. The effects were concentration dependent (10 to 100 μM) and appeared minutes after exposure.[101] Membrane properties of recorded neurons were not affected. The increased excitability revealed by these experiments appeared to be related to two facts: (1) that DHEAS inhibited a fast IPSP component of the evoked synaptic response and (2) the steroid produced a direct enhancement of EPSPs.[101]

The modulation of GABA$_A$ receptor complex by excitatory and inhibitory neurosteroids may be an important process by which these hormones regulate CNS both via endocrine and paracrine secretion. Neurosteroids are, as GCs,[49,52,56,102,103] also significantly involved in the regulation of CA^{2+} currents.[54]

V. EFFECTS OF STEROID HORMONES ON LTP

The hippocampal formation is a primary target for all steroid hormones and, in particular, the adreno-cortical steroids.[3,8] This structure is significantly involved in memory mechanisms. In turn, steroid hormones can affect memory processes. Two neural phenomena, kindling[104-105] and LTP, have been and are currently considered as putative mechanisms which could, in part at least, account for certain aspects of memory phenomena (see References 106 and 107). Kindling and LTP lead to an increase in efficiency of synaptic transmission. Although related, the two processes are by no means identical.[105,108]

Adrenalectomy markedly attenuates the amnesia which normally follows a retroactive kindled convulsion in an inhibitory aversion paradigm.[109] McIntyre[109] proposed "that brain seizure per se does not disrupt memory or retrieval, but rather modulates the adrenal system to promote its memorial effects."

In 1987, Dubrovsky et al.[110] reported the effects of an adrenal steroid metabolite, 5α-dihydrocorticosterone, on LTP. LTP was induced in the dentate gyrus, primed by stimulation of the perforant path. When injected 4 min prior to performing tetanic stimulations, this hormone significantly reduced the PS component of the response at all times.

Studies on the effects of stress on the induction of LTP also implicated adrenal steroids as modulators of the phenomenon.[38,111-114] In 1990, Diamond et al.[114] reported that exposure to a novel environment, a type of stress, interfered with primed burst potentiation (PB), a form of LTP.[115] Whereas LTP is commonly induced by 100 to more electrical pulses, PB potentiation can be induced by only 5 pulses patterned to mimic physiological activity which occurs in the hippocampus of behaving rats.[115] Shors et al.[70,112,113] reported attenuation of LTP with other types of stress. The first report of the direct effect of corticosterone injection on LTP was made in 1990.[8]

In ADX rats, i.v. injection of corticosterone dissolved in Nutralipid, 4 min prior to tetanus, significantly impaired the EPSP component of the evoked response immediately and 15 min after testing for LTP. Thereafter EPSP amplitudes were within normal values. Corticosterone significantly decreased the PS immediately after the train, the component remaining low 30 min after the train. A biphasic modulatory effect of corticosterone in hippocampal slices was showed by Rey et al.[47,48] With low concentrations, 0.5 nM, population spikes were greater than control. With corticosterone concentrations, 5 nM, population strikes decreased. The depressant effects of high corticosterone concentration on population strikes could be reversed by simultaneous incubation with RU486, a type II receptor antagonist. The high corticosterone concentration effects were significantly influenced by the CA^{2+} levels in the medium.[49]

The depressant effects of high corticosterone doses on LTP in the hippocampus were recently corroborated.[13] The relationship between corticosterone levels and LTP, however, is not a linear one. On the contrary, Bennet et al.[116] reported an inverted-U relationship between corticosterone and PB potentiation. Development of PB potentiation was suppressed both with very low and very high levels of plasma corticosterone. Intermediate levels of the hormone (11 to 20 µg/dl), on the other hand, produced high levels of potentiation on the CA1 population spike of the anesthetized rat. One possibility that emerged from experiments by Diamond et al.[12] is that corticosterone may have presynaptic effects. These authors found that with high corticosterone blood levels, post-tetanic potentiation (PTP), a presynaptically mediated short-term increase in synaptic efficacy, was blocked in the PB paradigm. Pavlides et al.,[13,117] working with the classical LTP paradigm in the dentate gyrus, did not detect changes in PTP with corticosterone. Data demonstrating presynaptic effects of steroids are available.[118]

With synthetic, selective type I and II steroid receptor activators and antagonists, their role in LTP was recently confirmed.[48,50,117] In ADX rats, administration of aldosterone, a specific type I agonist, produced a marked enhancement in long-term potentiation in comparison to either the ADX or sham-ADX controls. Administration of RU28318, a type I antagonist, which by itself had minimal effects, blocked the aldosterone enhancement. In contrast, administration of a specific type II agonist, RU28362, produced a marked decrease in the induction of long-term potentiation. The RU28362 effect was blocked by a prior injection of the type II antagonist, RU38486.

The relative content of 11β-hydroxysteroid dehydrogenase (11β-OHSD) in the hippocampus is low.[25] Thus, one of the essential steps to catalyze the reversible conversion of glucocorticoids to their 11 oxo metabolites is significantly diminished. As glucocorticoids show a high affinity for type I GRs, low levels of corticosterone will trigger effects mediated by these type I receptors.[18] Higher doses or concentration levels will activate type II receptors. These results appear consistent for different types of experiments using comparable electrophysiological indices.[18,50,117]

Activation of type I receptors not only enhanced but prolonged LTP. Aldosterone injection prolonged LTP duration when tested up to 48 h later in chronic ADX rats.[119] In this as well as other experiments by the Pavlides group, priming stimuli were effected at least more than 1 h after steroid injection to be certain that genomic effects could take place. Faster priming stimulation, 4 min after injection of the steroid, will, more likely, reflect nongenomic effects. Under these experimental conditions aldosterone significantly decreased development of EPSP over a 60-min period. The development of PS after aldosterone treatment did not differ significantly from that obtained with vehicle.[11]

Dubrovsky et al.[8-11] examined the effects of the mineralocorticoid hormones DOC and 18-OH-DOC on LTP. DOC significantly impaired the development of LTP and the impairment, observed in both components of the evoked potential (EPSP and population spike) increased with time in the EPSP, and was not significant after 30 min in the population strike. 18-OH-DOC is secreted by the rat adrenals in amounts exceeded only by corticosterone.[120] This mineralocorticoid injected 4 min prior to priming tetanization affected both components of the evoked potential responses in the dentate gyrus. The effects appeared immediately after testing and lasted for the 60-min recording period. The EPSP decreased significantly from control levels, reaching its lowest level at 60 min. The population strike was also significantly depressed after the train stimulation throughout the experiment.

In contrast, the 21-acetate of the hormone only moderately decreased the EPSPs and had no effect on the population strikes. The results suggest that the noted hormonal effects did not depend on the combination of the steroids with lipids on the cell wall, as 18-OH-DOC acetate has an enhanced lipophilicity imparted by masking the 21 hydroxyl group.[8,9]

Although a depressant steroid, tetrahydrodeoxycorticosterone (THDOC) marginally increased the amplitude of both the EPSP and population strike components of the dentate gyrus response during LTP. Hippocampal architectures would, in part at least, account for this phenomenon. Some of the excitatory (i.e., disinhibitory) effects of GABA on hippocampal slices may be due to the inhibition of GABAergic interneurons by GABAergic terminals.[121,122] Also, recent data show significant regional heterogeneities in the affinity of THDOC for its binding site; neuroactive steroid potently inhibits [35S] TBPS binding in the prefrontal cortex but not in the spinal cord.[123] Allotetrahydroprogesterone[9,11] decreased all recorded EPSPs and had no effect on the population strike development in comparison with vehicle.

Recently, it was shown that neurosteroid DHEAS enhances LTP development in the intact anesthetized rat.[124] Moreover, the enhancement was dose dependent, increasing with the doses used.

VI. SUMMARY

Electrophysiological methods allow for evaluation of effects manifested by changes in membrane and synaptic properties of neurons. At least two mechanisms appeared to mediate these effects: one restricted to the cell membrane, the second involving genomic processes.

A high degree of stereospecificity and marked functional changes with small configurational changes of the molecule characterize steroid hormones. Thus, for example, while β-estradiol acts as a very powerful estrogen, estradiol is almost totally devoid of activity.[74,86] The steroid corticosterone and its A-ring reduced metabolite may have contrasting effects on short-term effects and phenomenon,[42] i.e., immediate change in tonic firing rate of neurons. On the other hand, they behave like agonists in a long-term phenomenon, LTP.[8,110]

Glucocorticoid genomic effects are strongly determined by concentration factors. Low levels, due to the high affinity of corticosterone for the type I receptor, decreased accommodation and AHP, hence increased neuronal excitability.[18] On the other hand, by activating type II receptors, high levels decreased excitability.[18,52]

Gonadal steroids, estrogen and progesterone, showed a high degree of interdependence, priming effects, e.g., exposure to estrogens define intensity of progestin responses, and excitatory and depressant effects[74,76] in phenomena like seizures are counterbalanced.

The newly defined neurosteroid group, steroids producing *in situ* in the CNS and those resulting from metabolic degradation of parent compounds, P and DOC, also modulate CNS excitability.[29,82] They do it mainly through actions in the $GABA_A$ receptor complex and via control of membrane Ca^{2+} currents.[54,102] Extreme caution should be taken in extrapolating any one electrophysiological index and linearly correlating with behavioral phenomena which have, as a rule, multiple determinants.[9,71,125]

REFERENCES

1. Bullock, T. H., Orkland, R., and Grinnell, A., *Introduction to Nervous Systems,* W.A. Freeman & Co., San Francisco, 1977.
2. McEwen, B. S., Steroid hormones are multifunctional messengers to the brain, *Trends. Endocrinol. Metab.,* 2, 62, 1991.
3. Rees, H. D. and Gray, H. E., Glucocorticoids and mineralocorticoids: actions on brain and behavior, in *Peptides, Hormones and Behavior,* Nemeroff, C. B., Ed., Spectrum Publications, 1984, 579.
4. McEwen, B. S., Angulo, J., Cameron, H., Chao, H. M., Daniels, D., Gannon, M. N., Gould, E., Mendelson, S., Sakai, R., Spencer, R., and Woolley, C., Paradoxical effects of adrenal steroids on the brain; protection versus degeneration, *Biol. Psychiatry.,* 31, 177, 1992.
5. Sloviter, R., Valiquette, G., and Abrams, G., Selective loss of hippocampal granule cells in the mature rat brain after adrenalectomy, *Science,* 243, 535, 1989.
6. Sapolsky, R. M., Glucocorticoids, hippocampal damage and the glutamatergic synapse, *Prog. Brain Res.,* 86, 13, 1990.
7. Tsumoto, T., Long term potentiation and long term depression in the neocortex, *Prog. Neurobiol.,* 39, 209, 1992.
8. Dubrovsky, B., Filipini, D. L., Gijsbers, K., and Birmingham, M. K., Early and late effects of steroid hormones on the nervous system, in *Ciba Symposium 153: Steroids and Neuronal Activity,* Simmonds, M. A., Ed., John Wiley, Chichester, U.K., 1990, 240.
9. Dubrovsky, B., Gijsbers, K., Filipini, D. L., and Birmingham, M. K., Effects of adrenocortical steroids on long-term potentiation in the limbic system — basic mechanisms and behavioral consequences, *Cell Mol. Neurobiol.,* 13, 399, 1993.
10. Daw, N. W., Sato, H., Fox, K., Carmichael, T., and Giggerich, R., Cortisol reduces plasticity in the kitten visual cortex, *J. Neurobiol.,* 22, 158, 1991.
11. Filipini, D. L., Birmingham, M. K., Gijsbers, K., and Dubrovsky, B., Effects of adrenal steroid and their reduced metabolites on hippocampal long term potentiation, *J. Steroid Biochem. Mol. Biol.,* 40, 87, 1991.
12. Diamond, D. M., Fleshner, M., and Rose, Y., Psychological stress repeatedly blocks hippocampal primed burst potentiation in behaving rats, *Behav. Brain Res.,* 62, 1, 1994.
13. Pavlides, C., Watanabe, Y., and McEwen, B. S., Effects of glucocorticoids on hippocampal long-term potentiation, *Hippocampus,* 3, 183, 1993.
14. Bunge, M., On confusing "measure" with "measurement" in the methodology of behavioral science, in *The Methodological Unity of Science,* Bunge, M., Ed., Reidel Publishing Company, Dordrecht, Holland, 1973, 105.
15. Woodbury, D. M., Biochemical effects of adrenocortical steroids on the central nervous system, in *Handbook of Neurochemistry, Vol. VII,* Lajtha, A., Ed., Plenum, New York, 1972, 225.
16. Hall, E. D., Glucocorticoids effects on central nervous excitability and synaptic transmission, *Int. Rev. Neurobiol.,* 23, 165, 1982.
17. Joëls, M. and de Kloet, E. R., Effects of glucocorticoids and norepinephrine on the excitability in the hippocampus, *Science,* 245, 1502, 1989.
18. Joëls, M. and de Kloet, E. R., Control of neuronal excitability by corticosteroid hormones, *Trends Neurosci.,* 15, 25, 1991.
19. Ehlers, C. L., Chaplin, R. I., and Kaneko, W. M., Effects of chronic corticosterone treatment on electrophysiological measures in the rat, *Psychoneuroendocrinology,* 17, 691, 1992.
20. Heuser, G., Induction of anesthesia, seizures and sleep by steroid hormones, *Anesthesiology,* 28, 173, 1987.
21. File, S. E. and Simmonds, M. S., Myoclonic seizures in the mouse induced by alphaxolone and related steroid anaesthetics, *J. Pharm. Pharmacol.,* 40, 57, 1988.
22. Sandor, T. and Mehdi, Z., Steroids and evolution, in *Hormones and Evolution,* Barrington, E. J. W., Ed., Academic Press, New York, 1979, 1.
23. LeRoith, D., Shiloach, J., Roth, J., and Lesniak, M., Evolutionary origin of vertebrate hormones; substance similar to mammalian insulin are native to unicellular eukaryotes, *Proc. Natl. Acad. Sci. U.S.A.,* 77, 6184, 1980.
24. Funder, J. W., Pearce, P. J., Smith, R., and Smith, I., Mineralocorticoid action: target tissue specificity is enzyme, not receptor mediated, *Science,* 242, 583, 1988.
25. Sakai, R. R., Lakshini, V., Mouder, C., and McEwen, B., Immunocytochemical localization of 11β-hydroxysteroid dehydrogenase in hippocampus and other brain regions of the rat, *J. Neuroendocrinol.,* 4, 113, 1992.
26. Evans, R. M. and Arriza, J. L., A molecular framework for the actions of glucocorticoid hormones in the nervous system, *Neuron,* 2, 1105, 1989.
27. Celotti, F., Meliangi, R. C., and Martini, L., The 5 alpha-reductase in the brain: molecular aspects and relation to brain function, *Frontiers Neuroendocrinol.,* 13, 163, 1992.
28. Majewska, M. D., Steroid regulation of the GABA A receptor ligand binding, chloride transport and behavior, in *Ciba Symposium 153: Steroids and Neuronal Activity,* Simmonds, M. A., Ed., John Wiley, Chichester, U.K., 1990, 83.
29. Karavolas, H. J. and Hodges, D. R., Metabolism of progesterone and related steroids by neural and endocrine structures, in *Neurosteroids and Brain Function,* FIDIA Research Foundation Symposium Series Vol. 8, Costa, E. and Paul, S. M., Eds., Thieme Medical Publishers, New York, 1991, 135.
30. Paul, S. M. and Purdy, R. H., Neuroactive steroids, *FASEB,* 6, 2311, 1992.

31. Rupprecht, R., Reul, J. M. M. M., Trapp, T., van Steezsel, B., Wetzel, C., Damm, K., Zieglzänsberger, W., and Molsboer, F., Progesterone receptor-mediated effects of neuroactive steroids, *Neuron*, 11, 523, 1993.

32. Feldman, S., Electrophysiological effects of adrenocortical hormones on the brain, in *Steroid Hormone Regulation of the Brain*, Fuxe, K., Gustafson, J. A., and Wetterberg, J. A., Eds., Pergamon, Elmsford, NY, 1981, 175.

33. Dubrovsky, B., Illes, J., and Birmingham, M. K., Effects of 18-hydroxydeoxy-corticosterone on central nervous system excitability, *Experientia*, 42, 1027, 1986.

34. Dubrovsky, B., Williams, D., and Kraulis, I., Effects of deoxycorticosterone and its Ring A reduced metabolites on the central nervous system, *Exp. Neurol.*, 78, 728, 1982.

35. Holzbauer, M., Physiological aspects of steroids with anesthetic properties, *Med. Biol.*, 54, 227, 1976.

36. Joëls, M. and de Kloet, R. E., Corticosteroid actions on amino acid-mediated transmission in rat CA1 hippocampal cells, *J. Neurosci.*, 13, 4082, 1993.

37. Reiheld, C. T., Teyler, T. J., and Vardaris, R. M., Effects of corticosterone on the electrophysiology of hippocampal CA1 pyramidal cell *in vitro*, *Brain Res. Bull.*, 12, 349, 1984.

38. Teyler, T. J., Vardaris, R. M., Lewis, R., and Rawitch, A. D., Gonadal steroids: effects on excitability of hippocampal pyramidal cells, *Science*, 209, 1017, 1980.

39. Vidal, C., Jordan, W., and Zieglgansberger, W., Corticosterone reduces the excitability of hippocampal pyramidal cells *in vitro*, *Brain Res.*, 383, 54, 1986.

40. Henkin, R. I., The role of adrenal corticosteroids in sensory processes, in *Handbook of Physiology-Endocrinology*, Vol. VI, Blaschko, H., Sayers, G., and Smith, A. D., Eds., Williams & Wilkins, Baltimore, 1975, 209.

41. Pfaff, D. W., Silva, M. T. A., and Weiss, J. M., Telemetered recording of hormone effects on hippocampal neurons, *Science*, 172, 394, 1971.

42. Dubrovsky, B., Williams, D., and Kraulis, I., Effects of corticosterone and 5 α-dihydrocorticosterone on brain excitability in the rat, *J. Neurosci. Res.*, 14, 118, 1985.

43. Steiner, F. A., Ruf, K., and Akert, K., Steroid-sensitive neurones in rat brain: anatomical localization and responses to neurohumors and ACTH, *Brain Res.*, 12, 74, 1969.

44. Ben Barak, Y., Gutnick, M. J., and Feldman, S., Iontophoretically applied corticosteroids do not affect the firing of hippocampal neurons, *Neuroendocrinology*, 23, 248, 1977.

45. Mandelbrodt, I., Feldman, S., and Werman, R., Modification of responses to sensory and hippocampal stimuli in neurons of the rat mediobasal hypothalamus in the presence of iontophoretically applied cortisol, *Brain Res.*, 218, 115, 1981.

46. Mor, G., Saphier, D., and Feldman, S., Inhibition by corticosterone of paraventricular nucleus multiple unit activity responses to sensory stimuli in freely moving rats, *Exp. Neurol.*, 94, 391, 1986.

47. Rey, M., Carlier, E., and Soumireu-Mourat, B., Effects of corticosterone on hippocampal slice electrophysiology in normal and adrenalectomized BALB/c mice, *Neuroendocrinology*, 46, 424, 1987.

48. Rey, M., Carlier, E., and Soumireu-Mourat, B., Effects of RU 486 on hippocampal slice electrophysiology in normal and adrenalectomized BALB/c mice, *Neuroendocrinology*, 49, 120, 1989.

49. Talmi, M., Carlier, E., Rey, M., and Soumireu-Mourat, B., Modulation of the *in vitro* electrophysiological effect of corticosterone by extracellular calcium in the hippocampus, *Neuroendocrinology*, 55, 257, 1992.

50. Rey, M., Carlier, E., Talmi, M., and Soumireu-Mourat, B., *In vitro* functional differentiation of hippocampal corticosteroid mineralo- and glucocorticoid receptors, *CR Acad. Sci.*, 312, 247, 1991.

51. Zeise, M. L., Teschemacher, A., Arrigada, J., and Zieglgansberger, W., Corticosterone reduces synaptic inhibition in rat hippocampal and neocortical neurons *in vitro*, *J. Neuroendocrinol.*, 4, 107, 1992.

52. Kerr, S. D., Campbell, L. W., Hao, S. Y., and Landfield, P. W., Corticosteroid modulation of hippocampal potentials — increased effect with aging, *Science*, 245, 1505, 1989.

53. Arriza, J. L., Simerly, R. B., and Swanson, L., Adrenal mineralocorticoid as a mediator of glucocorticoid response, *Neuron*, 1, 887, 1988.

54. ffrench-Mullen, J. M. H., Danks, P., and Spence, K., Neurosteroids modulate calcium currents in hippocampal CA1 neurons via a pertussis toxin-sensitive G-protein-coupled mechanism, *J. Neurosci.*, 14, 1963, 1994.

55. Talmi, M., Carlier, E., and Soumireu-Mourat, B., Similar effects of aging and corticosterone treatment on mouse hippocampal function, *Neurobiol. Aging*, 14, 239, 1993.

56. Kerr, S. D., Campbell, L. W., Thibault, O., and Landfield, P. W., Hippocampal glucocorticoid receptor activation enhances voltage dependent Ca++ conductances: relevance to brain aging, *Proc. Natl. Acad. Sci. U.S.A.*, 89, 8527, 1992.

57. Karst, H. and Joëls, M., The induction of corticosteroid actions on membrane properties of hippocampal CA1 neurons requires protein synthesis, *Neuroscience Letters*, 130, 27, 1991.

58. Chen, Y. Z., Hua, S. Y., Wang, C. A., Wu, L. G., Gu, Q., and Xing, B. R., An electrophysiological study on the membrane receptor-mediated action of glucocorticoids in mammalian neurons, *Neuroendocrinology*, 53, 25, 1991.

59. Ke, F. C. and Ramirez, V. D., Binding of progesterone to nerve cell membranes of rat brain using progesterone conjugated to 125I Bovine Serum Albumin as a ligand, *J. Neurochem.*, 54, 467, 1990.

60. Csaba, G., Phylogeny and ontogeny of hormone receptors: the selection theory of receptor formation and hormonal imprinting, *Biol. Rev.*, 55, 47, 1980.

61. Schumacher, M., Rapid membrane effects of steroid hormones: an emerging concept in neuroendocrinology, *TINS*, 13, 359, 1990.

62a. Orchinik, M., Murray, T., and Moore, F. L., A corticosteroid receptor in neuronal membranes, *Science,* 252, 1848, 1991.

62. Baulieu, E. E., Steroid hormones in the brain: several mechanisms, in *Steroid Hormone Regulation of the Brain,* Fuxe, K., Gustafson, J. A., and Wetterberg, J. A., Eds., Pergamon, Elmsford, New York, 1981, 3.

63. Rose, J. D., Moore, F. L., and Orchinik, M., Rapid neurophysiological effects of corticosterone on medullary neurons: relationship to stress-induced suppression of courtship clasping in an amphibian, *Neuroendocrinology,* 57, 815, 1993.

64. Sadler, S. E. and Maller, J. L., Plasma membrane steroid hormone receptors, in *The Receptors, V. 1,* Conn, P. M., Ed., Academic Press, New York, 1984, 431.

65. Towle, A. C. and Sze, P. Y., Steroid binding to synaptic plasma membrane: differential binding of glucocorticoids and gonadal steroids, *J. Steroid Biochem.,* 18, 135, 1983.

66. Van den Berg, D. T. W. M., de Kloet, E. R., van Dijken, H. H., and de Jong, W., Differential central effects of mineralocorticoid and glucocorticoid agonists and antagonists on blood pressure, *Endocrinology,* 126, 118, 1990.

67. Selye, H., Confusion and controversy in the stress field, *J. Human Stress,* 1, 37, 1975.

68. Born, J., de Kloet, E. R., Wenz, H., Kern, W., and Fehm, H. L., Gluco- and mineralocorticoid effects on human sleep: a role for central corticosteroid receptors, *Am. J. Physiol.,* 260, 183, 1991.

69. Steiger, R., Rupprecht, R., Spengler, D., Guldner, J., Hemmeter, V., Rothe, B., Damm, K., and Molsboer, F., Functional properties of deoxycorticosterone and spirolactone, molecular characterization and effects on sleep-endocrine activity, *J. Psychiatr. Res.,* 27, 275, 1993.

70. Shors, T. J., Tocca, G., Patel, K. A., Baudry, M., and Thompson, R. F., Acute stress increases 3H-AMPA binding to the AMPA/quisqualate receptor in the hippocampus and the increase is not glucocorticoid-dependent, *Soc. Neurosci. Abstr.,* 17, 915, 1991.

71. Mason, J. W., Specificity in the organization of neuroendocrine response profiles, in *Frontiers in Neurology and Neuroscience Research,* Seeman, P. and Brown, G., Eds., University of Toronto Press, Toronto, 1974, 68.

72. Sapolsky, R. M. and Plotsky, P. M., Hypercortisolism and its possible neural bases, *Biol. Psychol.,* 27, 937, 1990.

73. Smith, S. S., Progesterone administration attenuates excitatory amino acid responses of cerebellar Purkinje cells, *Neuroscience,* 42, 309, 1991.

74. Smith, S. S., Female sex steroid hormones: from receptors to networks to performance — actions on the sensorimotor system, *Prog. Neurobiol.,* 44, 55, 1994.

75. Finn, D. A. and Gee, K. W., The influence of estrus on neurosteroid potency at the gamma-aminobutyric acid A receptor complex, *J. Pharmacol. Exp. Ther.,* 265, 1374, 1993.

76. Morrell, M. J., Hormones and epilepsy through the lifetime, *Epilepsia,* 33 (Suppl. 4), 49, 1992.

77. Backstrom, C. T., Landgren, S., Zetterlund, B., Blom, S., Dubrovsky, B., Bixo, M., and Sodergard, R., Effects of ovarian steroid hormones on brain excitability and their relation to epilepsy seizure variation during the menstrual cycle, in *Advances in Epileptology: XVth Epilepsy International Symposium,* Porter, R. J., Ed., Raven Press, New York, 1984, 269.

78. Logothetis, J., Harner, R., Morrell, F., and Torres, F., The role of oestrogen and catamenial exacerbations of epilepsy, *Neurology,* 9, 352, 1959.

79. Schmidt, P. J., Nieman, L. K., Grover, G. N., Muller, K. L., Merriam, G. R., and Rubinow, D. R., Lack of effect of induced menses on symptoms in women with premenstrual syndrome, *N. Engl. J. Med.,* 324, 1174, 1991.

80. Holzbauer, M., Birmingham, M. K., De Nicola, A. F., and Oliver, J. T., *In vivo* secretion of 3α-hydroxy-5α-pregnan-20-one, a potent anaesthetic steroid, by the adrenal gland of the rat, *J. Steroid Biochem.,* 22, 97, 1985.

81. Jung-Testa, I., Hu, Z., Baulieu, E. E., and Robel, P., Neurosteroids: biosynthesis of pregnenolone and progesterone in primary cultures of rat glial cells, *Endocrinology,* 125, 2083, 1989.

82. Majewska, M. D., Neurosteroids: endogenous bimodal modulators of the GABA A receptor. Mechanisms of action and physiological significance, *Prog. Neurobiol.,* 38, 379, 1992.

83. Landgren, S., Aasly, J., Backstrom, T., Dubrovsky, B., and Danielsson, J., The effect of progesterone and its metabolites on the interictal epileptiform discharge in the cat's cerebral cortex, *Acta Physiol. Scand.,* 131, 33, 1987.

84. Landgren, S., Selstam G., Aasly, J., and Danielsson, E., A method for recording effects of antiepileptic drugs in interictal discharge in the cat's cerebral cortex. Factors determining the distribution of external carotid artery infusion, *Acta Physiol. Scand.,* 128, 415, 1986.

85. Wong, M. and Moss, R. L., Long term and short term electrophysiological effects of estrogen on the synaptic properties of hippocampal CA1 neurons, *J. Neurosci.,* 12, 3217, 1992.

86. Foy, M. R. and Teyler, T. J., 7 α-Estradiol and 17 β-estradiol in hippocampus, *Brain Res. Bull.,* 10, 735, 1983.

87. Smith, S. S., Waterhouse, B. D., and Woodward, D. J., Sex steroid effects on extrahypothalamic CNS. I. Estrogen augments neuronal responsiveness to iontophoretically applied glutamate in rat cerebellum, *Brain Res.,* 422, 40, 1987.

88. Smith, S. S., Waterhouse, B. D., and Woodward, D. J., Locally applied estrogens potentiate glutamate-evoked excitation of cerebellar Purkinje ells, *Brain Res.,* 475, 272, 1988.

89. Minami, T., Oomura, Y., Nabekura, J., and Fukuda, A., 17 β-estradiol depolarization of hypothalamic neurons is mediated by cyclic AMP, *Brain Res.,* 519, 301, 1990.

90. Nabekura, J., Oomura, V., Minami, T., and Fukuda, A., Mechanism of the rapid effect of 17 β-estradiol on medial amygdala neurons, *Science,* 233, 226, 1986.

91. Wong, M. and Moss, R. L., Electrophysiological evidence for a rapid membrane action of the gonadal steroid 17 β-estradiol on CA1 pyramidal neurons of the rat hippocampus, *Brain Res.,* 543, 148, 1991.

92. Kelly, M. J., Moss, R. L., and Dudley, C. A., The specificity of the response of preoptic-septal area neurons to estrogen: 17 α-estradiol versus 17 β-estradiol and the response of extrahypothalamic neurons, *Exp. Brain Res.,* 30, 43, 1977.

93. Poulain, P. and Carette, B., Changes in the firing rate of single preoptic-septal neurons induced by direct applications of natural and conjugated steroids, in *Steroid Hormone Regulation of the Brain,* Fuxe, K., Gustafson, J. A., and Wetterberg, J. A., Eds., Pergamon, Elmsford, New York, 1981, 3.

94. Kelly, M. J., Loose, M. D., and Ronnekleiv, O. K., Estrogen suppresses μ-opioid- and GABA B-mediated hyperpolarization of hypothalamic arcuate neurons, *J. Neurosci.,* 12, 2745, 1992.

95. Schwartz-Giblin, S. and Pfaff, D. W., Ipsilateral and contralateral effects on cutaneous reflexes in a back muscle of the female rat: modulation by steroids relevant for reproductive behavior, *J. Neurophysiol.,* 64, 835, 1990.

96. Suza, S. and Sakuma, Y., Dihydrotestosterone-sensitive neurons in the male rat ventromedial hypothalamus, *Brain Res. Bull.,* 33, 205, 1994.

97. Corpechot, C., Robel, P., Axelson, M., Sjovall, J., and Baulieu, E. E., Characterization and measurement of dehydroepiandrosterone sulphate in the rat brain, *Proc. Natl. Acad. Sci. U.S.A.,* 78, 4704, 1981.

98. Lambert, J. J., Peters, J. A., Strugges, N. C., and Hales, T. G., Steroid modulation of the GABA A receptor complex: electrophysiological studies, in *Ciba Symposium 153: Steroids and Neuronal Activity,* Simmonds, M. A., Ed., John Wiley, Chichester, U.K., 1990, 83.

99. Kraulis, I., Foldes, G., Dubrovsky, B., Traikov, H., and Birmingham, M., Distribution, metabolism and biological activity of deoxycorticosterone in rat central nervous system, *Brain Res.,* 88, 1, 1975.

100. Majewska, M. D., Bluet-Pajot, M. T., Robel, P., and Baulieu, E. E., Pregnenolone sulfate antagonizes barbiturate-induced sleep, *Pharmacol. Biochem. Behav.,* 33, 701, 1989.

101. Harlan Meyer, J. and Gruol, D. L., Dehydroepiandrosterone sulfate alters synaptic potentials in area CA1 of the hippocampal slice, *Brain Res.,* 633, 253, 1994.

102. Elliot, E. M. and Sapolsky, R. M., Corticosterone impairs hippocampal neuronal calcium regulation — possible mediating mechanisms, *Brain Res.,* 602, 84, 1993.

103. Sze, P. Y. and Igbal, Z., Glucocorticoid action on depolarization dependent calcium influx in brain synaptosomes, *Neuroendocrinology,* 59, 457, 1994.

104. Racine, R., Kindling: the first decade, *Neurosurgery,* 3, 234, 1978.

105. Bliss, T. V. P. and Collingridge, G. L., A synaptic model of memory: long-term potentiation in the hippocampus, *Nature,* 361, 31, 1993.

106. Lynch, G. and Staubli, V., Possible contributions of long-term potentiation to the encoding and organization of memory, *Brain Res. Rev.,* 16, 204, 1991.

107. Madison, D. V., Malenka, R. C., and Nicoll, R. A., Mechanisms underlying long term potentiation of synaptic transmission, *Annu. Rev. Neurosci.,* 14, 379, 1991.

108. Cain, D. P., Long-term potentiation and kindling: how similar are the mechanisms?, *Trends Neurosci.,* 12, 6, 1989.

109. McIntyre, D. C., Kindling and memory. The adrenal system and the bisected brain, in *Limbic Mechanisms,* Livingstone, R. E. and Hornykiewcz, H., Eds., Plenum Press, NY, 1978, 496.

110. Dubrovsky, B., Liquornik, M., Noble, P., and Gijsbers, K., Effect of 5 α-dihydrocorticosterone on evoked potentials and long term potentiation, *Brain Res. Bull.,* 19, 635, 1987.

111. Foy, M. R., Stanton, M. E., Levine, S., and Thompson, R. F., Behavioral stress impairs long-term potentiation in rodent hippocampus, *Behav. Neural Biol.,* 48, 138, 1987.

112. Shors, T. J., Scib, T. B., Levine, S., and Thompson, R. F., Inescapable versus escapable shock modulates long-term potentiation in the rat hippocampus, *Science,* 244, 224, 1989.

113. Shors, T. J. and Thompson, R. F., Acute stress impairs (or induces) synaptic long-term potentiation (LTP) but does not affect paired-pulse facilitation in the stratum radiatum of rat hippocampus, *Synapse,* 11, 262, 1992.

114. Diamond, D. M., Bennett, M. C., Stevens, K. E., Wilson, R. L., and Rose, G. M., Exposure to a novel environment interferes with the induction of hippocampal primed burst potentiation, *Psychobiology,* 11, 273, 1990.

115. Rose, G. M. and Dunwiddle, T. V., Induction of hippocampal long term potentiation using physiologically patterned stimulation, *Neurosci. Lett.,* 69, 244, 1986.

116. Bennet, M. C., Diamond, D. M., Fleshner, M., and Rose, G. M., Serum corticosterone level predicts the magnitude of hippocampal primed burst potentiation and depression in urethane-anesthetized rats, *Psychobiology,* 19, 301, 1991.

117. Pavlides, C., Watanabe, Y., Magarinos, A. M., and McEwen, B. S., Opposing rules of type I and type II adrenal steroid receptors in hippocampal long term potentiation, *Neuroscience,* 68, 387, 1955.

118. Turner, J. P. and Simmonds, M. A., Modulation of the GABA A receptor complex by steroids in slices of rat cuneate nucleus, *Br. J. Pharmacol.,* 96, 409, 1989.

119. Pavlides, C., Kimura, A., Magarinos, A. M., and McEwen, B. S., Type I adrenal steroid receptors prolong hippocampal long term potentiation, *Neuro. Report,* 5,2673, 1994.

120. Birmingham, M. K. and Ward, P. J., The identification of the Porter Silber chromogen secreted by the rat adrenal, *J. Biol. Chem.,* 236, 1661, 1961.

121. Milner, T. A. and Bacon, C. E. D., GABAergic neurons in the rat hippocampal formation: ultrastructure and synaptic relationships with catecholaminergic terminals, *J. Neurosci.,* 9, 3410, 1989.

122. Soriano, E. and Frotscher, M., A GABAergic axoaxonic cell in the fascia dentata controls the main excitatory hippocampal pathway, *Brain Res.,* 503, 170, 1989.

123. Grobin, C. A., Roth, R. H., and Deutch, A. Y., Regulation of the prefrontal cortical system by the neuroactive steroid 3,21 dihydroxy-5α pregnane 20-one, *Brain Res.,* 578, 351, 1992.

124. Yoo, A., Harris, J., and Dubrovsky, B., Dehydroepiandrosterone sulfate increases PS in dentate gyrus LTP in intact rats, *Exp. Neurol.,* in press, 1995.

125. Bodnoff, S. R., Humphreys, A. G., Lehman, J. O., Diamond, D. M., Rose, G. M., and Meaney, M. J., Enduring effects of chronic corticosterone treatment on spatial learning, synaptic plasticity, and hippocampal neuropathology in young and mid-aged rats, *J. Neurosci.,* in press.

Chapter 6

Estrogen Effects on Neuronal Ultrastructure and Synaptic Plasticity

L. Leedom, L. M. Garcia-Segura, and F. Naftolin

CONTENTS

I. INTRODUCTION

Estrogen has profound effects on neuronal development including determining the number of neurons and their connections in several regions of the central nervous system (CNS).[1,2] These effects result in the sexual differentiation of the brain and spinal cord in rodents. In addition to these developmental actions, estrogen produces both reversible and irreversible alterations in the adult brain. This review addresses the effects of estrogen on neuronal cellular ultrastructure and synaptic plasticity in the adult CNS. The effects of estrogen have been most studied in the preoptic area, hypothalamic arcuate nucleus, hypothalamic ventromedial nucleus, amygdala, and hippocampus, all regions rich in estrogen receptors.[3] Estrogen action on the adult hypothalamus and limbic system results in the regulation of gonadotrophins and the expression of sexual behavior. Estrogen's neuroendocrine and behavioral actions are likely mediated through changes in protein synthesis and cellular ultrastructure which in turn produce changes in neurotransmitter levels and in the number and types of synapses.

II. ESTROGEN AND CELLULAR ULTRASTRUCTURE IN THE ADULT HYPOTHALAMIC ARCUATE NUCLEUS

Our laboratory has focused on the study of estrogen effects on the arcuate nucleus cellular structure and synaptic plasticity. This nucleus was chosen for study because of its central role in gonadotrophin regulation. In rats, there is a sexual dimorphism in luteinizing hormone (LH) and follicle-stimulating hormone (FSH) regulation in that changes in circulating estrogen can both decrease and increase pituitary LH and FSH secretion (negative and positive feedback, respectively) in females. In males only negative or reciprocal feedback is present.[4]

Hypothalamic mechanisms are thought to underlie the sexual dimorphism in gonadotrophin regulation.[4] During positive feedback in females, an estrogen-induced surge in blood LH and FSH results in ovulation.[4] If connections between the arcuate nucleus and anterior hypothalamus (where gonadotrophin-releasing hormone (GnRH) cell bodies are located) are disrupted, ovulation ceases, and animals enter constant estrus or diestrus.[5] The fact that cells of the arcuate nucleus have been demonstrated to contain estrogen receptors[3] and are functionally necessary for ovulation has led to the hypothesis that this nucleus is a site mediating estrogen-induced positive feedback.[4]

A. CYTOPLASMIC ELEMENTS

Electron microscopy of arcuate perikarya reveals a cellular organization unique to each day of the four-day estrous cycle.[6-9] Metestrus, the day following ovulation, is a period of relatively reduced hormonal and behavioral activity. Characteristics of arcuate nucleus neuronal perikarya include ample granulated secretory vesicles and rough and smooth endoplasmic reticulum. As the plasma estrogen level increases during diestrus, the number of secretory vesicles increases reaching a peak on the morning of proestrus.[7,8] Secretory vesicles disappear the afternoon of proestrus. The amount of rough and smooth endoplasmic reticulum and stacking of rough endoplasmic reticulum also increase through diestrus to diminish by the morning of proestrus.[7,8] On the morning of ovulation (estrus) the amount of endoplasmic reticulum is sparse.

Estrogen-sensitive, specialized structures appear and disappear during the cycle. The most prominent are the so-called whorl bodies[9] (Figure 1) or ribbon roles of smooth endoplasmic reticulum and the so-called cytoplasmic fibrillar bodies (CFBs) or nematosomes.[10] The whorl bodies are apparently extruded from the cisterns of the rough endoplasmic reticulum, forming into the shape of a hollow sphere or cap. The appearance of these structures is coincident with the fall of estradiol and rise of progesterone that accompany ovulation. Castration, which is also associated with decreased ovarian secretions, is accompanied by the appearance of whorl bodies in both male and female rats.[10] However, the time course following castration may be protracted over several weeks. Of great interest, whorl body-containing neuronal profiles have a one- to three-fold excess of axosomatic synapses in comparison to nonwhorl body profiles.[10] Compared to other arcuate nucleus neurons, whorl body-positive cells are also heavily labeled during immunostaining for glutamic acid decarboxylase (GAD), which synthesizes GABA.[6] Whorl bodies themselves are not labeled during immunostaining. Substance P-containing arcuate nucleus neurons have also been shown to undergo similar transformations during the rat estrous cycle.[11]

CFBs are nonmembrane-bound structures in the karyoplasm that contain massed electron-dense particles in circular form. They are not of nuclear origin.[10] CFBs become prominent during the period of disappearance of whorl bodies, for example, on estrogen treatment of castrates or one day following the proestrus peak of estrogen.[10] Upon pharmacologic estrogen treatment of two-week ovariectomized females, CFBs appear in the center of whorl bodies, raising the possibility that CFBs result from the estrogen-induced breakdown of the endoplasmic reticulum of whorl bodies. Although less is known about CFBs in this regard, it appears that CFB-containing cells are GABAergic and have decreased numbers of axosomatic synapses.

In summary, there are two apparently complementary intracellular estrogen-sensitive structures in arcuate GABAergic neurons. Thus far these have been found in GABAergic and substance P immuno-positive cells. Although their function remains unknown, they are associated with synaptic plasticity, *vidae infra,* and therefore are of great interest.

B. NUCLEAR ELEMENTS

Because estrogen's effects through receptor-mediated transcription and translation of genomic, i.e., nuclear information require passage of molecules across the nuclear membrane, our group has also studied neuronal nuclear membrane ultrastructure during each day of the estrous cycle.[12] Freeze-fracture replicas from hypothalamic punches from male rats and from met-, di-, pro-, and estrus morning females were examined. Replicated fracture faces from the nuclear membranes of randomly encountered cell bodies were photographed at $\times 27,000$ and the number of nuclear pores within a superimposed test square grid counted. Nuclear membranes from estrus and proestrus rats had significantly greater pore density and total pore number as compared to nuclear membranes from metestrus and diestrus rats. Nuclear pore density and total pore number in metestrus and diestrus females were not significantly different than in male control rats.[12] Further, in unpublished studies we have found that in estrus and proestrus rats, cell nuclei are enlarged, indicating that intracellular water may move into the nucleus during estrogen action.

C. NEURONAL MEMBRANE REMODELING DURING THE ESTROUS CYCLE

The perikaryal membrane represents the interface between the extracellular space (neuropil, glial processes, other cells, etc.) and the cellular machinery of the neuron. In freeze-fracture studies on the cerebellum we have shown that changes in the composition of postsynaptic membranes accompany

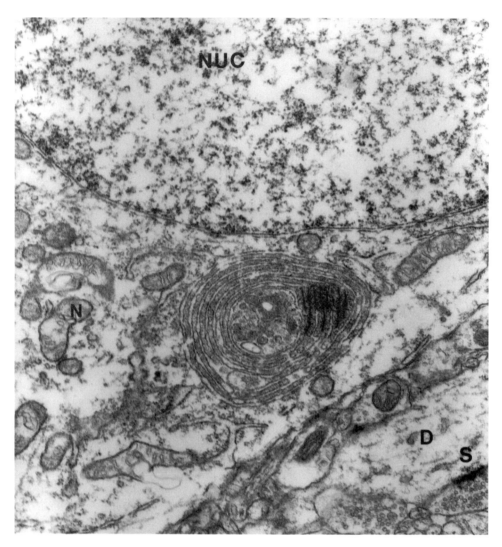

Figure 1 This electron micrograph shows a hypothalamic arcuate nucleus neuron (N) from an ovariectomized female rat beside a dendrite (D), the neuron's nucleus is labeled NUC, and a synapse on the dendrite is labeled S. Note the presence of a whorl body.

changes in synaptic dynamics. This fits with the general dogma that postsynaptic elements play a major role in determining presynaptic inputs.

Freeze-fracture permits visualization of components of membranes, including intramembranous domains of proteins. Using freeze-fracture it is possible to determine the diameter of intramembranous particles (IMP) and to classify them as small (less than 10 nM) and large (greater than 10 nM). Although these size differences have been attributed to functional implications, little is actually known of the identity of specific particles.

In early studies we found that there is a decreased number of IMPs in the neuronal membranes of males.[14] These studies demonstrated clear sexual dimorphism in perikaryal membranes, implying an effect from sex steroids.[14] In addition to determining that the IMP composition in the developing rat arcuate nucleus is related to perinatal sex steroid exposure,[15] we examined electron micrographs of freeze-fracture replicas in diestrus, proestrus, and estrus rats.[13] The IMP numerical density in the P face of the perikaryal plasma membrane was highest in diestrus when compared to proestrus and estrus. Differences were due to a significant decrease in the number of small IMPs in proestrus and estrus as compared to diestrus females. IMP density in dendritic shaft and spine membranes did not appear to fluctuate during the estrous cycle. As discussed below, the estrogen-induced decreases in neuronal IMP occur in perikaryal membrane locations which are also sites of synaptic plasticity.

The observed changes in arcuate nucleus neuronal IMP density during the estrous cycle are likely related to plasma estrogen levels. Immunoneutralization of cycling female rats with an antiestrogen antibody blocks the decrease in IMPs seen with estrus (unpublished data). We have also conducted freeze-fracture studies of arcuate nucleus neuronal membranes from high-dose estrogen-treated female rats.[16] Adult females showing at least two consecutive 4-day cycles were injected with estrogen valerate (EV) on the morning of diestrus (20 mg/kg). This protracted, high-dose estrogen treatment results in the failure of the estrogen-induced gonadotrophin surge, constant vaginal estrus, and elevated plasma estradiol levels similar to those of proestrus. Arcuate nucleus neuronal membranes were evaluated 3, 8, 16, and 32 weeks after injection. In perikaryal membranes, the nadir of density of P face small IMPs was found at 16 weeks. A similar pattern was observed in dendritic shafts but not in dendritic spines.[16]

The technique of fracture-labeling was used to identify the extracellular component of intramembranous particles affected by estrogen treatment.[16] Hypothalamic slices were incubated in sucrose to swell the extracellular spaces and further expose the neuronal glycocalyx, and concanavalin A was added to bind to oligosaccharides on extracellular glycoproteins. The concanavalin A was then labeled with horseradish peroxidase (HRP)-coated colloidal gold. It was found that the concanavalin A labeled particles closely corresponded to the small IMPs. This indicated that the small IMPs were likely intramembranous domains of proteins which extended extramembranous domains containing mannose-rich oligosaccharides.[17]

Having determined that estrogen can affect perikaryal membrane and glycocalyx composition, the mechanism for these effects was studied using the freeze-fracture[18] technique. The variation in IMP numbers during the estrous cycle and with EV treatment was associated with changes in the number of exo-endocytotic images in the plasma membrane of the perikaryon and preceded changes in synapses: the number of exo-endocytotic images was increased in proestrus morning females and in EV-treated females. We therefore proposed that endocytosis might play a role in the observed decrease in small IMPs. This hypothesis was tested *in vitro* using slices of unfixed arcuate nucleus incubated in artificial CSF.[18] HRP was used as a marker for endocytosis. Estradiol was added to the bath and the slices fixed at various intervals. The addition of estrogen to the bath resulted in an increase in the rate of endocytosis as measured by internalized HRP and an increase in the number of observed exo-endocytotic images. This effect was blocked by tamoxifen, indicating that it is estrogen receptor-dependent.

Because intramembranous proteins are involved in the formation of gap junctions, freeze-fracture has also been utlilized to study the effects of estrogen on gap junctions (Figure 2). Adult ovariectomized rats were treated with 17β-estradiol or oil vehicle and studied 2 days following injection.[19] Although gap junctions were relatively uncommon in both groups, their incidence was increased significantly in the estradiol-treated group. These results indicate that estradiol may increase both electrical and non-electrical coupling among arcuate nucleus neurons.[19]

D. ARCUATE NUCLEUS SYNAPTIC REMODELING DURING THE ESTROUS CYCLE

We have studied the number of synapses in the arcuate nucleus during the estrous cycle in order to determine whether the physiologic changes in circulating estrogen and gonadotrophin feedback control are associated with changes in synaptic connectivity in this nucleus.[20] Adult female Sprague Dawley rats demonstrating at least two consecutive four-day cycles were studied on metestrus, diestrus, proestrus, or estrus. Thin sections of arcuate nucleus were examined by transmission electron microscopy. All encountered neuronal cell bodies in a given section were photographed and a complete profile of each perikarya reconstructed. The length of perikaryal membrane studied, the length of each synaptic plate, and the percentage of perikaryal membrane covered by presynaptic terminals and in contact with glial processes were also determined. We found that the number of synapses was decreased in estrus when compared to the other days of the estrous cycle. There was a 30 to 50% reduction in axosomatic synapses in the 24 h between proestrus and estrus. Thereafter, synapses increased in number, recovering by metestrus. The reduction in axosomatic synapses in estrus was accompanied by a 17% increase in the percentage of neuronal membrane covered by glia from proestrus to estrus. Although we have extensively studied this neuroglial relationship, it is not possible at present to state whether the primary effect of estrogen is on the neurons or surrounding glia.[21]

These results indicated that a complete physiological cycle of denervation and reinnervation of arcuate neurons takes place in 48 h. We hypothesized that the increase in plasma estrogen seen in proestrus was

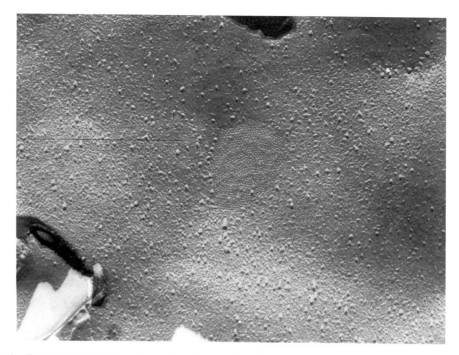

Figure 2 Freeze-fracture replica of arcuate nucleus perikaryal membrane E face, showing an oval gap junction amidst the flat lipid domain and dispersed intramembranous particles (original magnification approximately × 60,000).

related to the observed decrease in axosomatic synapses. In further support of this hypothesis, injection of adult female rats with EV results in a loss of axosomatic synapses which is observable by three weeks after injection and reaches a nadir by eight weeks after injection.[22] Finally, if the estrogen-induced gonadotrophin surge is blocked with an antiestrogen antibody, the loss of axosomatic synapses in the arcuate nucleus does not occur (unpublished data).

E. SUMMARY OF ESTROGEN-INDUCED CHANGES IN THE ADULT ARCUATE NUCLEUS

The rise in plasma estrogen between estrus and proestrus is accompanied by an increase in the protein synthetic machinery of the cell, the rough endoplasmic reticulum. Nuclear pores, which allow transport of mRNA from the nucleus to the cytoplasm, are increased in number. Rough and smooth endoplasmic reticulum are synthesized and their complexity peaks in diestrus. The prominence of the Golgi apparatus is greatest in proestrus as is the abundance of vesicles which presumably contain the post-Golgi processing products of the estrogen-induced increases in protein synthesis. The appearance and regression of the cellular organelles supporting protein synthesis indicates that the protein synthetic capability has been augmented and diminished during each estrous cycle.

Although the aforementioned studies have been conducted using rats; similar findings are being developed in primates. Ovariectomy of female monkeys results in the formation of whorl bodies in arcuate nucleus neurons.[23] Whorl bodies diminish and CFBs increase with estrogen treatment. Primate whorl body-containing neurons are also immunopositive for GAD,[22] as is the case in rats. Changes in arcuate nucleus neuronal IMPs and synaptic remodeling are also seen with estrogen treatment of female monkeys.[24] In both rodents and primates, estrogen-induced decreases in IMPs are correlated with reductions in axosomatic synapses. Thus, there is a temporal relationship between the number of IMPs and the number of axosomatic synapses (Figure 3).

Although estrogen usually causes augmentation of the protein synthetic apparatus in arcuate neurons, it is possible that the synthesis of some proteins is inhibited rather than stimulated by estrogen.[25,26] In a preliminary attempt to examine the role of protein synthesis in arcuate nucleus synaptic remodeling, we treated castrated female rats with cyclohexamide.[27] Cyclohexamide resulted in a significant inhibition of protein synthesis in the hypothalamus from 1 to 8 h following injection. By 9 h after cyclohexamide

- ESTROGEN | **+ ESTROGEN**

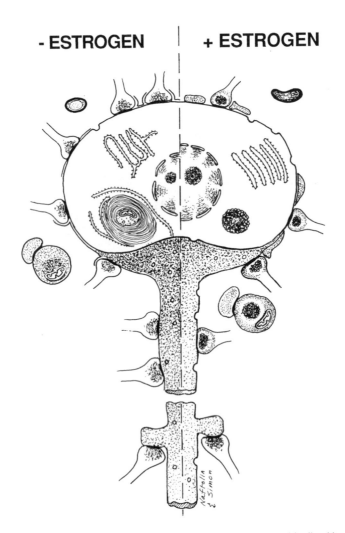

Figure 3 This schematic representation summarizes the effects of estrogen on idealized hypothalamic arcuate nucleus neurons. Castration (−estrogen) results in the appearance of whorl bodies, decreased nuclear pores, and nuclear size and increased numbers of synapses. With estrogen treatment whorl bodies disappear and nematosomes (CFBs) are seen along with stacked rough endoplasmic reticulum. Synapses are reduced in number. For details see text.

injection there was an observable decrease in small IMPs in arcuate neuronal perikarya. There was also a 41% decrease in axosomatic synapses and a 34% decrease in the percentage of neuronal membrane in contact with synaptic terminals. The decrease in the number of axosomatic synapses was accompanied by a prominent multilayered ensheathing of neuronal somas by thin astroglial processes. This effect of cyclohexamide was not observed in the cerebellum. Although the results of this study are interesting, they must be interpreted with caution given that cyclohexamide is a nonspecific agent whose use may also cause secondary effects.

III. ESTROGEN-INDUCED CHANGES IN THE PREOPTIC AREA

Witkin et al.[28] studied cellular structure and synaptology in GnRH immunoreactive neurons in the preoptic area and medial basal hypothalamus of the monkey using electron microscopy. Ovariectomized females were compared with "physiologically" estrogen replaced ovariectomized females. In ovariectomized females there was a significant increase in the apposition of glial processes to GnRH perikaryal membranes. This ensheathing was accompanied by a significant decrease in the synaptic inputs to the GnRH cells. These findings are consistent with the idea that following ovariectomy, the amount of presynaptic inhibition of GnRH cells decreases. It is possible that the mechanism for this effect of estrogen withdrawal lies in the cells innervating the GnRH neurons, because the GnRH neurons are not thought to possess significant estrogen receptors.[4]

The anteroventral periventricular nucleus (AVP) of the preoptic area is another region of the hypothalamus known to regulate gonadotrophins and ovulation. Lesion of this nucleus abolishes ovulation and produces constant estrus in female rats.[29] Langub et al.[29] have reported that the numbers of axosomatic synapses fluctuate in the AVP during the estrous cycle.

IV. ESTROGEN-INDUCED CHANGES IN CELL MORPHOLOGY IN THE VENTROMEDIAL HYPOTHALAMUS

A. CELLULAR ULTRASTRUCTURE

The ventromedial hypothalamus (VMH) is thought to play a crucial role in estrogen-induced female sexual behavior. For this reason other groups have studied estrogen-induced morphologic changes in this nucleus. Cohen and Pfaff[30] studied the ultrastructure of the ventrolateral and dorsomedial subdivisions of the VMH in ovariectomized and estrogen-treated ovariectomized rats. They noted increase and stacking of the rough endoplasmic reticulcum and a greater number of dense-core vesicles with estrogen treatment. In another study, ovariectomized rats were treated with estradiol and examined after 2 h.[31] Nucleolar, nuclear, and somal hypertrophy were seen with estrogen treatment as well as an increase in the amount of stacked rough endoplasmic reticulum. Another group found that a single dose of estradiol benzoate administered to ovariectomized rats was associated with an increase in the amount of endoplasmic reticulum, enlarged Golgi, and condensation of nucleolar material.[32]

B. SYNAPTIC ELEMENTS

Frankfurt et al.[33] reported that dendritic spine density in the VMH varied with the estrous cycle. The density of dendritic spines was found to be significantly greater at proestrus compared to diestrus. Castration resulted in a reduction in dendritic spine density which recovered after estrogen treatment. Estrogen treatment of castrated female rats also increased the number of axodendritic synapses per unit area in the VMH.[34]

V. CONCLUSIONS

We have reviewed changes in cell morphology and synaptic organization in estrogen-sensitive diencephalic regions during the estrous cycle and following estrogen treatment of castrated females. Our work has focused upon changes in the arcuate nucleus of rats and monkeys. In these species high levels of estrogen cause changes in postsynaptic neural membranes that are followed by the removal of axosomatic and axodendritic synapses. The main effect appears to be on inhibitory GABAergic synapses, although decreases in catecholaminergic synapses have been reported.[26] Estrogen-induced synaptic plasticity appears to be related to specific receptor-positive nuclei; whereas we found there is a decrease at midcycle (proestrus-estrus) in axosomatic synapses in the arcuate nucleus on the day spine synapses are reported to increase in the ventromedial nucleus. This may, however, be a characteristic of the nucleus rather than a differential effect of estrogen. For example, estrogen does promote dendritic spine growth and synapse formation in the hippocampus and amygdala, two other brain regions where spines are abundant.[35,36]

Due to space constraints we have not detailed estrogen effects on hypothalamic glia. However, it is apparent that glial cells are involved in synaptic plasticity induced by estrogen and other hormones. For example, synaptic changes in the arcuate nucleus are accompanied by morphologic alterations in glial cells. The surface density of cells immunoreactive for the specific astrocytic marker glial fibrillary acidic protein, the number of glial profiles in the neuropil, and the amount of perikaryal membrane covered by glia are all increased by estrogen. Finally, these changes are accompanied by estrogen-dependent gliosis.[37]

Another finding is the increase in the cellular rough endoplasmic reticulum and nuclear size by estrogen, indicating effects on protein synthesis. This was seen in both the arcuate and ventromedial nucleus. However, estrogen may differentially regulate protein synthesis in the ventromedial and arcuate nuclei. Jones et al.[38] demonstrated that in castrated female rats, a 6-h exposure to estrogen resulted in a 70% increase in product rRNA levels in the VMH and a 30% increase in the arcuate nucleus.

Many of the changes in cell morphology and synaptic pattern which occur following estrogen treatment likely result from the action of the steroid on brain proteins. Structural proteins, signal proteins, ion channels, cell adhesion molecules, neurotransmitter synthetic enzymes, and neuropeptides are among the likely targets for estrogenic regulation. The plethora of proteins involved makes their identification and study a technical challenge.[39] Future research will likely focus on the mechanism by which estrogenic regulation of proteins leads to the observed changes in morphology and synaptology and on how the observed changes in cellular ultrastructure relate to the observed changes in synaptology.

REFERENCES

1. Naftolin, F., Keefe, D., Apa, R., Palumbo, A., and Garcia-Segura, L.M., The apparent paradox of sexual differentiation of the brain, *Contrib. Gynecol. Obstet.,* 18, 24, 1989.
2. Arai, Y., Matsumoto, A., and Nishizuka, M., Gonadal control of synaptogenesis in the neuroendocrine brain, in *Endocrinology and Physiology of Reproduction,* Leung, P.C.K., Armstrong, D.T., Ruf, K.B., Moger, W.H., and Friesen, G.H., Plenum, New York, 1987.
3. Simerly, R.B., Chang, C., Muramatsu, M., and Swanson, L.W., Distribution of androgen and estrogen receptor mRNA-containing cells in the rat brain: an *in situ* hybridization study, *J. Comp. Neurol.,* 294, 76, 1990.
4. Horvath, T., Leedom, L., Garcia-Segura, L.M., and Naftolin, F., The effect of estrogen on synaptic remodeling in the rat hypothalamus, *Curr. Opin. Endocrinol., and Diabetes,* 2, 186, 1995.
5. Halasz, B. and Pupp, L., Hormone secretion of the anterior pituitary gland after physical disruption of all nervous pathways to the hypophysiotropic area, *Endocrinology,* 77, 553, 1965.
6. Leranth, C., Sakamoto, H., Maclusky, N.J., Shanabrough, M., and Naftolin, F., Estrogen responsive cells in the arcuate nucleus of the rat contain glutamic acid decarboxylase (GAD): an electron microscopic immunocytochemical study, *Brain Res.,* 331, 376, 1985.
7. Zambrano, D., The arcuate complex of the female rat during the sexual cycle, *Z. Zellforsch.,* 93, 560, 1969.
8. King, J.C., Williams, T.H., and Gerall, A.A., Transformations of hypothalamic arcuate nucleus, *Cell Tissue Res.,* 153, 497, 1974.
9. Naftolin, F., Leranth, C., and Garcia-Segura, L.M., Ultrastructural changes in hypothalamic cells during estrogen-induced gonadotropin feedback, *Neuroprotocols,* 1, 16, 1992.
10. Naftolin, F., Bruhlmann-Papazyan, M., Baetens, D., and Garcia-Segura, L.M., Neurons with whorl bodies have increased numbers of synapses, *Brain Res.,* 329, 289, 1985.
11. Tsuro, Y., Hisano, S., Okamura, Y., Tsukamoto, N., and Daikoku, S., Hypothalamic substance P-containing neurons. Sex dependent transformations associated with stages of the estrous cycle, *Brain Res.,* 305, 331, 1984.
12. Garcia-Segura, L.M., Olmos, G., Tranque, P., Aguilera, P., and Naftolin, F., Nuclear pores in rat hypothalamic arcuate neurons: sex differences and changes during the estrous cycle, *J. Neurocytol.,* 16, 469, 1987.
13. Garcia-Segura, L.M., Hernandez, P., Olmos, G., Tranque, P.A., and Naftolin, F., Neuronal membrane remodeling during the oestrus cycle: a freeze-fracture study in the arcuate nucleus of the rat hypothalamus, *J. Neurocytol.,* 17, 377, 1988.
14. Garcia-Segura, L.M., Baetens, D., and Naftolin, F., Sex differences and maturational changes in arcuate nucleus neuronal plasma membrane organization, *Dev. Brain Res.,* 19, 146, 1985.
15. Perez, J., Naftolin, F., and Garcia-Segura, L.M., Sexual differentiation of synaptic connectivity and neuronal plasma membrane in the arcuate nucleus of the rat hypothalamus, *Brain Res.,* 527, 116, 1990.
16. Olmos, G., Aguilera, P., Tranque, P., Naftolin, F., and Garcia-Segura, L.M., Estrogen-induced synaptic remodeling in adult rat brain is accompanied by the reorganization of neuronal membranes, *Brain Res.,* 425, 57, 1987.
17. Garcia-Segura, L.M., Perez, J., Tranque, P.A., Olmos, G., and Naftolin, F., Sex differences in plasma membrane concanavalin A binding in rat arcuate neurons, *Brain Res. Bull.,* 22, 651, 1989.
18. Garcia-Segura, L.M., Olmos, G., Tranque, P., and Naftolin, F., Rapid effects of gonadal steroids upon hypothalamic neuronal membrane ultrastructure, *J. Steroid Biochem.,* 27, 615, 1987.
19. Perez, J., Tranque, P.A., Naftolin, F., and Garcia-Segura, L.M., Gap junctions in the hypothalamic arcuate neurons of ovariectomized and estradiol-treated rats, *Neurosci. Lett.,* 108, 17, 1990.
20. Garcia-Segura, L.M., Luquin, S., Parducz, A., and Naftolin, F., Gonadal hormone regulation of glial fibrillary acidic protein immunoreactivity and glial ultrastructure in the rat neuroendocrine hypothalamus, *Glia,* 10, 59, 1994.
21. Olmos, G., Naftolin, F., Perez, J., Tranque, P.A., and Garcia-Segura, L.M., Synaptic remodeling in the rat arcuate nucleus during the estrous cycle, *Neuroscience,* 32, 663, 1989.
22. Garcia-Segura, L.M., Baetens, D., and Naftolin, F., Synaptic remodeling in arcuate nucleus after injection of estradiol valerate in adult female rats, *Brain Res.,* 366, 131, 1986.
23. Leranth, C., Shanabrough, M., and Naftolin, F., Estrogen induces ultrastructural changes in progesterone receptor-containing GABA neurons of the primate hypothalamus, *Neuroendocrinology,* 54, 571, 1991.
24. Naftolin, F., Leranth, C., Perez, J., and Garcia-Segura, L.M., Estrogen induces synaptic plasticity in adult primate neurons, *Neuroendocrinology,* 57, 935, 1993.

25. Rodriguez-Sierra, J.F., Heydorn, W.E., Creed, J.G., and Jacobowitz, D.M., Incorporation of amino acids into proteins of the hypothalamus of prepubertal female rats after estradiol treatment, Neuroendocrinology, 45, 459, 1987.

26. Jones, E.E. and Naftolin, F., Estrogen effects on the tuberoinfundibular dopaminergic system in the female rat rain, *Brain Res.,* 510, 84, 1990.

27. Perez, J., Naftolin, F., and Garcia-Segura, L.M., Cyclohexamide mimics effects of estradiol that are linked to synaptic plasticity of hypothalamic neurons, *J. Neurocytol.,* 22, 233, 1993.

28. Witkin, J.W., Ferin, M., Popilskis, S.J., and Silverman, A., Effects of gonadal steroids on the ultrastructure of GnRH neurons in the rhesus monkey: synaptic input and glia apposition, *Endocrinology,* 129, 1083, 1991.

29. Langub, C.M., Maley, B.E., and Watson, R.E., Estrous cycle-associated axosomatic synaptic plasticity upon estrogen receptive neurons in the rat preoptic area, *Brain Res.,* 641, 303, 1994.

30. Cohen, R.S. and Pfaff, D.W., Ultrastructural of neurons in the ventromedial nucleus of the hypothalamus in ovariectomized rats with or without estrogen treatment, *Cell Tissue Res.,* 217, 451, 1981.

31. Jones, K.J., Pfaff, D.W., and McEwen, B.S., Early estrogen-induced nuclear changes in rat hypothalamic ventromedial neurons: an ultrastructural and morphometric analysis, *J. Comp. Neurol.,* 239, 255, 1985.

32. Carrer, H.F. and Aoki, A., Ultrastructural changes in the hypothalamic ventromedial nucleus of ovariectomized rats after estrogen treatment, *Brain Res.,* 240, 221, 1982.

33. Frankfurt, M., Gould, E., Wooley, C.S., and McEwen, B.S., Gonadal steroids modify dendritic spine density in ventromedial hypothalamic neurons: a Golgi study in the adult rat, *Neuroendocrinology,* 51, 530, 1990.

34. Matsumoto, A. and Arai, Y., Male-female difference in synaptic organization of the ventromedial nucleus of the hypothalamus in the rat, *Neuroendocrinology,* 42, 232, 1986.

35. Nishizuka, M. and Arai, Y., Sexual dimorphism in synaptic organization in the amygdala and its dependence on neonatal hormone environment, *Brain Res.,* 212, 31, 1981.

36. Garcia-Segura, L.M., Chowen, J.A., Duenas, M., Torres-Aleman, I., and Naftolin, F., Gonadal steroids as promotors of neuroglial plasticity, *Psychoneuroendocrinology,* 19, 5, 445, 1994.

37. Gould, E., Wooley, C., Frankfurt, M., and McEwen, B.S., Gonadal steroids regulate dendritic spine density in hippocampal pyramidal cells in adulthood, *J. Neurosci.,* 10, 1286, 1990.

38. Jones, K., Harrington, C.A., Chikaraishi, D.M., and Pfaff, D.W., Steroid hormone regulation of ribosomal RNA in rat hypothalamus: early detection using *in situ* hybridization and precurseo-product ribosomal DNA probes, *J. Neurosci.,* 10, 1513, 1990.

39. McEwen, B.S., Coirini, H., Westlind, A., Frankfurt, M., Gould, E., Schumacher, M., and Wooley, C., Steroid hormones as mediators of neural plasticity, *J. Steriod Biochem.,* 39, 223, 1991.

Chapter 7

Modulation of Neurotrophins and Their Receptors by Adrenal Steroids

Gisela Barbany

CONTENTS

I. INTRODUCTION

The extensive neuronal degeneration that occurs in diseases such as Alzheimer's and Parkinson's diseases makes the identification and characterization of substances influencing neuronal survival interesting as possible alternative therapeutical approaches. One group of molecules influencing neuronal survival is the family of neurotrophic factors. Other substances, such as glucocorticoids have been shown to influence neuronal survival. Removal of glucocorticoids following adrenalectomy of newborn rats stimulates brain growth and cell proliferation.[1-3] Furthermore, glucocorticoids can damage the hippocampus when exposure is prolonged.[4] and impair the capacity of hippocampal neurons to survive various neurological insults.[5-7] Adrenalectomy has also been reported to protect hippocampal pyramidal cells from age-related loss.[8] However, glucocorticoids can also play a positive role for neuronal survival, since there is a selective loss of hippocampal granule cells in the mature rat brain after adrenalectomy, which can be prevented by adding corticosterone to the drinking water.[9] Furthermore, necrosis of granule cells seen in Addison´s disease may be due to the loss of adrenocortical function.[10]

Various mechanisms of action have been proposed to explain the neurodegenerative effects of glucocorticoids, such as enhancing glutamatergic signals[11] and impairment of glucose uptake.[12] A further possibility is that glucocorticoids influence neuronal survival by regulating expression of trophic factors.

II. THE NERVE GROWTH FACTOR FAMILY AND THEIR RECEPTORS

The prototype and best characterized member of the neurotrophin family is nerve growth factor (NGF). NGF was first discovered by Levi-Montalcini and Hamburger as a sarcoma-derived substance which supports survival of peripheral sympathetic and neural crest-derived sensory neurons.[13] NGF is synthesized at considerable distance from the cell body by peripheral tissues (referred to as targets). During development NGF is retrogradely transported from the target into the nerve terminal and up to the cell body,[14,15] and only those neurons receiving enough NGF will survive the period of neuronal death. NGF not only supports neuronal survival during the period of neuronal death, but also has been shown to influence a number of neuronal properties, such as neurotransmitter expression, acting as a differentiation factor.[16] Thus, NGF would match the number and properties of innervating neurons to the needs of the target tissue. In the periphery NGF is synthesized by non-neuronal cells,[17] opposite to the brain, where NGF is synthesized primarily by neurons.[18]

NGF belongs to a family of highly homologous proteins collectively known as the neurotrophins, which apart from NGF includes brain-derived neurotrophic factor (BDNF),[19] neurotrophin-3 (NT-3),[20-24] and neurotrophin-4 (NT-4).[25,26]

0-8493-7633-5/96/$0.00+$.50

114

All four members of the family are expressed in the brain although at different levels and with different regional distributions.[27] The highest levels of NGF mRNA correlate with innervation of cholinergic neurons in the brain.[28] NT-3 mRNA shows the most restricted distribution with higher levels in the hippocampus and cerebellum compared to other regions.[20,29] In contrast, BDNF[22,29] and NT-4[30] are widely distributed throughout the brain. The level of NT-4 in the brain is, however, significantly lower than the levels of the other neurotrophins.[30]

In the periphery the different neurotrophins support survival of specific and partially overlapping neuronal populations.[31,32] In the CNS the specific neurons responding to the different neurotrophins are not as well established. NGF has been shown to support basal forebrain cholinergic neurons.[33-35] BDNF has been shown to support the survival of retinal ganglion cells,[36] basal forebrain cholinergic neurons,[37] dopaminergic striatal cells,[38] and motoneurons.[39-41] Surprisingly, in BDNF knockouts, motoneurons and other neuronal populations are not particularly affected in the newborn animal, and in NGF knockouts basal forebrain cholinergic neurons are preserved at birth.[42] These results suggest that during development BDNF and NGF are not survival factors for these neurons.

NT-3 and NT-4 have been shown to stimulate the survival of embryonic locus coeruleus neurons in culture.[43] Moreover, NT-3 partially prevents the death of facial motoneurons in newborn rats,[40] and can prevent the death of adult central noradrenergic neurons exposed to 6-hydroxydopamine *in vivo*.[44] Hippocampal neurons respond to BDNF, NT-3, and NT-4.[45]

The neurotrophins exert their effects via a high affinity receptor, where a protein tyrosine kinase constitutes an essential component of the receptor.[46] The product of the tropomyosin receptor kinase (trk) proto-oncogene, a transmembrane protein of molecular weight 140 (p140^{trk}) is a functional receptor for NGF.[47,48] p140^{trk} (herein referred to as trkA) is a member of a transmembrane receptor family which also includes the products of the trkB (p145^{trkB}) and trkC (p145^{trkC}) genes. p145^{trkB} (TrkB) and p145^{trkC} (trkC) have recently been shown to be essential components of functional high affinity receptors for BDNF and NT-3, respectively.[49-51] trkB is also a functional receptor for NT-4 (Figure 1).[26,52] One puzzling feature is the existence of redundancy between the neurotrophins and their receptors. For instance, NT-3 binds preferentially to trkC,[49] but is also able to activate trkB (Figure 1).[53] Whether this redundancy has any biological significance is at present unknown.

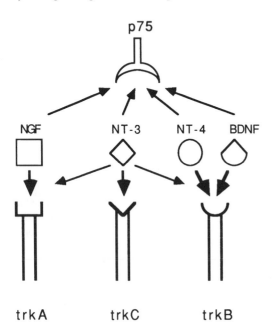

Figure 1 Schematic representation of the neurotrophins and their receptors. See text for details.

Neurotrophin binding to its respective tyrosine kinase receptor results in activation of its intrinsic tyrosine kinase activity (autophosphorylation) and signal transduction.[54,55]

The CNS patterns of expression of the different neurotrophin receptors are very different. TrkA expression is restricted to a few populations of CNS neurons.[56] On the contrary, trkB and trkC have broad and partially overlapping patterns of expression,[49,57,58] which suggests that many classes of neurons may respond to more than one neurotrophin. The finding that in many cases the same neuronal population expresses both neurotrophin and the receptor has led to the hypothesis that neurotrophins may act also in a paracrine or even autocrine manner, where the same population of cells expressing a particular neurotrophin also responds to it via a high affinity receptor.[59]

A complex feature of the trk receptors is the existence of several isoforms. Both trkB and trkC loci encode isoforms lacking the tyrosine kinase domain, the function of which still remains unclear.[60-63] The trkC locus encodes further multiple isoforms with an insert in the tyrosine kinase domain, also of unknown function.[60,61,64] Finally, a second isoform of trkA has been described that is apparently not expressed in the nervous system.[65]

In addition to binding to a receptor tyrosine kinase, the neurotrophins bind to a second transmembrane receptor, known as p75 or low affinity neurotrophin receptor (Figure 1).[20,25,66,67] p75 binds all neurotrophins with similar affinity, but does not seem to be a functional receptor in the absence of the trk receptor.[68] The function of p75 is still controversial.[54,55] Recently Lee et al. have shown that the presence of p75 is necessary for the developing nervous system, mice bearing a null mutation in the p75 gene have decreased pain sensitivity and cutaneous innervation.[69] Cultured neurons from these mice are less sensitive to NGF than wild-type neurons. How p75 enhances the survival response of sensory neurons to NGF is presently not understood.

III. EFFECTS OF GLUCOCORTICOIDS ON NEUROTROPHIN EXPRESSION IN CULTURE

L929 fibroblasts and primary brain cultures have been extensively used to study the effects of glucocorticoids on NGF regulation. The involvement of glucocorticoids in NGF regulation was first suggested by a report by Wion et al.,[70] where L929 cells, a mouse fibroblast cell line, were cultured in the presence of different concentrations of the synthetic glucocorticoid dexamethasone (DEX). The addition of DEX to the culture medium resulted in a dose-dependent decrease of NGF mRNA expression (Figure 2). The maximum reduction of NGF mRNA expression was elicited by 10^{-6} M DEX. This dose reduced NGF mRNA expression to less than 5% of the levels found in control cultures. The amount of NGF protein secreted by L929 fibroblasts was quantified in parallel cultures using double-site enzyme immunoassay. DEX treatment also reduced the NGF secreted to the medium to 25% of control cultures.

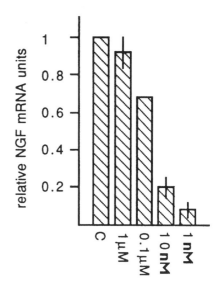

Figure 2 Densitometric analysis of the relative levels of NGF mRNA found in L929 cells cultured for 4 h with the indicated concentrations of dexamethasone. The value found in untreated cells was arbitrarily set to one. (From Wion, D. et al., *Exp. Cell Res.*, 162, 562, 1986. With permission.)

The rat endogenous glucocorticoid, corticosterone, showed the same effect on NGF mRNA expression in L929 cells, reducing NGF mRNA to 10% of control levels,[71] and, moreover, inhibited the induction of NGF mRNA elicited by phorbol ester.[72] In the presence of the transcription inhibitor actinomycin D corticosterone showed no effect on NGF mRNA decay,[72] suggesting that corticosterone reduces NGF mRNA levels by inhibiting NGF gene transcription and not by affecting NGF mRNA stability. The inhibition of NGF mRNA expression by corticosterone is likely to be mediated via the glucocorticoid receptor, since corticosterone showed no inhibitory effect on NGF induction by phorbol ester in the presence of the glucocorticoid antagonist RU486.[72] Interleukin-1, a cytokine secreted by activated macrophages, is a potent inducer of NGF mRNA expression.[73,74] Using primary sciatic fibroblasts Lindholm et al. found that DEX also inhibits the interleukin-1-mediated increase of NGF mRNA levels.[75]

Primary cultures from the hippocampus express NGF both in glial cells and neurons.[74] DEX treatment decreased NGF basal levels as well as the increase originated by interleukin-1 treatment in these cultures. From this study it is not possible to establish if DEX decreased NGF mRNA expression in glial cells, neurons, or both. Also, in iris transplants and iris primary fibroblast cultures DEX reduced NGF mRNA expression as well as the amount of NGF protein secreted to the medium.[76] In conclusion, DEX decreases both basal and induced NGF mRNA and protein expression in fibroblast cell line L929 and in primary cultures.

Glucocorticoids, upon binding to their receptor travel to the nucleus, where the complex binds to the so-called glucocorticoid responsive element (GRE) in the promoter region of the responsive genes, enhancing or repressing transcription. A possible explanation for the observed effects of DEX on NGF regulation is that DEX interferes with the NGF gene activity. This hypothesis was tested by Lindholm et al.[75] using a vector containing the NGF promoter region linked to a chloramphenicol acetyltransferase (CAT) reporter gene transfected into sciatic fibroblasts or NIH 3T3 cells. Addition of DEX reduced the basal CAT activity and also blocked the induction originated by serum or interleukin-1, which suggests that DEX is acting on the NGF promoter blocking transcription. Deletion experiments in the NGF promoter revealed a region between bases −437 and −275 necessary for DEX-induced suppression of NGF expression. Surprisingly, this region of the NGF promoter does not show any homology to the consensus sequence of the GRE.[77] Other genes, such as the proliferin gene, have previously also been shown to respond to glucocorticoids without carrying a classical GRE.[78]

In conclusion, DEX has a negative regulation on NGF gene expression in fibroblasts, which seems to be mediated via the glucocorticoid receptor.

The effects of glucocorticoids on NGF expression are, however, not only negative. Lindholm et al.[79] found that in primary hippocampal cultures DEX has differential effects depending on the cell type considered. DEX $(0.5 \times 10^{-6} M)$ inhibited the stimulatory effect of TGF-β1 on NGF mRNA expression in astrocytes, however, it increased NGF mRNA expression in cultured hippocampal neurons. The increase was maximal after 6 h and thereafter the NGF mRNA levels declined to reach control levels by 24 h. In hippocampal cultures DEX treatment did not affect the level of the other neurotrophin studied, BDNF. The increase of NGF mRNA in hippocampal neurons appears to be mediated via the glucocorticoid receptor, since the glucocorticoid antagonist RU38486 inhibited the positive effect of DEX on NGF mRNA. This inhibition was, however, incomplete.[79] Similar results were obtained by Leach Scully and Otten[80] when they incubated immortalized mouse hippocampal neurons with $10^{-6} M$ DEX. Interestingly, the time course of the NGF mRNA increase differed in hippocampal neurons derived from embryonic or postnatal tissue. In the postnatally derived cell line the increase was considerably greater and lasted for a longer period than in the embryonic line. These two cell lines also differ in the response of NT-3 mRNA to DEX. In embryonic cells NT-3 mRNA raises significantly with DEX treatment, whereas no change is seen in the postnatally derived cell line.[80] These results suggest that the changes originated by glucocorticoids in neurotrophin mRNA expression may be developmentally regulated.

Sapolsky et al.[7] have shown that glucocorticoids can enhance neuronal degeneration *in vitro*. It is, however, unlikely that the increase in NGF mRNA seen after DEX treatment is a response to neuronal damage, since the NGF increase is so rapid (detectable after 1 or 2 h) before damage due to glucocorticoids exposure is apparent.

Thus, the effect of DEX on NGF regulation seems to be cell specific. This is not unique for NGF regulation by DEX. It has been previously shown that DEX has a cell-specific effect on the collagenase gene, increasing or decreasing its expression depending on the cell type considered.[81-83] The stimulatory or inhibitory effect of DEX on this gene depends on the cellular context, in particular on the interaction with the Jun/Fos family of transcription factors.[84] These transcription factors are known to form homo- or heterodimers with other members of the family and bind to a specific recognition site, termed AP1 site, in the promoter region of the corresponding gene.[85] It is of interest to note the existence of a functional AP1 site in the first intron of the NGF gene between positions +33 and +50 as revealed by DNase I footprint analysis.[86] In the rat astrocytic cell line C6 NGF mRNA expression is transiently stimulated by cyclic AMP and this stimulation is preceded by a transient increase in c-fos mRNA level, suggesting an involvement of this transcription factor in neurotrophin gene regulation.[87] In fibroblasts carrying an exogenous c-fos construct under control of the metallothionin promoter, induction of the exogenous c-fos gene is followed with some delay by an increase in NGF mRNA.[86] Thus, an interaction between the glucocorticoid receptor and the Fos/Jun family of transcription factors, in a way similar to the collagenase gene, could explain the differential effects of DEX on the NGF gene in different cell types.

IV. EFFECTS OF DEX ADMINISTRATION ON NEUROTROPHIN EXPRESSION *IN VIVO*

The finding showing that glucocorticoids have a profound influence on neuronal survival of hippocampal neurons prompted a series of studies to explore the possibility that changes in neurotrophins could underlie the observed effects of glucocorticoids on neuronal survival. Several groups have studied the effects of DEX on neurotrophin and particularly NGF expression *in vivo*. After a single injection of DEX NGF mRNA has been found to increase in the hippocampus,[79,88,89] in the cerebral cortex,[88,90] and in the septum.[89] This increase is most likely due to an upregulation of NGF mRNA in neurons. The time course of this induction shows a maximal increase of 2.5-fold, that was reached 3 to 4 h after the injection and NGF mRNA remained at this elevated level at 10 h (Figure 3). Between 10 and 24 h NGF mRNA returned to control levels.[88,89] In the hippocampus and the cortex NGF mRNA had decreased below control levels 24 h after the injection and remained low at the latest time point analyzed 48 h (60% of control); (Figure 3).[88,89] NT-3 mRNA showed a similar pattern after DEX injection. An initial increase, which was maximal 1(hippocampus) and 2 h after the injection (cortex). The increase was followed by a decrease that resulted in levels lower than control levels 24 and 48 h after injection (Figure 3).[88] In contrast, BDNF mRNA showed no significant changes in the first hours following DEX injection but decreased to approximately 60% of control after 48 h (Figure 3).[88]

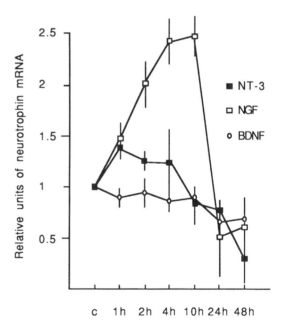

Figure 3 Time course of DEX-induced changes in neurotrophin mRNA expression in the cortex. Poly(A)+ RNA from the indicated time points after dexamethasone injection was electrophoresed in a formaldehyde-containing agarose gel, blotted onto nitrocellulose, and hybridized to probes specific for NGF, BDNF, and NT-3. Autoradiograms from two independent experiments were then analyzed by densitometric scanning and the level of neurotrophin mRNA was calculated relative to the level of α-actin mRNA. Data are expressed as percentage of control values. (From Barbany, G. and Persson, H., *Eur. J. Neurosci.*, 4, 396–403, 1992. By permission of Oxford University Press.)

Studies analyzing NGF protein after DEX treatment showed a small but significant elevation in a similar time-dependent fashion as seen for the mRNA. The peak elevation was observed at 12 h (hippocampus) and at 18 h (septum) after the injection.[89] The relative short interval between maximal increases in NGF mRNA and protein suggests that the NGF is synthesized locally. Retrograde transport requires up to 24 h to occur[91] and is thus unlikely to contribute to the increase. A similar elevation was seen in cortical NGF protein after subcutaneous DEX administration.[90]

Because DEX treatment influences NGF expression in different brain regions, it was of interest to assess whether there are any changes in NGF content following circadian changes in corticosterone. No significant change was seen in NGF protein in the hippocampus of rats killed at different times of the day.[79] However, in the septum the NGF protein content was slightly lower in the early morning, suggesting that the NGF protein transported to the septum may vary with the diurnal alterations of corticosterone in serum.

Sciatic nerve lesion has been shown to upregulate synthesis of NGF in proximal and distal nerve segments.[92] Pretreatment with DEX completely suppresses the lesion-mediated increase in sciatic NGF mRNA. In contrast, no effect was seen on the low basal levels of NGF mRNA in the intact sciatic nerve.[75]

A number of stimuli have been shown to upregulate NGF and BDNF mRNAs in hippocampus and cortex. Chronic seizures induced by electrolytic lesion[93,94] or by electrical kindling stimulation[95] lead to marked increases of NGF and BDNF mRNAs in the hippocampus and parietal and piriform cortices. Administration of the excitotoxin kainic acid also causes an increase of NGF and BDNF mRNAs in these brain regions.[96-98] DEX treatment can also modulate this increase. The combined administration of DEX and kainic acid with 2-h intervals results in increased levels of NGF mRNA in the dentate gyrus compared to kainic acid only (Figure 4).[99] This additive effect of DEX and kainic acid implies that the two treatments elicit increases of NGF mRNA by two different mechanisms. The kainic-acid mediated increase appears to be due to glutamate release and activation of the excitatory amino acid receptor of the non-methyl-D-aspartate subtype,[98] while the effect of DEX could be due to the presence of a glucocorticoid responsive element in the NGF promoter.[75] Surprisingly, when kainic acid was administered 24 h after DEX there was still a potentiation of kainic acid increase in NGF mRNA (Figure 4), even though DEX alone produces a decrease in NGF mRNA after 24 h.[88] One possible explanation could be changes in neuronal excitability due to DEX treatment, rendering neurons more susceptible to kainic acid. Supporting this hypothesis several groups have reported electrophysiological changes in the hippocampus in response to adrenal steroids.[100,101] The levels of BDNF mRNA in the cerebral cortex or hippocampus did not differ significantly between animals receiving both treatments with 2-h intervals between them or only kainic acid (Figure 4). Thus, there is no additive effect for BDNF mRNA regulation between DEX and kainic acid. This result is consistent with the finding that BDNF mRNA does not change in response to DEX alone after 4 h.[88] In contrast, when kainic acid was given 24 h after DEX injection, the increase of BDNF mRNA was significantly attenuated in both the dentate gyrus (Figure 4) and the piriform cortex, which is consistent with the decrease of BDNF mRNA seen 24 h after DEX treatment only.[88] In agreement with this result Cosi et al.[102] found that DEX treatment of hippocampal neurons in culture blunted the increase of BDNF mRNA elicited by high potassium or kainic acid.

Taken together these results suggest that different mechanisms underlie the increases of BDNF and NGF mRNAs seen after kainic acid exposure, which are modulated by DEX in different ways.

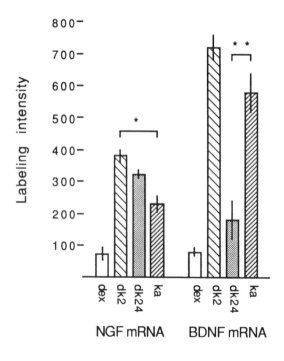

Figure 4 Relative amounts of NGF and BDNF mRNAs in the dentate gyrus after dexamethasone (n = 6), kainic acid only (n = 6) and dexamethasone plus kainic acid within 2- (n = 6) and 24-h (n = 4) intervals. Optical density values were determined using a microcomputer imaging device and converted to relative levels of radioactivity using autoradiographic ^{14}C scales as external standards. Data are presented as means ± S.E.M. Statistical analysis was performed using ANOVA followed by Scheffe's F-test. Asterisks indicate statistically significant differences between kainic acid treatment and animals receiving dexamethasone plus kainic acid ($*p$ < 0.01; $**p$ <0.001). (Reprinted from Barbany, G. and Persson, H., *Neuroscience*, 54, 909–922, 1993. With kind permission from Elsevier Science Ltd, The Boulevard, Langford Lane, Kindlington OX5 1GB, U.K.)

V. ADRENALECTOMY AND NEUROTROPHINS

The removal of glucocorticoids by adrenalectomy (ADX) results in a marked increase in cell death in the granule cell layer of the dentate gyrus as demonstrated by a dramatic raise in the number of pyknotic cells[103] and by extensive neuronal loss.[9]

The effect of ADX on NGF protein content in the hippocampus was studied by means of a biological assay, where hippocampal extracts were tested for their neurite-promoting activity on superior cervical ganglion neurons.[104] Twelve days after ADX NGF biological activity in the hippocampal extract was reduced to less than half of the activity of sham-operated animals. NGF immunoreactivity in the hippocampus was also decreased after ADX, particularly in the CA2 and CA3 regions and in the dentate gyrus.[104] The decrease in NGF protein was most likely a result of reduced transcription, since NGF mRNA levels decrease in the hippocampus three days after ADX (Figure 5).[88,105] The levels of the mRNAs for BDNF and NT-3 are also reduced as a result of ADX (Figure 5).[88] The mRNA expression for these three neurotrophins is also reduced in the cerebral cortex after ADX. These results suggest that adrenal steroids are required to maintain normal levels of neurotrophin expression in the rat brain. This hypothesis is further strengthened by the fact that the decrease induced by ADX is reversed when the animals received glucocorticoid replacement after ADX (Figure 5).[88] In a different study Chao et al.,[106] using *in situ* hybridization, found no evidence for any changes in hippocampal expression of the mRNAs for BDNF and NT-3 7 days after ADX. One possible explanation is that the effects of ADX on the levels of BDNF and NT-3 may be transient and that 7 days after ADX the expression of these two neurotrophins is restored to basal levels.

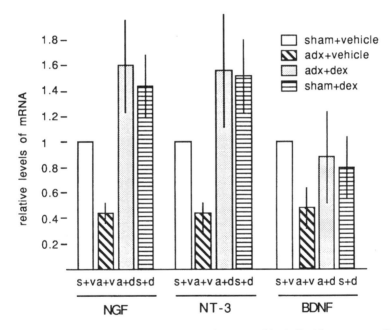

Figure 5 Densitometric analysis of the relative amounts of neurotrophins in the hippocampus after adrenalectomy. Densitometric scanning was performed on autoradiograms from three different experiments. The levels of neurotrophin mRNAs were then calculated relative to the level of α-actin mRNA. The values are expressed in arbitrary units with the level of mRNA in sham-operated, vehicle-injected animals set to 1.0. (s + v) sham operation plus vehicle; (a + v) adrenalectomy plus vehicle; (a + d) adrenalectomy plus dexamethasone; (s+d) sham operation plus dexamethasone. (From Barbany, G. and Persson, H., *Eur. J. Neurosci.*, 4, 396–403, 1992. By permission of Oxford University Press.)

Depleting catecholamine stores with reserpine treatment has been shown to induce NGF mRNA and protein in the cerebral cortex but not in other brain areas in 3-week-old rats.[107] Reserpine increases corticosterone levels via release of ACTH,[108] which could explain the increase of NGF mRNA after reserpine treatment. Supporting this hypothesis reserpine fails to increase NGF mRNA in ADX animals.[107] It remains to be clarified, however, why reserpine increased NGF mRNA in the cerebral cortex but not in the hippocampus, which is an important target site for glucocorticoid action. One possibility may be differences in the development of glucocorticoid receptors in these two brain regions. In fact, the cytosol binding capacity for glucocorticoids reaches adult levels in the hippocampus around four weeks of age,[109] suggesting that at three weeks hippocampal glucocorticoid receptors, although present, may not be fully functional.

ADX not only decreases basal neurotrophin mRNA expression, but can also affect the kainic acid-elicited increases in NGF and BDNF. Quantitative *in situ* hybridization showed that ADX attenuated the kainic acid-mediated increase of BDNF in the dentate gyrus as well as in the CA1 region of the hippocampus (Figure 6) and in the frontoparietal and piriform cortices (Figure 6).[99] The increase of NGF mRNA elicited by kainic acid in the dentate gyrus was almost completely abolished in ADX animals. Thus, adrenal steroids seem not only necessary to maintain basal expression of neurotrophins, but also for the neurons to respond with increased levels of neurotrophin mRNA after kainic acid treatment. However, an indirect action cannot be excluded, since it has been shown that ADX attenuates the stress-induced elevations in extracellular glutamate concentrations in the hippocampus.[110]

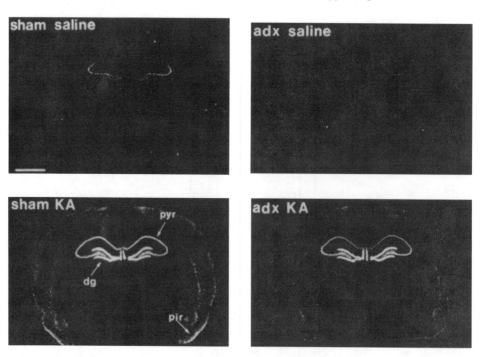

Figure 6 Expression of BDNF mRNA in the brain of ADX rats after kainic acid treatment. Prints of autoradiograms from sham animals receiving saline, sham animals plus kainic acid, ADX animals plus kainic acid, and ADX animals plus saline. Scale bar is 2.3 mm. (ADX) adrenalectomized; (dg) dentate gyrus; (KA) kainic acid, (pyr) pyramidal cell layer; (pir) piriform cortex. (Reprinted from Barbany, G. and Persson, H., *Neuroscience,* 54, 909-922, 1993. With kind permission from Elsevier Science Ltd., The Boulevard, Langford Lane, Kindlington OX5 1GB, U.K.)

VI. NEUROTROPHIN RECEPTORS AND ADRENAL STEROIDS

Glucocorticoids and NGF have antagonistic effects on neuroendocrine precursor cells. The choice of developing into adrenal chromaffin cells or sympathetic neurons is in part determined by the relative levels of glucocorticoids and NGF.[111] This observation prompted a series of studies to determine whether glucocorticoids influenced the expression of the NGF receptor, p75. Tocco et al.[112] found that DEX treatment of PC12 causes a decrease in the number of low affinity binding sites already detectable after 6 h and that was maximal after 3 days of treatment. This result was confirmed by Foreman et al.[113] who showed that treatment of PC12 cells with 1 μM DEX for 3 days resulted in decreased NGF receptor protein as measured by cross-linking and subsequent immunoprecipitation of the ^{125}I-NGF/NGFR complexes with a p75 specific monoclonal antibody or NGF polyclonal antisera. Subsequently, Yakovlev et al.[114] analyzed in PC12 and

C6 2B cells (a glioma cell line expressing p75) whether DEX affected p75 transcription. After incubation with DEX p75 mRNA expression decreases in both cell lines in a dose-dependent manner.[114] Moreover, DEX also blocked the increase in p75 mRNA originated by NGF treatment of PC12 cells.[113]

In vivo one single s.c. injection of DEX failed to change p75 mRNA in the septum. However, when the injection was repeated 24 h after the first one, a 60% decrease in p75 mRNA content was seen in the septum.[114] Aloe showed that following ADX p75 immunoreactivity increases in the septum. To determine whether increased p75 immunoreactivity was a result of increased septohippocampal retrograde transport, radiolabeled monoclonal Ab against p75 was injected into the hippocampus and the distribution of radiolabeled cells in the septum analyzed 24 h later. The number of radiolabeled neurons in the septum of ADX rats was greater than in control rats, suggesting that ADX could increase septohippocampal retrograde transport.[104] Taken together these results suggest that glucocorticoids regulate p75 expression and possibly function both *in vitro* and *in vivo*.

In contrast to p75, glucocorticoids did not change the expression level of trkB or trkC mRNAs in the rat brain. ADX by itself did not have any effect on trkB and trkC mRNAs expression in the hippocampus[99,106] or cerebral cortex.[99] However, ADX attenuated the kainic acid increase of trkB mRNA in the rat brain. One possible explanation for this unexpected result is decreased levels of BDNF protein in ADX animals treated with kainic acid compared to kainic acid alone, rather than a direct effect of glucocorticoids on trkB mRNA expression.

DEX injection has been shown to induce trk phosphorylation in the septum.[89] This induction of trk phosphorylation is not likely to be a direct effect of DEX, but rather mediated by NGF. The time course showed that trk phosphorylation was detectable 12 h after DEX treatment, which parallels the induction of NGF protein in the septum by DEX (Figure 7). This result opens up the possibility that the elevation of NGF mRNA and, subsequently, of NGF protein caused by DEX is sufficient to evoke an NGF-mediated biological response.

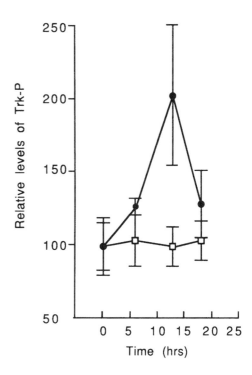

Figure 7 Time course of trk phosphorylation in the septum and hippocampus after dexamethasone injection. Dexamethasone was injected i.p. at a dose of 20 mg/kg and the animals were sacrificed 6, 12, and 18 h later. Graph of cumulative densitometric values (mean ± standard deviation). Statistically significant ($p < 0.05$) difference from control value for corresponding brain region. (From Saporito, M. S. et al., *Mol. Pharmacol.*, 45, 395–401, 1994. With permission.)

VII. CONCLUSIONS

Neurotrophin expression in the rat brain is regulated by adrenal steroids. Normal levels of circulating glucocorticoids appear to be necessary to maintain basal levels of trophic factor expression in several brain regions. The presence of adrenal steroids also seems necessary for the neurons to respond with increased neurotrophin mRNA expression to kainic acid. The effect of DEX on NGF expression is apparently cell specific; decreasing NGF mRNA expression in fibroblasts and astrocytes, while increasing

NGF mRNA in neurons. The increase of NGF protein caused by DEX treatment seems sufficient to initiate NGF-mediated biological responses. Moreover, the expression and possibly the function of the neurotrophin low affinity receptor p75 appear to be regulated by adrenal steroids. In contrast, trk receptor expression does not respond to glucocorticoids. Taken together these results open up the possibility that the effects of glucocorticoids on neuronal survival could, at least in part, be mediated via altered levels of neurotrophins. Moreover, they provide evidence that it is feasible to pharmacologically induce NGF *in vivo* using glucocorticoids.

REFERENCES

1. Devenport, L. D. and Devenport, J. S., The effects of adrenal hormones on brain and body size, *Physiol. Psychol.,* 10, 399, 1982.
2. Meyer, J. S., Early adrenalectomy stimulates subsequent growth and development of the rat brain, *Exp. Neurol.,* 82, 432, 1983.
3. Yehuda, R., Fairman, K. R., and Meyer, J. S., Enhanced brain proliferation following early adrenalectomy in rats, *J. Neurochem.,* 53, 241, 1989.
4. Sapolsky, R., Krey, L., and McEwen, B., Prolonged glucocorticoid exposure reduces hippocampal neuron number: implications for ageing, *J. Neurosci.,* 5, 1222, 1985.
5. Sapolsky, R. and Puslinelli, W., Glucocorticoids potentiate ischemic injury to neurons: therapeutic implications, *Science,* 229, 1397, 1985.
6. Sapolsky, R., Glucocorticoid toxicity in the hippocampus: temporal aspects of synergy with kainic acid, *Neuroendocrinology,* 43, 440, 1986.
7. Sapolsky, R. M., Glucocorticoids, hippocampal damage and the glutamatergic synapse, *Prog. Brain Res.,* 86, 13, 1990.
8. Landsfield, P., Baskin, R., and Pitler, T., Brain-ageing correlates: retardation by hormonal-pharmacological treatments, *Science,* 214, 581, 1981.
9. Sloviter, R. S., Valiquette, G., Abrams, G. M., Ronk, E. C., Sollas, A. L., Paul, L. A., and Neubort, S., Selective loss of hippocampal granule cells in the mature rat brain after adrenalectomy, *Science,* 243, 535, 1989.
10. Maehlen, J. and Torik, A., Necrosis of granule cells of hippocampus in adrenocortical failure, *Acta Neuropathol.,* 80, 85, 1990.
11. Armanini, M. P., Hutchins, C., Stein, B. A., and Sapolsky, R. M., Glucocorticoid endangerment of hippocampal neurons is NMDA-receptor dependent, *Brain Res.,* 532, 7, 1990.
12. Horner, H. C., Packan, D. R., and Sapolsky, R. M., Glucocorticoids inhibit glucose transport in cultured hippocampal neurons and glia, *Neuroendocrinology,* 52, 57, 1990.
13. Levi-Montalcini, R. and Hamburger, V., Selective growth stimulating effects of mouse sarcoma on the sensory and sympathetic nervous system of the chick embryo, *Cancer Res.,* 14, 49, 1951.
14. Levi-Montalcini, R. and Angeletti, P. U., Nerve growth factor, *Physiol. Rev.,* 48, 534, 1968.
15. Thoenen, H. and Barde, Y. A., Physiology of nerve growth factor, *Physiol. Rev.,* 60, 1284, 1980.
16. Mobley, W. C., Rutkowski, J. L., Tennekoon, G. I., Gemski, J., Buchanan, K., and Johnston, M. V., Nerve growth factor increases choline acetyltransferase activity in developing basal forebrain neurons, *Mol. Brain Res.,* 1, 53, 1986.
17. Bandtlow, C. E., Heuman, R., Schwab, M. E., and Thoenen, H., Cellular localization of nerve growth factor synthesis by *in situ* hybridization, *EMBO J.,* 6, 891, 1987.
18. Ayer-LeLievre, C., Olson, L., Ebendal, T., Seiger, Å., and Persson, H., Expression of the β-nerve growth factor gene in hippocampal neurons, *Science,* 240, 1339, 1988.
19. Barde, Y.-A., Edgar, D., and Thoenen, H., Purification of a new neurotrophic factor from mammalian brain, *EMBO J.,* 1, 549, 1982.
20. Ernfors, P., Ibañez, C. F., Ebendal, T., Olson, L., and Persson, H., Molecular cloning and neurotrophic activities of a protein with structural similarities to nerve growth factor: developmental and topographical expression in the brain, *Proc. Natl. Acad. Sci. U.S.A.,* 87, 5454, 1990.
21. Kaisho, Y., Yoshimura, K., and Nakahama, K., Cloning and expression of a cDNA encoding a novel human neurotrophic factor, *FEBS Lett.,* 266, 187, 1990.
22. Hohn, A., Leibrock, J., Bailey, K., and Barde, Y. A., Identification and characterization of a novel member of the nerve growth factor/brain-derived neurotrophic factor family, *Nature,* 344, 339, 1990.
23. Maisonpierre, P. C., Belluscio, L., Squinto, S., Ip, N. Y., Furth, M. E., Lindsay, R. M., and Yancopolous, G. D., Neurotrophin-3: a neurotrophic factor related to NGF and BDNF, *Science,* 247, 1446, 1990.
24. Rosenthal, A., Goeddel, D. V., Nguyen, T., Lewis, M., Shih, A., Laramee, G. R., Nikolics, K., and Winslow, J. W., Primary structure and biological activity of a novel human neurotrophic factor, *Neuron,* 4, 767, 1990.
25. Hallbök, F., Ibañez, C. F., and Persson, H., Evolutionary studies of the nerve growth factor family reveal a novel member abundantly expressed in *Xenopus* ovary, *Neuron,* 6, 845, 1991.
26. Berkemeier, L. R., Winslow, J. W., Kaplan, D. R., Nikolics, K., Goeddel, D. V., and Rosenthal, A., Neurotrophin-5: a novel neurotrophic factor that activates trk and trkB, *Neuron,* 7, 857, 1991.
27. Persson, H., Neurotrophin production in the brain, *Neurosciences,* 5, 227, 1993.

28. Korsching, S., Auburger, G., Heumann, R., Scott, J., and Thoenen, H., Levels of nerve growth factor and its mRNA in the central nervous system of the rat correlate with cholinergic innervation, *EMBO J.,* 4, 1389, 1985.

29. Ernfors, P., Wetmore, C., Olson, L., and Persson, H., Identification of cells in rat brain and peripheral tissues expressing mRNA for members of the nerve growth factor family, *Neuron,* 5, 511, 1990.

30. Timmusk, T., Belluardo, N., Metsis, M., and Persson, H., Widespread and developmentally regulated expression of neurotrophin-4 mRNA in rat brain and peripheral tissues, *Eur. J. Neurosci.,* 60, 287, 1994.

31. Eide, F. F., Lowenstein, H., and Reichardt, L. F., Neurotrophins and their receptors — current concepts and implications for neurologic disease, *Exp. Neurol.,* 121, 200, 1993.

32. Korsching, S., The neurotrophic factor concept: a re-examination, *J. Neurosci.,* 14, 1542, 1993.

33. Hefti, F., Nerve growth factor promotes survival of septal cholinergic neurons after fimbrial transections, *J. Neurosci.,* 6, 2155, 1986.

34. Williams, L. R., Varon, S., Peterson, G., Wictorin, K., Fischer, W., Björklund, A., and Gage, F. H., Continuous infusion of nerve growth factor prevents basal forebrain neuronal death after fimbria fornix transection, *Proc. Natl. Acad. Sci. U.S.A.,* 83, 9231, 1986.

35. Kromer, L. F., Nerve growth factor treatment after brain injury prevents neuronal death, *Science,* 235, 214, 1987.

36. Johnson, J. E., Barde, Y.-A., Schwab, M., and Thoenen, H., Brain-derived neurotrophic factor supports the survival of cultured rat retinal ganglion cells, *J. Neurosci.,* 6, 3031, 1986.

37. Alderson, R. F., Alterman, A. L., Barde, Y.-A., and Lindsay, R. M., Brain-derived neurotrophic factor increases survival and differentiated functions of rat septal cholinergic neurons in culture, *Neuron,* 5, 297, 1990.

38. Knusel, B., Winslow, J., Rosenthal, A., Burton, L., Seid, D., Nikolics, K., and Hefti, F., Promotion of central cholinergic and dopaminergic neuron differentiation by brain-derived neurotrophic factor but not neurotrophin 3, *Proc. Natl. Acad. Sci. U.S.A.,* 88, 961, 1991.

39. Yan, Q., Elliot, J., and Snider, W. D., Brain-derived neurotrophic factor rescues spinal motoneurons from axotomy-induced cell death, *Nature,* 360, 753, 1992.

40. Sendtner, M., Holtzman, B., Kolbeck, R., Thoenen, H., and Barde, Y.-A., Brain-derived neurotrophic factor prevents the death of motoneurons in newborn rats after brain section, *Nature,* 360, 757, 1992.

41. Oppenheim, R. W., Qin-Wei, Y., Prevette, D., and Yan, Q., Brain-derived neurotrophic factor rescues developing avian motoneurons from cell death, *Nature,* 360, 755, 1992.

42. Snider, W. D., Functions of the neurotrophins during nervous system development: what the knockouts are teaching us, *Cell,* 77, 627, 1994.

43. Friedman, W. J., Ibañez, C. F., Hallbök, F., Persson, H., Cain, L. D., Dreyfus, C. F., and Black, I. B., Differential actions of neurotrophins in the locus coeruleus and basal forebrain, *Exp. Neurol.,* 119, 72, 1993.

44. Arenas, E. and Persson, H., Neurotrophin-3 prevents the death of adult central noradrenergic neurons *in vivo, Nature,* 367, 368, 1994.

45. Ip, N. Y., Li, Y., Yancopoulos, G. D., and Lindsay, R. M., Cultured hippocampal neurons show responses to BDNF, NT-3 and NT-4, but not NGF, *J. Neurosci.,* 13, 3394, 1993.

46. Meakin, S. O. and Shooter, E. M., Molecular investigations on the high-affinity nerve growth factor receptor, *Neuron,* 6, 153, 1991.

47. Klein, R., Jing, S. Q., Nanduri, V., O'Rourke, E., and Barbacid, M., The trk proto-oncogene encodes a receptor for nerve growth factor, *Cell,* 65, 189, 1991.

48. Kaplan, D. R., Hempstead, B. L., Martin-Zanca, D., Chao, M. V., and Parada, L. F., The trk proto-oncogene product: a signal transducing receptor for nerve growth factor, *Science,* 252, 554, 1991.

49. Lamballe, F., Klein, R., and Barbacid, M., TrkC, a new member of the family of tyrosine protein kinases, is a receptor for neurotrophin-3, *Cell,* 66, 967, 1991.

50. Squinto, S. P., Stitt, T. N., Aldrich, T. H., Davis, S., Bianco, S. M., Radziejewski, C., Glass, D. J., Masiakowski, P., Furth, M. E., Valenzuela, D. M., DiStefano, P. S., and Yancopoulos, G. D., trkB encodes a functional receptor for brain-derived neurotrophic factor and neurotrophin-3 but not for nerve growth factor, *Cell,* 65, 885, 1991.

51. Soppet, D., Escandon, E., Maragos, J., Middlemas, D. S., Reid, S. W., Blair, J., Burton, L. E., Stanton, B. R., Kaplan, D. R., Hunter, T., Nikolics, K., and Parada, L. F., The neurotrophic factors brain-derived neurotrophic factor and neurotrophin-3 are ligands for the trkB tyrosine kinase receptor, *Cell,* 65, 895, 1991.

52. Ip, N. Y., Ibañez, C. F., Nye, S. H., McClain, J., Jones, P. F., Gies, D. R., Belluscio, L., Le Beau, M. M., Espinosa, III, R., Squinto, S. P., Persson, H., and Yancopoulos, G., Mammalian neurotrophin-4: structure, chromosomal localization, tissue distribution and receptor specificity, *Proc. Natl. Acad. Sci. U.S.A.,* 89, 3060, 1992.

53. Klein, R., Nanduri, V., Jing, S., Lamballe, F., Tapley, P., Bryant, S., C., C.-C., Jones, K. R., Reichardt, L. F., and Barbacid, M., The trkB tyrosine kinase is a receptor for brain-derived neurotrophic factor and neurotrophin-3, *Cell,* 66, 395, 1991.

54. Barbacid, M., Nerve growth factor: a tale of two receptors, *Oncogene,* 8, 2033, 1993.

55. Chao, M. V., Neurotrophin receptors: a window into neuronal differentiation, *Neuron,* 9, 583, 1992.

56. Holtzman, D. M., Li, Y., Parada, L. F., Kinsman, S., Chen, C.-K., Valetta, J. S., Zhou, J., Long, J. B., and Mobley, W. C., p140trk mRNA marks NGF-responsive forebrain neurons: evidence that trk gene expression is induced by NGF, *Neuron,* 9, 465, 1992.

57. Klein, R., Martin-Zanca, D., Barbacid, M., and Parada, L. F., Expression of the tyrosine kinase receptor gene trkB is confined to the murine embryonic and adult nervous system, *Development,* 109, 845, 1990.

58. Merlio, J., Ernfors, P., Jaber, M., and Persson, H., Molecular cloning of rat trkC and identification of cells expressing mRNAs for members of the trk family in the rat central nervous system, *Neuroscience,* 51, 513, 1992.

59. Merlio, J.-P., Ernfors, P., Kokaia, Z., Middlemas, D. S., Bengzon, J., Kokaia, M., Smith, M.-L., Siesjö, B. K., Hunter, T., Lindvall, O., and Persson, H., Increased production of the trkB protein tyrosine kinase receptor after brain insults, *Neuron,* 10, 151, 1993.

60. Tsoulfas, P., Soppet, D., Escadon, E., Tesarollo, L., Mendoza-Ramirez, J.-L., Rosenthal, A., Nikolics, K., and Parada, L. F., The rat trkC locus encodes multiple neurogenic receptors that exhibit differential response to neurotrophin-3 in PC12 cells, *Neuron,* 10, 975, 1993.

61. Valenzuela, D. M., Maisonpierre, P. C., Glass, D. J., Rojas, E., Nuñez, L., Kong, Y., Gies, D. R., Stitt, T. N., Ip, N. Y., and Yancopoulos, G. D., Alternative forms of trkC with different functional capabilities, *Neuron,* 10, 963, 1993.

62. Middlemas, D. S., Lindberg, R. A., and Hunter, T., trkB, a neural receptor protein-tyrosine kinase: evidence for a full-length and two truncated receptors, *Mol. Cell Biol.,* 11, 143, 1991.

63. Klein, R., Conway, D., Parada, L. F., and Barbacid, M., The trkB tyrosine protein kinase gene codes for a second neurogenic receptor that lacks the catalytic kinase domain, *Cell,* 61, 647, 1990.

64. Lamballe, F., Tapley, P., and Barbacid, M., TrkC encodes multiple neurotrophin-3 receptors with distinct biological properties and substrate specificities, *EMBO J.,* 12, 3083, 1993.

65. Barker, P. A., Lomen-Hoerth, C., Gensch, E. M., Meakin, S., Glass, D. J., and Shooter, E. M., Tissue-specific alternative splicing generates two isoforms of the trkA receptor, *J. Biol. Chem.,* 268, 15150, 1993.

66. Rodriguez-Tebar, A., Dechant, G., and Barde, Y.-A., Binding of brain-derived neurotrophic factor to the nerve growth factor receptor, *Neuron,* 4, 487, 1990.

67. Rodriguez-Tebar, A., Dechant, G., Götz, R., and Barde, Y.-A., Binding of neurotrophin-3 to its neuronal receptors and interactions with nerve growth factor and brain-derived neurotrophic factor, *EMBO J.,* 11, 917, 1992.

68. Meakin, S. O. and Shooter, E. M., The nerve growth factor family of receptors, *TINS,* 15, 323, 1992.

69. Lee, K.-F., Bachman, K., Landis, S., and Jaenisch, R., Dependence on p75 for innervation of some sympathetic targets, *Science,* 263, 1447, 1994.

70. Wion, D., Houlgatte, R., and Brachet, P., Dexamethasone rapidly reduces the expression of β-NGF gene in mouse L-929 cells, *Exp. Cell Res.,* 162, 562, 1986.

71. Siminoski, K., Murphy, R. A., Rennert, P., and Heinrich, G., Corticosterone, testosterone and aldosterone reduce levels of nerve growth factor messenger ribonucleic acid in L-929 fibroblasts, *Endocrinology,* 121, 1432, 1987.

72. D´Mello, S. R. and Heinrich, G., Multiple signalling pathways interact in the regulation of nerve growth factor production in L929 fibroblasts, *J. Neurochem.,* 57, 1570, 1991.

73. Lindholm, D., Heumann, R., Meyer, M., and Thoenen, H., Interleukin-1 regulates synthesis of nerve growth factor in non-neuronal cells of rat sciatic nerve, *Nature,* 330, 658, 1987.

74. Friedman, W. J., Lärkfors, C., Ayer-LeLievre, C., Ebendal, T., Olson, L., and Persson, H., Regulation of β-nerve growth factor expression by inflammatory mediators in hippocampal cultures, *J. Neurosc. Res.,* 27, 374, 1990.

75. Lindholm, D., Hengerer, B., Heuman, R., Carroll, P., and Thoenen, H., Glucocorticoid hormones negatively regulate nerve growth factor expression *in vivo* and in cultured rat fibroblasts, *Eur. J. Neurosci.,* 2, 795, 1990.

76. Houlgatte, R., Wion, D., and Brachet, P., Levels of nerve growth factor secreted by rat primary fibroblasts and iris transplants are influenced by serum and glucocorticoids, *Dev. Brain Res.,* 47, 171, 1989.

77. Strähle, U., Klock, G., and Schütz, G., A DNA sequence of 15 base pairs is sufficient to mediate both glucocorticoid and progesterone induction of gene expression, *Proc. Natl. Acad. Sci. U.S.A.,* 84, 7871, 1987.

78. Mordacq, J. C. and Linzer, D. I. H., Co-localization of elements required for phorbol ester stimulation and glucocorticoid repression of proliferin gene expression, *Genes Dev.,* 3, 760, 1989.

79. Lindholm, D., Castrén, E., Hengerer, B., Zafra, F., Berninger, B., and Thoenen, H., Differential regulation of nerve growth factor (NGF) synthesis in neurons and astrocytes by glucocorticoid hormones, *Eur. J. Neurosci.,* 4, 404, 1992.

80. Leach Scully, J. and Otten, U., Glucocorticoid modulation of neurotrophin expression in immortalized mouse hippocampal neurons, *Neurosci. Lett.,* 155, 11, 1993.

81. Jonat, C., Rahmsdorf, H. J., Park, K., Cato, A. C. B., Gebel, S., Ponta, H., and Herrlich, P., Antitumor promotion and antiinflammation: down modulation of AP-1 (Fos/Jun) activity by glucocorticoid hormone, *Cell,* 62, 1189, 1990.

82. Schüle, R., Rangarajan, P., Kliever, S., Ransone, L. J., Bolado, J., Yang, N., Verma, I. M., and Evans, R. M., Functional antagonism between oncoprotein c-jun and the glucocorticoid receptor, *Cell,* 62, 1217, 1990.

83. Yang-Yen, H.-F., Chambard, J.-C., Sun, H.-L., Smeal, T., Schmidt, T. J., Drouin, J., and Karin, M., Transcriptional interference between c-jun and the glucocorticoid receptor: mutual inhibition of DNA binding due to direct protein-protein interaction, *Cell,* 62, 1205, 1990.

84. Diamond, M. I., Miner, J. N., Yoshinaga, S. K., and Yamamoto, K. R., Transcription factor interactions: selectors of positive or negative regulation from a single DNA element, *Science,* 249, 1266, 1990.

85. Morgan, J. I. and Curran, T., Stimulus transcription coupling in the nervous system: involvement of the inducible proto-oncogenes fos and jun, *Annu. Rev. Neurosci.,* 14, 421, 1991.

86. Hengerer, B., Lindholm, D., Heumann, R., Ruther, U., Wagner, E. F., and Thoenen, H., Lesion-induced increase in nerve growth factor mRNA is mediated by c-fos, *Proc. Natl. Acad. Sci. U.S.A.,* 87, 3899, 1990.

87. Mocchetti, I., De Bernardi, M. A., Szekely, A. M., Alho, H., Brooker, G., and Costa, E., Regulation of NGF biosynthesis by β-adrenergic receptor activation in astrocytoma cells: a potential role of c-Fos protein, *Proc. Natl. Acad. Sci. U.S.A.,* 86, 3891, 1989.

88. Barbany, G. and Persson, H., Regulation of neurotrophin mRNA expression in the rat brain by glucocorticoids, *Eur. J. Neurosci.,* 4, 396, 1992.

89. Saporito, M. S., Brown, E. R., Hartpeuce, C. K., Wilkox, H. M., Robins, E., Vaught, J. L., and Carswell, S., Systemic dexamethasone administration increases septal trk autophosphorylation in adult rats via an induction of nerve growth factor, *Mol. Pharmacol.,* 45, 395, 1994.

90. Fabrazzo, M., Costa, E., and Mocchetti, I., Stimulation of nerve growth factor biosynthesis in developing rat brain by reserpine: steroids as potential mediators, *Mol. Pharmacol.,* 39, 144, 1991.

91. Distefano, P. S. B., Friedman, C., Rasziejewski, C., Alexander, P., Boland, C., Schick, M., Lindsay, M., and Wiegand, S. G., The neurotrophins BDNF, NT-3 and NGF display distinct patters of retrograde transport in peripheral and central neurons, *Neuron,* 8, 983, 1992.

92. Heumann, R., Lindholm, D., Bandtlow, C., Meyer, M., Radeke, M. J., Misko, T. P., Shooter, E., and Thoenen, H., Differential regulation of mRNA encoding nerve growth factor and its receptor in rat sciatic nerve during development, degeneration, and regeneration: role of macrophages, *Proc. Natl. Acad. Sci. U.S.A.,* 84, 8735, 1987.

93. Gall, C. M. and Isackson, P. J., Limbic seizures increase neuronal production of messenger RNA for nerve growth factor, *Science,* 245, 758, 1989.

94. Isackson, P. J., Huntsman, M. M., Murray, K. D., and Gall, C. M., BDNF mRNA expression is increased in adult rat forebrain after limbic seizures: temporal patterns of induction distinct from NGF, *Neuron,* 6, 937, 1991.

95. Ernfors, P., Bengzon, J., Kokaia, Z., Persson, H., and Lindvall, O., Increased levels of messenger RNAs for neurotrophic factors in the brain during kindling epileptogenesis, *Neuron,* 7, 165, 1991.

96. Ballarin, M., Ernfors, P., Lindefors, N., and Persson, H., Hippocampal damage and kainic acid injection induce a rapid increase in mRNA for BDNF and NGF in the rat brain, *Exp. Neurol.,* 114, 35, 1991.

97. Gall, C., Murray, K., and Isackson, P. J., Kainic acid-induced seizures stimulate increased expression of nerve growth factor mRNA in rat hippocampus, *Mol. Brain Res.,* 9, 113, 1991.

98. Zafra, F., Hengerer, B., Leibrock, J., Thoenen, H., and Lindholm, D., Activity dependent regulation of BDNF and NGF mRNAs in the rat hippocampus is mediated by non-NMDA glutamate receptors, *EMBO J.,* 9, 3545, 1990.

99. Barbany, G. and Persson, H., Adrenalectomy attenuates kainic acid-elicited increases of messengerRNAs for neurotrophins and their receptors in the rat brain, *Neuroscience,* 54, 909, 1993.

100. Kerr, D. S., Campbell, L. W., Hao, S.-Y., and W., L. P., Corticosteroid modulation of hippocampal potentials: Increased effect with ageing, *Science,* 245, 1505, 1989.

101. Doi, N., Miyahara, S., and Hori, N., Glucocorticoids promote localized activity of rat hippocampal pyramidal neurons in brain slices, *Neurosci. Lett.,* 123, 99, 1991.

102. Cosi, C., Spoerri, P. E., Comelli, M. C., Guidolin, D., and Skaper, S., Glucocorticoids depress activity-dependent expression of BDNF mRNA in hippocampal neurons, *Neuroreport,* 4, 527, 1993.

103. Gould, E., Wooley, C. S., and McEwen, B. S., Short-term glucocorticoid manipulations affect neuronal morphology and survival in the adult dentate gyrus, *Neuroscience,* 37, 367, 1990.

104. Aloe, L., Adrenalectomy decreases nerve growth factor in young adult rat hippocampus, *Proc. Natl. Acad. Sci. U.S.A.,* 86, 5636, 1989.

105. Follesa, P. and Mocchetti, I., Regulation of basic fibroblasts growth factor mRNA by b-adrenergic receptor activation and adrenal steroids in rat central nervous system, *Mol. Pharmacol.,* 43, 132, 1993.

106. Chao, H. M. and McEwen, B. S., Glucocorticoids and the expression of mRNAs for neurotrophins, their receptors and GAP-43 in the rat hippocampus, *Mol. Brain Res.,* 26, 271, 1994.

107. Fabrazzo, M., Costa, E., and Mocchetti, I., Stimulation of nerve growth factor biosynthesis in developing rat brain by reserpine. steroids as potential mediators, *Mol. Pharmacol.,* 39, 144, 1991.

108. Westerman, E. O., Maikel, R. P., and Brodie, B. B., On the mechanism of pituitary-adrenal stimulation by reserpine, *J. Pharmacol. Exp. Ther.,* 138, 208, 1962.

109. Clayton, C. J., Grosser, B. I., and Stevens, W., The ontogeny of cortisone and dexamethasone receptors in the rat brain, *Brain Res.,* 134, 445, 1977.

110. Lowy, M. T., Gault, L., and Yamamoto, B. K., Adrenalectomy attenuates stress-induced elevations in extracellular glutamate concentrations in the hippocampus, *J. Neurochem.,* 61, 1957, 1993.

111. Landis, S. C. and Patterson, P. H., Neural crest lineages, *Trends Neurosci.,* 4, 172, 1981.

112. Tocco, M. D., Contreras, M. L., Koizumi, S., Dickens, G., and Guroff, G., Decreased levels of nerve growth factor receptor on dexamethasone-treated PC12 cells, *J. Neurosci. Res.,* 20, 411, 1988.

113. Foreman, P. J., Taglialatela, G., Jackson, G. R., and Perez-Polo, J. R., Dexamethasone blocks nerve growth factor induction of nerve growth factor receptor mRNA in PC12 cells, *J. Neurosci. Res.,* 31, 52, 1992.

114. Yakovlev, A. G., De Bernardi, M. A., Fabrazzo, M., Brooker, G., Costa, E., and Mocchetti, I., Regulation of nerve growth factor receptor mRNA content by dexmethasone: *in vitro* and *in vivo* studies, *Neurosci. Lett.,* 116, 216, 1990.

Chapter 8

Sexual Dimorphism in the CNS and the Role of Steroids

Antonio Guillamón and Santiago Segovia

CONTENTS

I. INTRODUCTION: THE ORGANIZATIONAL CONCEPT

The functions of gonadal steroids are intimately related to physiological and behavioral mechanisms that control reproduction, a process of great diversity in nature.[1] In the phylum vertebrata, societies present different degrees of complexities in which species-specific strategies for mating and rearing have evolved to maximize individual reproductive success and the survival of offspring until maturity.

As reviewed by Crews,[2] different mechanisms trigger sex determination in vertebrates. In mammals, birds and other gonochoristic vertebrates, specific chromosomes inherited from parents at the time of fertilization determine gonadal sex. Recently, the *sex-determining region of the Y chromosome* (SRY) in mice[3] and man[4] was identified. However, sex determination can be guided by temperature in some reptiles and apparently by social stimuli in some fishes.[2]

The role of gonadal steroids in differentiating male and female mammals at the morphological, physiological, and behavioral levels has been guided from thoughts arising from embryological and behavioral studies.

Working with sexually indifferent rabbit embryos, Jost[5,6] showed that an indifferent urogenital tract goes through male differentiation if a testis develops and female differentiation if an ovary develops. Embryos castrated before sexual differentiation develop as phenotypic females. These experiments demonstrate that the induced phenotype in mammals is male and testicular secretions are necessary for male development.[7] Thus, it is widely assumed that the male is the organized sex, while the female is the neutral sex; the opposite is believed to occur in the class aves.[2]

Phoenix et al.[8] studied the sexual behavior of male and female guinea pigs born of mothers treated with testosterone propionate (TP) during pregnancy. TP females were less likely to show lordosis and were more likely to display mounting behavior in adulthood than were control animals when both were gonadectomized and treated with the appropriate sex hormones. The authors suggested that an androgen administered prenatally has an *organizing* action on the neural tissues mediating sexual behavior and during adulthood gonadal hormones are *activational*. Although this hypothesis found some early resistance (see Baum[9] for historical comments), it has been of great heuristic value for guiding research on sexual dimorphism in the CNS and the differentiation of coital behavior in mammals.[10] It is amply supported by experimental evidence; however, some constraints[11] have been brought up and caution should be taken before generalizing a broad concept of organization to the phylum vertebrata.[2]

Gonadal steroids acting perinatally, prepuberally, and in adulthood initially provide the necessary structural differentiation in all the systems involved in reproduction (including the CNS) and later provide the gender-specific functioning of these systems. During development gonadal steroids probably shape a profile in male and female mammals that could place them into two different neurobehavioral dimensions.[12] This chapter examines the role which gonadal steroids play in the development of sex differences in the mammalian CNS. Those interested in aves should refer to the excellent reviews[13,14] in the literature.

II. SEXUAL DIMORPHISM IN MAMMALIAN CNS

A. FUNCTIONAL APPROACH

Because the brain regions involved in reproductive physiology and behavior are sexually dimorphic, sexual dimorphism is better understood from a functional approach.[15-18] The sexual dimorphism in the vomeronasal system[18] (VNS) and some motor nuclei in the spinal cord, the spinal nucleus of the bulbocavernosus system,[19,20] (SNBS), provides a useful framework.

Reproductive behaviors are influenced by the endocrinological state in conjunction with a variety of sensory stimuli, in particular olfactory stimuli seem to play an important role in eliciting endocrinological and behavioral responses related to reproduction. The VNS[21-35] (Figure 1) mediates the effects of pheromones, which are involved in eliciting and maintaining masculine and feminine sexual[32-34,36-52] and parental care[32-34,53-61] behaviors, and primer pheromone mechanisms which affect puberty,[32-34,62-64] estrous cycle,[32-34,65-68] and gestation.[32-34,69-72] At the same time, the SNBS innervates penile muscles involved in male copulatory reflexes (Figure 2).[73,74] Thus, the study of development of sex differences in the VNS and the SNBS will facilitate demonstration of how gonadal steroids sexually differentiate (1) a sensorial organ (the vomeronasal), (2) the complex circuitry (VNS) that supports important physiological and behavioral aspects of reproductive behaviors, and (3) the ejaculatory reflexes (SNBS).

B. SEXUAL DIMORPHISM IN THE VOMERONASAL SYSTEM

1. Sexual Dimorphism in the Vomeronasal Organ

The vomeronasal organ (VNO), a chemosensorial structure located in the anteroventral portion of the nasal cavity, is present in most vertebrate species[32,33] including man.[75] It conveys chemical stimuli related to reproductive and other social behaviors.[34]

Gonadal steroids present shortly after birth develop a sexually dimorphic VNO in the rat.[76] As shown in Table 1, males present higher morphometric measures than females with respect to the VNO; the overall volume, neuroepithelium volume, and number of bipolar neurons are all larger in males. These sex differences are abolished and inverted when they are orchidectomized and females are androgenized (using TP) on the day of birth (D1) and studied in adulthood (D90). Gonadectomized males do not differ from control females, but do differ from control males, while D1 androgenized females present greater morphometric measures than control females, but not compared to control males. The nuclei of the bipolar neurons also present sex diferences (Table 1). They are larger in female than in male rats and gonadal steroids induce these sex differences.

The role which gonadal steroids play in maintaining sex differences in the VNO in adulthood has been examined.[77] When male and female rats are gonadectomized at the age of three months and sacrificed when they are six months old, a decrease is produced in the neuroepithelial height of both sexes with respect to control animals. The treatments affect the nuclear size in females to a greater extent than in males, since the females experience a 20% decrease in nuclear area while the reduction in area in males is 5%. No ultrastructural cellular change was observed in this study.

These results indicate that gonadal steroids act shortly after birth (at least on D1) to develop sex differences in the VNO and are necessary later in both sexes to maintain the normal morphometric parameters of this organ. Sexual dimorphism in the VNO may suggest the existence of sex differences in coding stimuli related to reproductive behaviors.[76] Nevertheless, the picture might actually be considerably more complex since studies have shown the heterogeneity of vomeronasal receptor cells and the specific nature of projections from subclasses of vomeronasal nerve fibers to the accessory olfactory bulb.[78,79]

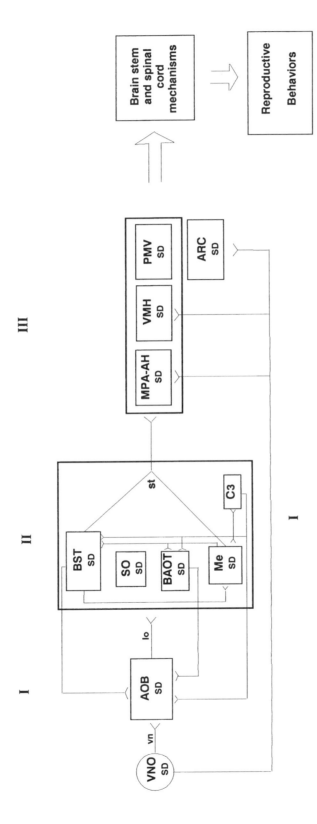

Figure 1 Anatomical and functional model of the vomeronasal system in rodents. I, II, and III: primary, secondary, and tertiary vomeronasal projections. SD: sex differences; AOB: accessory olfactory bulb; ARC: arcuate nucleus; BAOT: bed nucleus of the accessory olfactory tract; BST: bed nucleus of the stria terminalis; C₃: posteromedial cortical amygdaloid nucleus; Me: medial amygdaloid nucleus; MPA-AH: medial preoptic area-anterior hypothalamic continuum; PMV: premammillary nucleus, ventral part; SO: supraoptic nucleus; VMH: ventromedial hypothalamic nucleus; VNO: vomeronasal organ; lo: lateral olfactory tract; st: stria terminalis; and vn: vomeronasal nerve. (Modified from Segovia, S. and Guillamón, A., *Brain Res. Rev.*, 18, 51, 1993. With permission.)

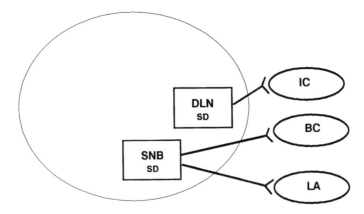

Figure 2 Spinal nucleus of the bulbocavernosus system. L5–L6: 5th and 6th spinal cord segments; SD: sex differences; DLN: dorsolateral nucleus; SNB: spinal nucleus of the bulbocavernosus; BC: bulbocavernosus muscle; IC: ischiocavernosus muscle; and LA: levator ani muscle.

Table 1 Sexual Dimorphism in the Vomeronasal Organ and Effects of Early Postnatal Male Orchidectomy and Female Androgenization on Its Development

	Males		Females	
	Control	Orchidectomized	Control	Androgenized
VNO volume (mm³)	1.7263 ± 0.071[aa]	1.0779 ± 0.036[bb]	1.0811 ± 0.068	1.6631 ± 0.125[b]
Neuroepithelium volume (mm³)	0.1838 ± 0.013[a]	0.1141 ± 0.011[b]	0.1134 ± 0.012	0.2023 ± 0.011[bb]
Number of bipolar neurons	37961 ± 720[a]	31560 ± 720[b]	30175 ± 1154	38819 ± 1576[b]
Nuclear size of the bipolar neurons (μm²)	22.88 ± 0.007[aa]	23.73 ± 0.010[bb]	23.52 ± 0.009	22.94 ± 0.006[bb]

Note: Data show means ± S.E.M. Student's *t*-test for *post hoc* comparisons.

Comparisons between control groups: [a]$p <0.005$; [aa]$p <0.001$.

Comparisons with respect to the control group of the same genetic sex: [b]$p <0.005$; [bb]$p <0.001$.

From Segovia, S. and Guillamón, A., *Brain Res. Rev.,* 18, 51, 1993. With permission.

2. Sexual Dimorphism in the Accessory Olfactory Bulb

The vomeronasal nerve projects to the accessory olfactory bulb (AOB; Figure 1), a structure embedded in the dorsocaudal portion of the olfactory bulb related to reproductive physiology and behavior in both male and female rats.[38,45,80-84]

The rat AOB is sexually dimorphic (Figure 3). Males show greater morphometric values (Table 2) in relation to total volume, the volume of each of its layers, the number of mitral and light and dark granule cells, the size of mitral somata, and the number of denditric branches growing out from the main dendritic stalks in these cells.[85-89] Male orchidectomy and female androgenization on D1 supresses and inverts all these sex differences except for the dark granule cells in the females (Table 2).[85-89]

There is evidence that AOB granule cells have two neurogenetic periods.[90,91] The first takes place prenatally (embryonic (E) E17 to 19 days after conception) and the second occurs postnatally (between D1 to D20), but proliferation for the latter is more intense between D6 to D8. Thus, it has been suspected that AOB dark granule cells might actually originate postnatally and it has been hypothesized[88] that these cells could have a different critical period for the organizing effects of androgens from that of the light granule cells. In fact, androgenization is effective in masculinizing females when it is performed on D14 or D1 plus D14, but not only on D1 (Figure 4). Therefore, the organizing effects of androgens involved in sexual differentiation of the AOB light and dark granule cells depend on different periods of maximal susceptibility which are related to the neurogenetic timing of these two particular types of cells.[88]

A developmental study[92] of the AOB from D1 to D60 has shown that the AOB volume increases 600% from D1 to D18 and then declines, so that by D60 it is 66% of its maximum value. The number of mitral cells also declines from D18 while the total number of granule cells increases from D1 to D18 but does not change significantly through D60.

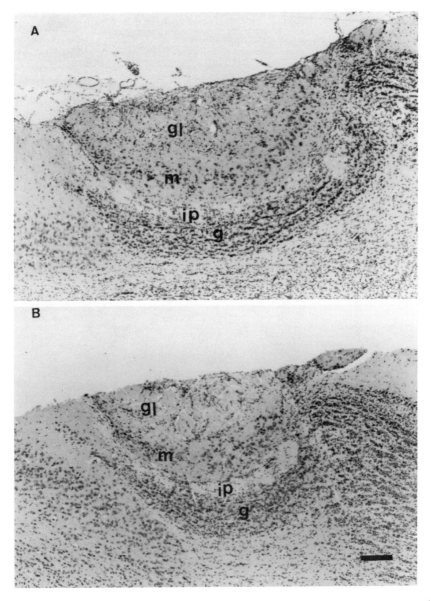

Figure 3 A sagittal section of the accessory olfactory bulb showing sex differences. g: granular layer; gl: glomerular-external plexiform layer; ip: inner plexiform layer; and m: mitral cells layer. (A) Male; (B) female. Bar 200 µm. (Courtesy of Dr. C. Pérez-Laso.)

Studies have also been carried out on the effects of androgens on the development of the AOB. Roos et al.[86,93] have shown that sex differences, measured taking into account the surface area of the AOB, appear as early as D7. They have also studied the effects of TP treatment from D20 onward on AOB development in rats castrated at birth or on D20. TP fails to restore AOB development in animals castrated at birth but completely restores it in those castrated at D20. The authors suggest that secretion of androgen during the perinatal and prepubescent periods appears to be critical for male AOB development.

Because the mammalian brain metabolizes testosterone to 17β-estradiol (E2) and 5α-androstan-17β-ol-3-one (DHT)[94-96] and E2 is known to be responsible for the masculinization of some VNS structures (the sexually dimorphic nucleus of the preoptic area (SDN-POA),[97-99] the medial amygdaloid nucleus (Me),[100] and the bed nucleus of the accessory olfactory tract (BAOT))[101] further investigation[102] has been made into the effects of single administrations of estradiol benzoate (EB) and DHT on males castrated at birth. D1 castration decreases mitral cells layer volume and the number of mitral cells of the male

Table 2 Sexual Dimorphism in the Accessory Olfactory Bulb and Effects of Early Postnatal Male Orchidectomy and Female Androgenization on Its Development

	Males		Females	
	Control	Orchidectomized	Control	Androgenized
Layer volume (mm³)				
VN-glomerular	0.036 ± 0.010	0.017 ± 0.008[bb]	0.034 ± 0.020	0.052 ± 0.020
Mitral cells	0.065 ± 0.010[aaaaa]	0.034 ± 0.010[bbb]	0.038 ± 0.003	0.072 ± 0.030[b]
Plexiform	0.032 ± 0.006[aaa]	0.020 ± 0.006[b]	0.020 ± 0.006	0.036 ± 0.013[b]
Granular	0.073 ± 0.020[aaaaa]	0.039 ± 0.010[bbbb]	0.041 ± 0.007	0.063 ± 0.020[b]
Total volume (mm³)	0.211 ± 0.050[aaa]	0.113 ± 0.030[bbbb]	0.137 ± 0.040	0.231 ± 0.090[b]
Number of neurons				
Mitral	5646 ± 536.36[aaaa]	3910 ± 258.86[b]	3850 ± 339.56	7410 ± 966.16[bb]
Light granule	53340 ± 3543.71[aa]	35208 ± 3604.61[bbb]	35936 ± 3917.16	47260 ± 3861.12[b]
Dark granule	8448 ± 1309.06[aa]	3634 ± 479.19[bbb]	3910 ± 502.91	3642 ± 360.82
Mitral cells				
Somatic area (μm²)	188.87 ± 6.872[aa]	158.07 ± 7.201[bb]	158.15 ± 6.346	176.46 ± 5.589[b]
Number of branches	43.075 ± 3.131[a]	30.500 ± 2.456[bb]	34.325 ± 2.137	65.100 ± 5.825[bbb]

Note: Data show means ± S.E.M. Mann-Whitney test for volume *post hoc* comparisons: Student's *t*-test for neuron number, somatic area, and number of branches *post hoc* comparisons.

Comparisons between control groups: [a]p <0.05; [aa]p <0.01; [aaa]p <0.025; [aaaa]p <0.005; [aaaaa]p <0.001.

Comparisons with respect to the control group of the same genetic sex: [b]p <0.05; [bb]p <0.01; [bbb]p <0.005; [bbbb]p <0.001.

From Segovia, S. and Guillamón, A., *Brain Res. Rev.*, 18, 51, 1993. With permission.

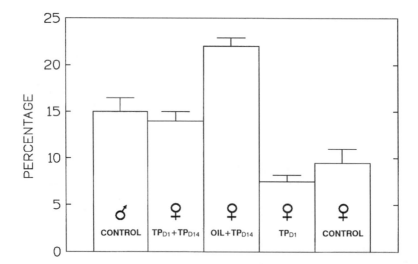

Figure 4 Percentage of AOB dark granule cells in relation to the total number (light plus dark) granule cells. ♂: male; ♀: female; ♀ TP_{D1} + TP_{D14}: androgenized females on D1 and D14; ♀ oil + TP_{D14}: females treated with oil (vehicle) the day of birth (D1) and androgenized on D14; ♀ TP_{D1}: females androgenized on D1; and control: vehicle-treated female. (Modified from Segovia, S. and Guillamón, A., *Brain Res. Rev.*, 18, 51, 1993. With permission.)

AOB. This effect is counteracted by a single injection of EB on D1, but administration of DHT also on D1 is unable to restore the normal morphological pattern. In addition, long-term administration of DHT (from D6 to D20) to males produces a significant decrease in AOB volume with respect to intact and control (vehicle) males and it abolishes sex differences compared to intact females.[103]

These experiments suggest that the hormone E2 (but not DHT) is involved in masculinization of the male rat AOB. Moreover, AOB labeled cells expressing mRNA encoding androgen and estrogen receptors have been shown to exist.[104] Thus, intracellular aromatization of testosterone to estradiol might explain the mechanism which masculinizes the male AOB. However, to our knowledge, no studies to show aromatase activity in the AOB have been undertaken to date.

It has been suggested[86,93] that the secretion of androgens during the perinatal and prepubescent periods could be critical for male AOB development. However, the presence of EB on D1 is sufficient to restore a masculine AOB in males orchidectomized at birth[102] and TP administered on D1 masculinizes the female AOB.[87-89] It seems that the action of androgens for AOB masculinization is more critical at the early postnatal than at the prepubescent period.

Moreover, evidence indicates that gonadal steroids could modulate the GABA-A/benzodiazepine receptor Cl-channel complex (GABA/BDZ receptor complex[105-107]). Perinatal administration of diazepam (DZ) has been shown to alter development of sex differences in the AOB[108,109] since prenatal (E14 to 21), postnatal (D1 to D16), and pre- and postnatal (E14 to E21 and D1 to D16) administration of DZ produces a drop in the volume of the mitral cell layer and in the number of mitral and dark and light granule cells in the male AOB, but does not affect these parameters in the female.[108,109] Thus, the perinatal exposure to DZ on the part of males produces the same effects as early postnatal (D1) gonadectomy.[108,109] These results and others[105-107] indicating that sex steroids exert membrane effects on neurotransmitter receptors have prompted the hypothesis[18] that organization of sex differences of the brain might not be mediated only by the genomic action of gonadal steroids alone, but by changes taking place in the membrane as well.

3. Sexual Dimorphism in the Bed Nucleus of the Accessory Olfactory Tract

The BAOT is a forebrain group of cells associated with the accessory olfactory tract (Figure 1). BAOT lesions facilitate maternal behavior in both virgin female[60] and male[61] rats, and electrical stimulation increases the sensitization latency for this behavior (Pacheco and Rosenblatt, personal communication).

The BAOT undergoes sexual differentiation (Figure 5).[110] Table 3 shows that male rats have greater BAOT volumes, neuronal numbers, and neuron/glia ratios than females, but no sex difference has been found in terms of the number of glial cells. Androgens play an important role in inducing sexual dimorphism in this nucleus since male orchidectomy and female androgenization (TP) on D1 reverse sex differences in the BAOT.[110]

Sex differences in the neuron/glia ratio deserve some comments because males show larger ratios than control female rats (Table 3). Similar findings have been reported for the cortex,[111] but the mechanism(s) that controls sex differences in this parameter is unknown. In primary cultures of fetal rat hypothalamic cells it has been shown that aromatase is located primarily in neurons while 5α-reductase is confined mainly to non-neuronal cells.[112] This distribution might contribute to explain sex differences in neuron/glia ratio.

Studies have examined the role of E2 and DHT in the development of sex differences in the BAOT.[101,113] Orchidectomized males injected on D1 with a single dose of EB show volumes and neuron numbers similar to those of control males (Figure 6); however, orchidectomized males and orchidectomized males given a single dose of DHT on D1 show decreases in BAOT volume and neuron number with respect to control males.[101] Thus, it seems that E2 (but not DHT) plays an important role in the masculinization of this structure. This conclusion is supported by a study[114] showing that BAOT cells are targets for E2. Aromatization activity has also been reported in the amygdaloid complex probably including the BAOT.[115]

Long-term postnatal administration of DHT (D6 to D20) to male pups produces a decrease in BAOT volume to such an extent that sex differences are no longer seen.[113] When this experiment was carried out, females were also treated long term (D6 to D20) with cyproterone acetate (CA). CA caused an increment in BAOT volume in such a manner that no sex differences were seen between the CA females and the vehicle-treated males.[113] The effect(s) of DHT on male BAOT volume was interpreted taking into account the inhibitory effects of the hormone on secretion of the LH and the subsequent decrease in circulating levels of T.[113] However, CA treatment induced a masculinization of the female BAOT volume and suggests that it might be possible that the presence of DHT in this period (D6 to D20) is responsible for feminization of the female rat.[113]

4. Sexual Dimorphism in the Bed Nucleus of the Stria Terminalis

The bed nucleus of the stria terminalis (BST) is a complex olfactory structure (Figure 1) composed of four main divisions that can be distinguished along an anterior-posterior gradient.[116-119] They are the

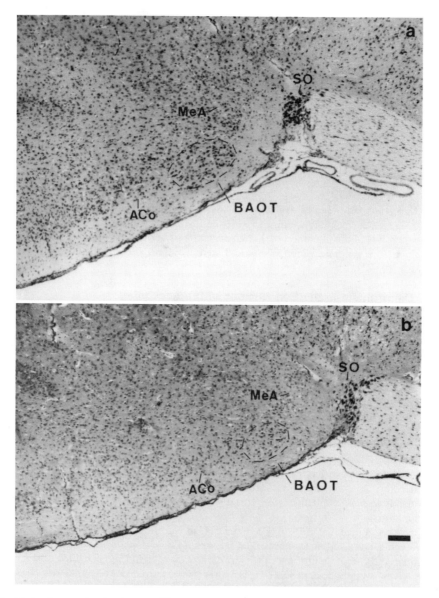

Figure 5 Photomicrographs showing sex differences in the bed nucleus of the accessory olfactory tract (BAOT). (A) Male. (B) Female. ACo: anterocortical amygdaloid area; MeA: medial amygdaloid nucleus, anterior part; and SO: supraoptic nucleus. Bar: 100 µm. (Courtesy of Dr. P. Collado.)

medial (BSTM), lateral (BSTL), ventral (BSTV), and intermediate (BSTI) divisions. The BSTM is a component of the VNS and the BSTL belongs to the main olfactory system (MOS).[18]

The BST does not present sexual dimorphism in its overall volume,[119] however, sex differences can be observed in the volume of some divisions (Table 4). The volume of the medial posterior subdivision (BSTMP) is larger in male compared to female rats.[119] Analogous sex differences have also been found in the guinea pig[120] and in human beings.[121] Sex differences in the rat BSTMP are due to the existence of sexual dimorphism in the "encapsulated" region of this subdivision. These differences are caused by the action of gonadal hormones shortly after birth since male orchidectomy and female androgenization (both on D1) invert sex differences (Table 4).[119]

Female rats present greater measures for volume in the medial anterior region than males (Table 4).[119] Male gonadectomy on D1 increases the BSTMA volume to a size similar to that of control females but androgenization (D1) has no effect on females (Table 4).[119]

Table 3 Sexual Dimorphism in the Bed Nucleus of the Accessory Olfactory Tract and Effects of Early Postnatal Male Orchidectomy and Females Androgenization on Its Development

	Males		Females	
	Control	Orchidectomized	Control	Androgenized
Volume (mm³)	0.030 ± 0.001	0.027 ± 0.0016[bb]	0.022 ± 0.001[aaaa]	0.030 ± 0.0017[bbbb]
Number of neurons	1690 ± 158.7	872 ± 44.50	1033 ± 104.8[aa]	1571 ± 158.4[bb]
Number of glial cells	473 ± 66.01	685 ± 101.3[bbb]	389 ± 32.4	420 ± 46.08
Neuron/glia ratio	3.81 ± 0.35	1.40 ± 0.18[aaaaa]	2.65 ± 0.16[a]	3.91 ± 0.48

Note: Data show means ± S.E.M. Student's *t*-test for *post hoc* comparisons.

With respect to the control males: [a]p <0.02; [aa]p <0.005; [aaa]p <0.002; [aaaa]p <0.0005; [aaaaa]p <0.00025.

With respect to the control females: [b]p <0.05; [bb]p <0.02; [bbb]p <0.009; [bbbb]p <0.0005.

From Segovia, S. and Guillamón, A., *Brain Res. Rev.,* 18, 51, 1993. With permission.

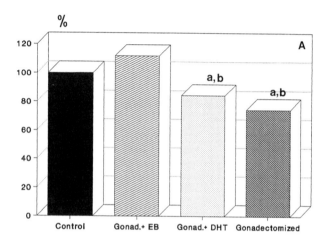

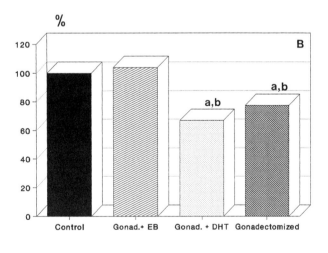

Figure 6 Hormonal manipulations in males on the day of birth (D1) and their differentiating effects on the volume (A) and number of neurons (B) of the bed nucleus of the accessory olfactory tract. Control: vehicle (oil) treated males. Gonad. + EB: D1 gonadectomized males injected with a single dose of estradiol benzoate; Gonad. + DHT: D1 gonadectomized males injected with a single dose of dihydrotestosterone; and Gonadectomized: D1 gonadectomized males. (A) p <0.01 with respect to control males. (B) p <0.01 with respect to gonadectomized and EB-treated males. Control = 100%. (Modified from Collado, P. et al., *Dev. Brain Res.,* 75, 285, 1993. With permission.)

Sex differences can also be found in the neuronal population of the BSTMP, BSTMA, and the anterior region of the lateral division of the BST (BSTLA; Table 4).[122,123] Males have more neurons than females in the "encapsulated" region of the BSTMP.[122,123] Gonadectomy on D1 decreases the number of neurons in males so that they no longer differ from control females, while female androgenization (D1) increases the number of neurons to levels similar to control males (Table 4).[122,123]

Table 4 Sexual Dimorphism in the Bed Nucleus of the Stria Terminalis and Effects of Early Postnatal Male Orchidectomy and Female Androgenization on Its Development

	Males		Females	
	Control	Orchidectomized	Control	Androgenized
BST total volume (mm³)	1.078 ± 0.069	1.148 ± 0.072	1.145 ± 0.072	1.266 ± 0.047[bbb]
BSTMP volume (mm³)	0.389 ± 0.019	0.303 ± 0.004[bbb]	0.327 ± 0.020[a]	0.441 ± 0.031
BSTMA volume (mm³)	0.264 ± 0.017	0.399 ± 0.021[bbbb]	0.362 ± 0.034[aa]	0.393 ± 0.020
Number of BSTMP neurons	39769 ± 2037.4	33662 ± 670.4[bb]	30117 ± 1218.4[aaa]	40548 ± 3443.5[bb]
Number of BSTLA neurons	3828 ± 252.5	4776 ± 515.2[b]	5434 ± 95.7[aaaa]	4577 ± 210.2[bbbbbb]

Note: Data show means ± S.E.M. Student's *t*-test for *post hoc* comparisons.

Comparisons between control groups: [a]$p <0.05$; [aa]$p <0.025$; [aaa]$p <0.0025$; [aaaa]$p <0.0005$.

Comparisons with respect to the control group of the same genetic sex: [b]$p <0.07$; [bb]$p <0.025$; [bbb]$p <0.01$; [bbbb]$p <0.0025$; [bbbbb]$p <0.001$; [bbbbbb]$p <0.0005$.

From Segovia, S. and Guillamón, A., *Brain Res. Rev.*, 18, 51, 1993. With permission.

Females have a greater number of neurons in the BSTMA and BSTLA than males (Table 4).[122,123] Orchidectomy of males on D1 increases the number of neurons in these regions to such an extent that the males do not differ from control females and androgenization (D1) of females decreases the number of neurons in the BSTLA to levels similar to those of control males. However, female androgenization on D1 has no effect on the number of neurons in the BSTMA.[122,123]

All these results indicate two main aspects in relation to the sex differences seen in the BST. First, the BST is not a homogeneous structure in terms of sex differences because two patterns can be observed. In some regions males show greater morphometric values than females (BSTMP)[119] while in others,[119,122,123] (BSTMA and BSTLA) the exact opposite is true. Second, because male gonadectomy on D1 *increases* the number of neurons in the BSTMA and the BSTLA[122,123] and female androgenization (D1) *decreases* the number of neurons in the BSTLA, it has been suggested that androgens play an "inhibitory" role in the development of sexual dimorphism.[122,123]

5. Sexual Dimorphism in the Medial Amygdala

The Me is a sexually dimorphic nucleus[100] implicated in reproductive behavior[39,54] and its neurons are targets for androgens and estrogens.[104,114] Overal Me volume is larger in male than in female rats,[100] and this can be seen beginning on D21 and is clearly evident on D30. This sex difference can be abolished by treating the female with estradiol from D1 to D30.[100] The volumes of Me neuron nuclei are larger in females than in males, orchidectomy of males abolishes this difference.[124]

The synaptic organization of the Me is also sexually dimorphic.[125-127] The total number of shaft synapses in adult male rats is significantly larger than in normal females. Administration of TP to females on D5 causes a marked increase in shaft synapses, while neonatal orchidectomy produces the opposite effect in males. Sex differences have also been reported in terms of dendritic spine synapses in the ventral molecular layer of the Me.[125-127]

6. Sexual Dimorphism in the Medial Preoptic Area

The medial preoptic area (MPA) is fully implicated in many aspects of reproductive physiology and behavior.[46,49,128,129] The neurons of this region have androgen and estrogen receptors[104] and concentrated gonadal steroids.[130,131] The MPA receives vomeronasal input directly from the VNO[30] and through the stria terminalis from the Me and BST (BSTMA and BSTMP).[116]

Dörner and Staudt[132] first showed sex differences in the MPA when they found sexual dimorphism in the volume of the neuronal nuclei of the MPA-anterior hypothalamus (AH) continuum. Female rats show a larger nuclear cell volume than males but males orchidectomized on D1 increase their nuclear cell volume. The effects of D1 male orchidectomy can be counteracted by administration of TP on D3.

Gorski and colleagues[97,98] have reported another example of sexual dimorphism in the MPA. It refers to a small group of cells in the male MPA which shows a higher stain density than the corresponding group in female rats. This group of cells was named the SDN-POA and it shows the following differentiating parameters: (1) its volume is several times larger in male than in female rats; (2) males present larger neurons; and (3) the ratio of cells inside to outside, or neuronal density, is greater in males than in females. The SDN-POA has been observed in other species[120,133-136] including man.[137-139]

The SDN-POA has attracted a considerable amount of research[17] but before studying the effects of gonadal hormones on the development of its sex differences, we must deal with its anatomical boundaries. Since the first description of the SDN-POA, its anatomical identity has undergone several modifications. First, it was described as being medially located in the preoptic area and as having a larger volume in male than in female rats.[97,98] Later, it was said to correspond to the central part of the medial preoptic nucleus (MPNc).[140] More recently, it has been described as a heterogeneous cellular group[141,142] (also larger in males than in females) composed of cells belonging to three cytoarchitectonic divisions: the MPNc, the anteroventral MPN (MPNav), and the medial part of the MPN (MPNm) *excluding* the MPNav and the MPNc (the so-called MPNm-excl). The MPNav is present in males but virtually absent in females.[141]

The SDN-POA hormone-sensitive period seems to begin prenatally[143] on E18 and end postnatally[144] on D5. D1 male castration decreases SDN-POA volume,[97] but it is counteracted by a single injection of TP on D2,[145] administration of TP on days 2, 3, 4, or 5,[144] or gonadal graft reimplantation.[145] Although postnatal treatment with TP increases the SDN-POA volume in the female rat,[97,145] only pre- and postnatal treatment with TP is able to abolish sex differences.[146,147]

These experiments show that androgens act perinatally to masculinize the SDN-POA. However, there is substantial evidence pointing out that, in fact, SDN-POA masculinization depends on estrogens, because perinatal treatment of males with the estrogen antagonist tamoxifen decreases male SDN-POA volume to levels shown by females[148,149] and pre- and postnatal treatment of females with the synthetic estrogen diethylstilbestrol increases female volume, thereby abolishing sexual dimorphism.[148-150] Furthermore, when an aromatase inhibitor (CGS) was administered to mothers in the late gestational age and to neonates from D1 to D14 (sacrificed on D30), the CGS-treated males showed a decrease in SDN-POA area compared to that of the control females.[151] Recently, McCarthy et al.[152] infused antisense oligodeoxynucleotides to the estrogen receptor messenger RNA, a scrambled nucleotide sequence, or a vehicle in the hypothalamus of three-day-old female pups (normal and slightly androgenized). The authors found that infusion of antisense DNA to the estrogen receptor mRNA on D3 permanently modified the volumes of the SDN-POA and parastrial nucleus (see below). The SDN-POA of the androgenized females was about 35% larger than that of the normal females. It was significantly smaller in the antisense oligo-infused androgenized females than in the two control groups (scrambled and vehicle), and its size was comparable to that of normal females. All these experiments suggest that estrogen converted from androgen plays an important function in the development of sexual dimorphism of the SDN-POA, and that aromatase activity in the perinatal period is important in the process.[151]

Experiments in which female rats are postnatally treated with tamoxifen are of theoretical value in understanding female brain differentiation. Döhler and colleagues[153] have shown that newborn female rats treated with tamoxifen suffer a reduction in the SDN-POA volume of about 50% with respect to control females and that they show permanent anovulatory sterility. Thus, the development of the female morphological pattern in the SDN-POA requires some (low) exposure to estrogens.[153] Taking into account that estrogens are necessary for development of both the male and female SDN-POA, Döhler et al.[17,153,154] have developed a "quantitative" model of estrogenic induction of brain differentiation in males and females.

The human SDN-POA also appears to be sexually dimorphic since it is two times greater in volume[137,138] and has more cells in men compared to women.[137] This sexual dimorphism seems to be independent of sex steroids in the adult, because adult women suffering from the virilization syndrome do not differ from normal women.[138] Swaab and Hofman[139] have also reported that cells reach peak value around 2 to 4 years after birth. Sexual differentiation takes place after this point due to a decrease in the number of cells in women.[139] Moreover, women and men have different periods of cell loss.[137-139]

The anteroventral periventricular nucleus (AVPv), which is implicated in controlling the release of gonadotropin,[155] also presents sex differences.[142,156] Bloch and Gorski[142] have found that the volume of this nucleus is greater in female than in male rats.

The female AVPv contains significantly more tyrosine hydroxylase (TH) stained perikarya and a greater density of TH-stained fibers than that of the male rat.[157] *In situ* hybridization histochemistry has shown[158] that the female AVPv has over three times as many TH mRNA-containing cells as the comparable male structure. This sex difference appears to be dependent on perinatal levels of gonadal steroids, and treatment of newborn females with testosterone decreases this parameter. It could be suggested, as for the BSTMA[122] and BSTLA,[123] that androgens might play an "inhibitory" role in sexual differentiation of the AVPv.

The development of sex differences in the AVPv has recently been reviewed by Arai and colleagues.[159] A significant decrease in female AVPv can be achieved through several types of treatment: (1) administration of TP on E17 or E21; (2) injections of T from D1 to D7; and (3) administration of estradiol from D1 to D5. However, treatment from D1 to D5 with T combined with tamoxifen or DHT alone does not affect development of the normal female pattern. Neonatal castration does not increase the AVPv volume in the male rat either.

Sex differences in the number of AVPv neurons have been shown to be detectable as early as E21.[159] When pregnant female rats were injected with TP from E14 to E18 and the fetuses were studied at E21, the incidence of pycnotic cells in the TP females and oil-treated males significantly increases with respect to the control (oil) females. Consequently, the authors[159] suggested that testicular or exogenous androgen might stimulate cell death.

The paraestrial nucleus (PS) has been considered a part of the MPA[140] but in terms of sex differences in the overall preoptic area (POA), it has been suggested[160] more recently that the striatal preoptic area (STPOA)-PS continuum might be considered a discrete area located dorsally within the POA. The PS shows sex differentiation, because its volume is larger in female than in male rats.[160] Early postnatal (D1) treatments of orchidectomy of males and androgenization of females reverse the normal pattern of sex differences.[152,160] Because orchidectomy significantly increases the PS volume in males and D1 androgenization has exactly the opposite effect on females, it has been suggested[12,18,122,123,160] that testicular androgen secretion might be an "inhibitor." If this is true, this unexpected function of androgens (similar to that seem in BSTMA, BSTLA, and AVPv) has theoretical implications (see below) for understanding sexual differentiation of the brain.[12,18]

7. Sexual Dimorphism in Hypothalamic Nuclei that Receive Vomeronasal Input

The ventromedial hypothalamic nucleus (VMH) and ventral part of the premammillary nucleus (PMV) receive vomeronasal input[116,161-164] and the former also receives it directly from the VNO.[30] The two nuclei are implicated in reproductive physiology and behavior[49,129,165,166] and they have receptors for gonadal steroids.[104,131,167]

The volume of the VMH is considerably larger in male than in female rats.[168] Orchidectomy on D1 reduces VMH volume in males to that of the control female rats, however, the same treatment on D7 does not produce this result.[168]

The nuclei of VMH neurons are larger in size in female than in male rats and D1 castration of males significantly increases this parameter.[129,169] In addition, sex differences related to the regional distribution of E-concentrating neurons have also been found as part of the pattern of VMH synaptic connectivity. Males show larger numbers of shaft and spine synapses in the ventrolateral region of the VMH than do female rats, but no differences have been seen in the dorsomedial region.[170] Castration of males on D1 reduces the number of these synapses to a level similar to that of control females. At the same time, neonatal administration of TP to female rats increases the number of synapses observed in the same region. This group of females did not differ from the control males.[170]

Synaptogenesis in the VMH takes place up to D45. Sex differences have been reported in the synaptic organization of the neuropil of the ventrolateral subdivision of the VMH.[171] The numerical densities of spine and shaft synapses in the adult male are normally higher than in the female. Neonatal treatment of female rats with testosterone increases the density of axodendritic synapses to levels similar to males and administration of tamoxifen to newborn males significantly reduces the density of spine synapses to levels comparable to normal females.[171]

Sex differences have also been established in the PMV because the nuclei of the neurons in this nucleus are significantly larger in size in male than in female rats.[129] PMV neurons concentrate estradiol and testosterone[167] and show regional sex differences in terms of concentration of estradiol.[167] Many E-concentrating neurons are located in and around the PMV, but few labeled neurons are found in the dorsal region.[167]

The arcuate nucleus, which is involved in controlling growth hormone, LH, and prolactin secretion, receives direct input from the VNO[30] and, as García-Segura et al.[172] have shown, it is an excellent model for studying the role of gonadal hormones in synaptic plasticity. Development of sex differences in this nucleus have been described with respect to axosomatic synapses, with female rats showing greater numbers of these after D10 than males. Treatment of females with TP on D5 decreases the number to such an extent that sex differences are no longer seen.[173]

Using anterograde and retrograde tract tracing techniques Smithson et al.[35] have recently shown that in the rat there might be a direct projection from the AOB to the supraoptic nucleus (SO). The SO shows sex differences.[174-181] Madeira et al.[174] have studied the SO in rats at several ages (2, 6, 12, and 18 months) and have found an age-dependent increase in neuron size involving primarily the vasopressin neurons. These age-related changes are significantly greater in males than in females in terms of SO volume, the cross-sectional areas of neuronal somata and nuclei, and neuron volume. The same group has also reported[175] ultrastructural sex differences and a morphometric analysis has revealed the presence of more glial fibrillary acidic protein (GFPA) immunostaining in the male SO.[176]

To our knowledge, no experiments have examined the effects of gonadal steroids on the development of sex differences in the rat SO. It should also be kept in mind that no estrogen receptor mRNA-containing cells have been found in it and the androgen receptors have only been labeled weakly.[104] Recently, it has been proposed[182] that estradiol affects the oxytocinergic system via an indirect route since the oxytocin mRNA content in the SO neurons of ovariectomized female rats increases after short- (2 days) and long-term (2 months) treatment with the hormone. However, ovariectomy of prepubertal female rats induces neuronal degeneration in the SO and this effect can be counteracted by treatment with EB.[183] In other species (the guinea pig) SO oxytocin-immunoreactivity cells revealed estrogen receptor immunoreactivity, and it has been suggested that the hormone controls the oxytocin system by acting directly on the magnocellular neurons.[184]

C. THE SPINAL NUCLEUS OF THE BULBOCAVERNOSUS SYSTEM

The SNBS (Figure 2) encompasses the spinal nucleus itself (SNB) and its striated target musculature.[185] Four muscles make up this musculature: the medial and lateral bulbocavernosus (BC), the levator ani (LA), and the ischiocavernosus (IS); in rodentia these muscles are only present in males.[19] The BC and LA muscles are implicated in the penile reflex necessary for male copulatory behavior.[74]

The motoneurons of the SNB, located in the lumbar spinal cord (L5 and L6), innervate the BC and LA muscles, while a subpopulation of motoneurons of the spinal dorsolateral nucleus (DLN) sends fibers to innervate the IC.[185] Both the SNB and the DLN show sex differences.[19,20]

The SNB is diminished or absent in normal female rats and males with testicular feminization mutation (Tfm).[185] Females have only one third the number of SNB motoneurons found in males and their motoneurons are one half size larger.[186] These sex differences seem to depend on perinatal androgens.

In adult male rats SNB motoneurons concentrate T and DHT, but not E2.[185] Moreover, genetic male rats treated prenatally with flutamide and gonadectomized at birth develop a feminine number and size of SNB motoneurons and they do not have BC and LA muscles.[187] Administration of flutamide during pregnancy (from E11 to E21) also reduces the number of SNB neurons in gonadally intact male rats, but the same treatment does not affect the number of SNB neurons in the female.[188] In addition, prenatally (last week of pregnancy) or neonatally (D1 to D5) TP administration increases the number of motoneurons in the female SNB.[189]

Studies have researched the effects of reduced and aromatized metabolites of T on the development of SNB sex differences. Prenatal administration of DHT does not masculinize the female SNB, even though the perineal muscles persist.[190] Perinatal administration of E2 does not masculinize the number of neurons in the SNB either.[191,192] However, females perinatally treated with E2 and DHT develop a masculine SNB.[192]

As mentioned above, the DLN also presents sex differences because males have a greater number of motoneurons than females,[193] as is the case with the SBN. The development of this difference depends on perinatal androgens. Females treated with a single dose of TP on D2 show a significant increase in the number of DLN motoneurons with respect to control females[193] and prenatal administration of DHT (propionate) masculinizes the DLN.[190] Moreover, administration of flutamide during the last 11 days of pregnancy decreases the number of DLN neurons in the gonadally intact male, but it does not affect the female.[188]

The development of sex differences in the SNBS seems to occur in two periods. During the last week of pregnancy and shortly after birth (D1 to D5) androgenization of females masculinizes the number of SNB neurons, and a few days later (D7 to D11) it increases their size but does not affect their number.[189] Furthermore, the sex difference found in terms of the size of SNB motoneurons is controlled in adulthood by androgens. Adult male gonadectomy causes a decrease in size of the motoneurons without altering the neuron number; this effect on neuron size can be reversed through administration of TP.[186]

Androgens induce sex differences in the SNB by promoting neuronal survival or by preventing natural neuronal death in males.[194] During the prenatal period, both male and female rats show similar increases in SNB neuron number[194] and all of the later SNB neurons appear before gestational days 14 to 15.[195] This suggests that androgens are not involved in the neurogenesis of SNB neurons[194,195] and that sex differences develop shortly after birth when natural neuronal death occurs, with females showing more SNB cell death than males.[194] However, the effects of androgens preventing natural neuronal death in the SNB could be due to acting on the musculature innervated by this nucleus.[196] It is also known that androgens regulate the size of the rat BC and LA muscles locally.[197] The ciliary neurotrophic factor (CNTF) is involved in preventing SNB cell death and the calcitonin gene-related peptide (CGRP) seems to play a functional role.[198]

Sex differences have also been found in the spinal cord of humans and dogs. Specifically, males show significantly more motoneurons than females in Onuf's nucleus, the nucleus that innervates the perineal muscles involved in copulatory behavior.[199]

In rodents, the SNB receives projections from the hypothalamic PVN and lesions in the PVN impair the penile reflex.[200] Retrograde tracers injected in the lower lumbar spinal cord enable observation of labeled cells in the dorsal and lateral parvocellular subnuclei of the PVN and the lateral and dorsal hypothalamus, but not in the MPA.[200] The PVN is the main source of oxytocin[201] in the CNS and oxytocin is known to be implicated in male sexual responses.[202] Oxytocin and neurophysin inmunoreactives are present in fibers originating in the PVN, and these fibers can be found in the rostrocaudal area of the SNB.[200] At this point it is worth noting that: (1) the PVN is innervated by fibers from sexually dimorphic nuclei that receive vomeronasal input, such as the Me, VMH, BST, and the lateral and medial POAs[203,205] and (2) the PVN also receives projections from the locus coeruleus (LC)[203] for which sex differences have been found, because female rats show greater morphological parameters (volume and neuron number) than males.[206-208]

III. THE ROLE OF STEROIDS IN INDUCING TWO PATTERNS OF SEX DIFFERENCES IN THE CNS

The mechanisms by which gonadal hormones control the development of sex differences in the CNS must account for the existence of two morphological patterns; one in which males present larger volumes and numbers of neurons than females (Table 5) and the exact opposite, in which females show larger volumes and greater numbers of neurons than males (Table 6). The existence of these two forms of CNS sex differences precludes any simple explanation of hormonal mechanisms implicated in their development. However, the easiest way to approach this complex problem is to analyze each pattern separately and then determine whether both patterns share some common mechanism(s).

The mechanisms promoting greater volumes and/or neuron numbers in the male compared to the female seem initially to be androgen dependent for all the structures listed in Table 5. This dependence on androgens is evidenced by the fact that early postnatal gonadectomy to males *decreases* the number of neurons and/or volumetric measures, while early postnatal or perinatal androgenization of females *increases* these same measures. The group of structures listed in Table 5 contains three different subgroups: (1) structures in which masculinization might depend on the aromatization of testosterone to estradiol such as AOB,[102] BAOT,[101] SDN-POA,[148-152] and Me;[100] (2) nuclei in which masculinization seems to depend on androgen with no need for aromatization of T to E2, as is the case with the SNB[191,192] and DLN;[188] and (3) structures which, to our knowledge, the aromatization hypothesis has not been tested, such as the BSTMP and VMH.

The aromatization hypothesis has been widely tested in the SDN-POA, which belongs to the first group above. There is a decrease in the normal male morphological parameter when the SDN-POA is perinatally treated with: (1) an antiestrogen[148,149] (tamoxifen); (2) an intrahypothalamic infusion of antisense[152] DNA to ER mRNA; or (3) an aromatase inhibitor.[151] On the other hand, females present a masculinized SDN-POA with pre- and postnatally administered diethylstilbestrol.[147,150] Thus, there is evidence that masculinization of the SDN-POA in the male is due to the aromatization of testoterone to estradiol. This mechanism of male masculinization could also be suggested for the AOB,[102] the BAOT,[101] and the Me.[100] A similar mechanism of masculinization could be suggested for the third group of structures (BSTMP and VMH) because these nuclei show steroid receptors[104,167] and aromatase activity,[115,209] and because neonatal orchidectomy decreases their numbers of neurons and/or volumes.

Table 5 Morphological Patterns ♂ > ♀

	Androgen experimental manipulations				Estrogen experimental manipulations		
Structure	**Neonatal male gonadectomy**	**Neonatal female androgenization**	**Neonatal female CA treatment**	**Perinatal male Fl treatment**	**Neonatal perinatal ♀Gx + EB**	**Neonatal female Tx treatment**	**Testing aromatization hypothesis**
VNO	$\downarrow$	$\uparrow$					
AOB	$\downarrow$	$\uparrow$			$\uparrow$		+
BAOT	$\downarrow$	$\uparrow$	$\uparrow$		$\uparrow$		+
BSTMP	$\downarrow$	$\uparrow$					
SDN-POA	$\downarrow$	$\uparrow$			$\uparrow$	$\downarrow$	+++
Me	$\downarrow$	$\uparrow$					
VMH	$\downarrow$	$\uparrow$					
SNB		$\uparrow$		$\downarrow$	—		—
DLN		$\uparrow$		$\downarrow$			—

Note: $\uparrow$: Morphometric increment; $\downarrow$: morphometric decrement; —: no effect; +: positive; +++: highly positive; CA: cyproterone acetate; Fl: flutamide; Tx: tamoxifen; EB: estradiol benzoate; and Gx: gondectomized. See text for references.

142

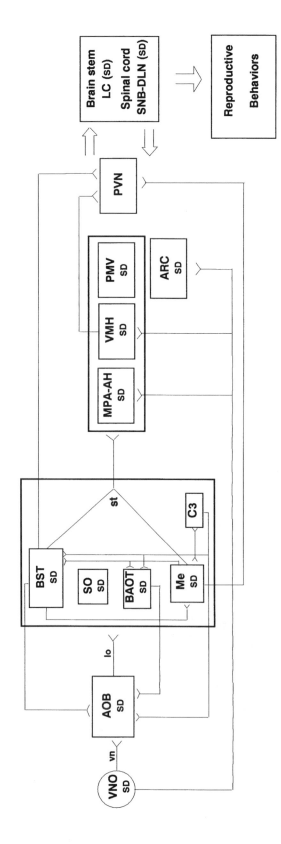

Figure 7 A morphological and functional (motivational) approach showing two sexually dimorphic networks (VNS-SNBS) that control sex differences in reproductive behaviors. LC: locus coeruleus and PVN: paraventricular nucleus of the hypothalamus. For the remaining nuclei see Figures 1 and 2.

Table 6 Morphological Patterns ♀ > ♂

Structure	Neonatal male gonadectomy	Neonatal female androgenization
BSTMA	↑	—
BSTLA	↑	↓
AVPV	—	↓
PS	↑	↓
ARC		↓
LC	—	↓

Note: ↑ Morphometric increment; ↓ morphometric decrement; — no effect. See text for references.

In the SNB and DLN nuclei, from the second group above, male masculinization seems to depend only on androgens, because prenatal administration of flutamide,[188] prenatal administration of flutamide followed by neonatal gonadectomy,[187] or prenatal stress[210] decreases the number of neurons in the SNB. Conversely, prenatal or neonatal administration of TP increases the number of neurons in the female SNB.[189] However, perinatal[191,192] EB, or prenatal[190] DHT treatments are ineffective in masculinizing the female SNB. In relation to the DLN, prenatal administration of flutamide is known to decrease the number of neurons in the male DLN[188] while an increase is seen in the androgenized female.[193]

Several points should be stressed with respect to the SNBS as a whole. First, the critical period of action of gonadal steroids occurs perinatally. Second, estradiol has not been found to affect the development of sex differences in the system. Third, it has been demonstrated that sex differences are established by postnatally differential neuronal death.[194] Androgens seem to prevent natural neuronal death acting not only on the SNB neurons but also on the target musculature of this nucleus.[196,197]

Do gonadal steroids play any role in the normal feminization of the female in the pattern in which males present greater morphological measures than females? This question is of important theoretical value, because the brain is thought to be inherently female unless male differentiation is superimposed by androgens or estrogens.[17] Döhler et al.[148] have shown that postnatal treatment of the female with tamoxifen decreases the SDN-POA volume with respect to control females, and McCarthy et al.[152] have shown that normal females treated by infusing antisense DNA to ER mRNA in the hypothalamus also present a significant reduction in SDN-POA volume compared to control groups. Thus, the normal development of the female SDN-POA involves differentiation that is regulated by exposure to estrogen acting at the estrogen receptor.[211] This picture might be more complex because BAOT volume increases in females treated with cyproteron acetate from D6 to D20,[113] suggesting that the presence of DHT for this period is responsible for the feminization in the female rat.[113] It is also known that the prenatal administration of flutamide causes female offspring to show higher levels of sexual receptivity (than control females) in adulthood after ovariectomy and administration of threshold doses of EB and progesterone.[212] All these studies could indicate that androgen and estrogen might play an active role in the differentiation of the feminine phenotype in the morphological pattern in which males present greater morphological measures than females.

The second pattern of CNS sex differences, in which females present greater morphological measures (volume and/or neuron number) than males (Table 6), has been discovered more recently and the mechanisms involved must still be demonstrated. However, it will not be possible to understand sex differences in the CNS without describing these mechanisms, because apparently they pose some constraints to generalizing the mechanisms already accepted for explaining the opposite pattern (see above and Table 5).

How can it be explained that female neonatal androgenization produces a *decrease* in volume and/or number of neurons in some structures (i.e., BSTLA, AVPv, PS, and LC) while *increasing* these essential parameters in others (see Table 5)? Furthermore, how can we explain the fact that male neonatal gonadectomy is able to *increase* the neuron number of some structures (Table 6) while the same treatment produces the exact opposite effect in others (Table 5)? Interestingly, the same treatment (presence or absence of the same hormones) produces different effects depending on two variables: genetic sex and CNS nucleus.

From the available data it is possible to put forth some hypotheses and ideas. First, it has been suggested that the greater measures presented by females with respect to males could be explained by an ovary-dependent mechanism(s).[12,18,119,123,160,206,207] Thus, estrogen (except in the SNBS) might promote greater neuron numbers in both patterns of sex differences. Second, the lesser measures of males in relation to females might be the consequence of an "inhibitory" effect[12,18,123,160] of androgens because neonatal gonadectomy of the male *increases* volume and/or neuron number in BSTMA,[119,122] BSTLA,[123] and PS[160] and pre- and/or postnatal androgenization of the female *decreases* the volume and/or number of neurons in BSTLA,[123] AVPv,[159] PS,[160] LC,[206] and DLN.[190]

After the work of Phoenix et al.,[8] the nongenomic effects of neuroactive steroids were associated to their "activational" effects that mediate reproductive functions in adulthood. However, there is evidence indicating that neurotransmitters might actually be involved in the development of CNS sex differences, because perinatal administration of drugs that affect adrenergic, serotoninergic, cholinergic, dopaminergic (see Döhler[17]), and GABAergic[108,213,214] systems are able to alter the development of sex differences in the brain and reproductive behavior. At the same time it is well known that the neuronal membrane is a target for steroids that induce changes in ion channels.[106,107,215] The existence of neurosteroids synthesized in brain tissue[216,217] and the observation that perinatal administration of agonist and antagonist substances to neurotransmitter receptors can modify development of sex differences in the brain have led us to suggest[18] that nongenomic effects of steroids are also involved in the development of CNS sex differences. If that is true, the synapse might be the final target for both the genomic and nongenomic actions of steroids which "organize" the masculine and feminine phenotypes.

IV. CONCLUSIONS

For many years research on sex differences in the CNS and behavior has been guided by the "organizational-activational" paradigm and has mainly focused on the morphological pattern in which males present greater measures than females.[218-220] The pattern has acted as a mechanism to confirm the paradigm. However, weak aspects have emerged in dealing with new data produced during the last decade.[11,12,18,221,222]

The main constraints of the paradigm are

1. The existence of two opposite morphological patterns of sex differences in the CNS (males greater than females and females greater than males) in contrast with the single pattern (males greater than females) considered in the paradigm.
2. Literature (see previous sections) that indicates the female phenotype is achieved through the active role of gonadal steroids, contrary to the idea that only the male is actively differentiated by gonadal steroids.
3. In the original paradigm steroid membrane action was activational and was reserved to explain the role of steroids seen in the adult; however, as has been shown membrane processes might participate in promoting sex differences as well (see previous sections).
4. Literature[221,223] indicating the existence of brain structural changes induced by genomic actions in adulthood.
5. The existence of morphological and functional sex differences arising in the absence of gonadal steroids.[224]
6. Experiments that show the same gonadal steroid acting in opposite ways depending on the genetic sex and CNS nucleus (see Tables 5 and 6).
7. The existence of sex differences caused by a possible "domino effect" in CNS nuclei that do not concentrate gonadal steroids.[105,222]
8. Experiments indicating sex differences in the brain controlling sexually dimorphic reproductive functions affect complex neural networks[12,18] (see Figures 1 and 7).

These considerations, and surely more to come, will drive future research on how sex differences in the brain are built.

ACKNOWLEDGMENTS

The work of the authors is supported by DYGICYT grants PB93-0291-C03-01 (A.G.) and PM91-0207 (S.S.). We are grateful to Drs. A. Del Abril, M. R. De Blas, and P. Collado for their comments on the first draft of this chapter. Dr. F. Claro, Ms. C. Garcia-Malo de Molina, Ms. Mary Litzler, and Ms. R. Sanchez helped us with the editorial aspects of the work.

REFERENCES

1. Blackwelder, R. E. and Shepherd, B. A., *The Diversity of Animal Reproduction*, CRC Press, Boca Raton, FL, 1981.
2. Crews, D., The organizational concept and vertebrates without sex chromosomes, *Brain. Behav. Evol.*, 42, 202, 1993.
3. Gubbay, J., Collignon, J., Capel, B., Economov, A., Münsterberg, A., Vivian, N., Goodfellow, P., and Lovell-Badge, R., A gene mapping to the sex-determining region of the Y chromosome is a member of a novel family of embryonically expressed genes, *Nature*, 346, 245, 1990.
4. Sinclair, A. H., Berta, P., Palmer, M. S., Hawkins, J. R., Griffith, B. L., Smith, M. J., Foster, J. W., Frischauf, A. M., Lovell-Badge, R., and Goodfellow, P., A gene from the human sex determining region encodes a protein with homology to a conserved DNA-binding motif, *Nature*, 346, 240, 1990.
5. Jost, A., Problems in fetal endocrinology: the gonadal and hypophyseal hormones, *Recent Prog. Horm. Res.*, 8, 379, 1953.
6. Jost, A., A new look at the mechanisms controlling sexual differentiation in mammals, *Johns Hopkins Med. J.*, 130, 38, 1972.
7. George, F. W. and Wilson, J. D., Sex determination and differentiation, in *The Physiology of Reproduction*, Knobil, E. and Neill, J. D., Eds., Raven Press, New York, 1988, Vol. 1, 3.
8. Phoenix, C. H., Goy, R. W., Gerall, A. A., and Young, W. C., Organizing action of prenatally administered testosterone propionate on the tissues mediating mating behavior in the female guinea pig, *Endocrinology*, 65, 369, 1959.
9. Baum, M. J., Frank Beach's research on the sexual differentiation of behavior and his struggle with the "organizational" hypothesis, *Neurosci. Biobehav. Rev.*, 14, 201, 1990.
10. Baum, M. J., Differentiation of coital behavior in mammals: a comparative analysis, *Neurosci. Biobehav. Rev.*, 3, 265, 1979.
11. Arnold, A. P. and Bredlove, S. M., Organizational and activational effects of sex steroids on brain and behavior: a reanalysis, *Horm. Behav.*, 19, 469, 1985.
12. Guillamón, A. and Segovia, S., Sexual dimorphism in the accessory olfactory system, in *The Development of Sex Differences and Similarities in Behavior*, Haug, M., Whalen, R. E., Aron, C., and Olsen, K. L., Eds., Kluwer Academic Publishing, Dordrecht, 1993, 363.
13. Konishi, M., Birdsong: from behavior to neuron, *Annu. Rev. Neurosci.*, 8, 125, 1985.
14. Nottebohm, F., Nottebohm, M. E., and Crane, L., Developmental and seasonal changes in canary song and their relation to changes in the anatomy of song-control nuclei, *Behav. Neural Biol.*, 46, 445, 1986.
15. Kelley, D. B., Sexually dimorphic behaviors, *Annu. Rev. Neurosci.*, 11, 225, 1988.
16. De Vries, G. J., Sex differences in neurotransmitters systems, *J. Neuroendocrinology*, 2, 1, 1990.
17. Döhler, K. D., The pre- and postnatal influence of hormones and neurotransmitters on sexual differentiation of the mammalian hypothalamus, *Int. Rev. Cytol.*, 131, 1, 1991.
18. Segovia, S. and Guillamón, A., Sexual dimorphism in the vomeronasal pathway and sex differences in reproductive behaviors, *Brain Res. Rev.*, 18, 51, 1993.
19. Breedlove, S. M., Steroid influences on the development and function of a neuromuscular system, *Prog. Brain Res.*, 61, 147, 1984.
20. Tobet, F. A. and Fox, T. O., Sex differences in neuronal morphology influenced hormonally throughout life, in *Handbook of Behavioral Neurobiology: Sexual Differentiation*, Vol. 11, Gerall. A. A., Moltz, H., and Ward, I. L., Eds., Plenum Press, New York, 1992, 41.
21. Raisman, G., An experimental study of the projection of the amygdala to the accessory olfactory bulb and its relationship to the concept of a dual olfactory system, *Exp. Brain Res.*, 14, 395, 1972.
22. Winans, S. S. and Scalia, F., Amygdaloid nucleus: new afferent input from the vomeronasal organ, *Science*, 170, 330, 1970.
23. Scalia, F. and Winans, S. S., The differential projections of the olfactory bulb and accessory olfactory bulb in mammals, *J. Comp. Neurol.*, 161, 31, 1975.
24. Cowan, W. M., Raisman, G., and Powell, T. P., The connections of the amygdala, *J. Neurol. Neurosurg. Psychiatry*, 28, 137, 1965.
25. Heimer, L., The olfactory connections of the diencephalon in the rat, *Brain. Behav. Evol.*, 6, 484, 1972.
26. Powell, T. P., Cowan, W. M., and Raisman, G., The central olfactory connections, *J. Anat.*, 99, 791, 1965.
27. Scott, J. W. and Leonard, C. H., The olfactory connections of the lateral hypothalamus in the rat, mouse, and hamster, *J. Comp. Neurol.*, 141, 331, 1971.
28. Barber, P. C. and Raisman, G., An autoradiographic investigation of the projection of the vomeronasal organ to the accessory olfactory bulb in the mouse, *Brain Res.*, 81, 21, 1974.
29. Barber, P. C. and Raisman, G., Replacement of receptor neurons after section of the vomeronasal nerves in the adult mouse, *Brain Res.*, 147, 297, 1978.
30. Larriva-Sahd, J., Rondán, A., Orozco-Estevez, H., and Sánchez-Robles, M. R., Evidence of a direct projection of the vomeronasal organ to the medial preoptic nucleus and hypothalamus, *Neurosci. Lett.*, 163, 45, 1993.
31. Broadwell, R. D., Olfactory relationships of the telencephalon and diencephalon in the rabbit. I. An autoradiographic study of the efferent connections of the main and accessory olfactory bulbs, *J. Comp. Neurol.*, 163, 329, 1975.

32. Halpern, M., The organization and function of the vomeronasal system, *Annu. Rev. Neurosci.*, 10, 325, 1987.
33. Wysocki, C. J., Neurobehavioral evidence for the involvement of the vomeronasal system in mammalian reproduction, *Neurosci. Biobehav. Rev.*, 3, 301, 1979.
34. Wysocki, C. J. and Meredith, M., The vomeronasal system, in *Neurobiology of Taste and Smell*, Finger, T. E. and Silver, W. L., Eds., John Wiley, New York, 1987, 125.
35. Smithson, K. G., Weiss, M. L., and Hatton, G. I., Supraoptic nucleus afferents from the accessory olfactory bulb: evidence from anterograde and retrograde tract tracing in the rat, *Brain Res. Bull.*, 29, 209, 1992.
36. Commins, D. and Yahr, D., Lesions of the sexually dimorphic area disrupt mating and marking in male gerbils, *Brain Res. Bull.*, 13, 185, 1984.
37. Emery, D. E. and Sachs, B. D., Copulatory behavior in male rats with lesions in the bed nucleus of the stria terminalis, *Physiol. Behav.*, 17, 803, 1976.
38. Kelche, C. and Aron, C., Olfactory cues and accessory olfactory bulb lesion: effects on sexual behavior in the cyclic female rat, *Physiol. Behav.*, 33, 45, 1984.
39. Lehman, M. N., Winans, S. S., and Powers, J. B., Medial nucleus of the amygdala mediates chemosensory control of male hamster sexual behavior, *Science*, 210, 557, 1980.
40. Mackay-Sim, A. and Rose, J. D., Removal of the vomeronasal organ impairs lordosis in female hamsters: effect is reversed by luteinising hormone-releasing hormone, *Neuroendocrinology*, 42, 489, 1986.
41. Claro, F., Segovia, S., Guillamón, A., and Del Abril, A., Lesions in the medial posterior region of the BST impair sexual behavior in sexually experienced and inexperienced male rats, *Brain Res. Bull.*, 36, 1, 1995.
42. Malsbury, C. W., Kow, L. M., and Pfaff, D. W., Effects of medial hypothalamic lesions on the lordosis response and other behaviors in female golden hamsters, *Physiol. Behav.*, 19, 223, 1977.
43. Meredith, M., Vomeronasal organ removal before sexual experience impairs male hamster mating behavior, *Physiol. Behav.*, 36, 737, 1986.
44. Meredith, M., Marques, D. M., O'Connell, R. J., and Stern, F. L., Vomeronasal pump: significance for male hamster sexual behavior, *Science*, 207, 1224, 1980.
45. O'Connell, R. J. and Meredith, M., Effects of volatile and non-volatile chemical signals on male sex behavior mediated by the main and accessory systems, *Behav. Neurosci.*, 98, 1984, 1083.
46. Pfaff, D. W. and Schwartz-Giblin, S., Cellular mechanisms of female reproductive behaviors, in *The Physiology of Reproduction*, Knobil, E. and Neill, J. D., Eds., Raven Press, New York, 1988, Vol. 2, 1487.
47. Powers, B. and Valenstein, E. S., Sexual receptivity: facilitation by medial preoptic lesions in female rats, *Science*, 175, 1003, 1972.
48. Powers, J. B., Fields, R. B., and Winans, S. S., Olfactory and vomeronasal system participation in male hamster attraction to female vaginal secretions, *Physiol. Behav.*, 22, 77, 1979.
49. Sachs, B. D. and Meisel, R. L., The physiology of male sexual behavior, in *The Physiology of Reproduction*, Knobil, E. and Neill, J. D., Eds., Raven Press, New York, 1988, Vol. 2, 1393.
50. Saito, T. R. and Moltz, H., Copulatory behavior of sexually naive and sexually experienced male rats following removal of the vomeronasal organ, *Physiol. Behav.*, 37, 507, 1986.
51. Saito, T. R. and Moltz, H., Sexual behavior in the female rat following removal of the vomeronasal organ, *Physiol. Behav.*, 38, 81, 1986.
52. Winans, S. S. and Powers, J. B., Olfactory and vomeronasal deafferentiation of male hamsters: histological and behavioral analyses, *Brain Res.*, 126, 325, 1977.
53. Fleming, A., Vaccarino, F., Tambosso, L., and Chee, P. H., Vomeronasal and olfactory system modulation of maternal behavior in rat, *Science*, 203, 372, 1979.
54. Fleming, A., Vaccarino, F., and Luebke, C., Amygdaloid inhibition of maternal behavior in the nulliparous female rat, *Physiol. Behav.*, 25, 731, 1980.
55. Marques, D. M., Roles of the main olfactory and vomeronasal systems in the responses of the female hamster to young, *Behav. Neural Biol.*, 26, 311, 1979.
56. Menella, J. A. and Moltz, H., Infanticide in the male rat: the role of the vomeronasal organ, *Physiol. Behav.*, 42, 303, 1988.
57. Numan, M., Maternal behavior, in *The Physiology of Reproduction*, Knobil, E. and Neill, J. D. Eds., Raven Press, New York, 1988, Vol. 2, 1569.
58. Saito, T. R., Induction of maternal behavior in sexually unexperienced male rats following removal of the vomeronasal organ, *Jpn. J. Vet. Sci.*, 48, 1029, 1986.
59. Saito, T. R., Kamata, K., Nakamura, M., and Inaba, M., Maternal behavior in virgin female rats following removal of the vomeronasal organ, *Zool. Sci.*, 5, 1141, 1988.
60. Del Cerro, M. C. R., Izquierdo, M. A. P., Collado, P., Segovia, S., and Guillamón, A., Bilateral lesions of the bed nucleus of the accessory olfactory tract facilitate maternal behavior in virgin female rats, *Physiol. Behav.*, 50, 67, 1991.
61. Izquierdo, M. A. P., Collado, P., Segovia, S., Guillamón, A., and Del Cerro, M. C. R., Maternal behavior induced in male rats by bilateral lesions of the bed nucleus of the accessory olfactory tract, *Physiol. Behav.*, 52, 707, 1992.
62. Lomas, D. E. and Keverne, E. B., Role of the vomeronasal organ and prolactin in acceleration of puberty in female mice, *J. Reprod. Fertil.*, 66, 101, 1982.

63. Sánchez-Criado, J. E., Involvement of the vomeronasal system in the reproductive physiology of the rat, in *Olfaction and Endocrine Regulation*, Breihpohl, W., Ed., IRL Press, London, 1982, 209.

64. Vandenbergh, J. G., Pheromones and mammalian reproduction, in *The Physiology of Reproduction*, Knobil, E. and Neill, J. D., Eds., Raven Press, New York, 1988, Vol. 2, 1679.

65. Johns, M. A., Feder, H. M., Komisaruk, B. R., and Mayer, A.D., Urine-induced reflex ovulation in anovulatory rats may be a vomeronasal effect, *Nature*, 272, 446, 1978.

66. Mora, O. A., Sánchez-Criado, J. E., and Guisado, S., Role of the vomeronasal organ on the estral cycle reduction by pheromones in the rat, *Rev. Esp. Fisiol.*, 41, 305, 1985.

67. Reynolds, J. and Keverne, E. B., The accessory olfactory system and its role in the pheromonally mediated supression of oestrus in grouped mice, *J. Reprod. Fertil.*, 57, 31, 1979.

68. Sánchez-Criado, J. E., Blockade of the pheromonal effects in rat by central deafferentiation of the accessory olfactory system, *Rev. Esp. Fisiol.*, 35, 137, 1979.

69. Bellringer, J. F., Pratt, H. P. M., and Keverne, E. B., Involvement of the vomeronasal organ and prolactin in pheromonal induction of delayed implantation in mice, *J. Reprod. Fertil.*, 59, 223, 1980.

70. Keverne, E. B., Pheromonal influences on the endocrine regulation of reproduction, *TINS*, 66, 381, 1983.

71. Keverne, E. B. and De la Riva, C., Pheromones in mice: reciprocal interactions between the nose and brain, *Nature*, 296, 148, 1982.

72. Rajendren, G. and Dominic, C. J., Effect of bilateral transection of the lateral olfactory tract on the male-induced implantation failure (the Bruce effect) in mice, *Physiol. Behav.*, 36, 587, 1986.

73. Sachs, B. D., Role of rat's striated penile reflexes, copulation, and the induction of pregnancy, *J. Reprod. Fertil.*, 66, 433, 1982.

74. Hart, B. L. and Melese-D'Hospital, P. Y., Penile mechanisms and the role of the striated penile muscles in penile reflex, *Physiol. Behav.*, 31, 807, 1983.

75. Stensaas, L. J., Lavker, R. M., Monti-Bloch, L., Grosser, B. I., and Berliner, D. L., Ultrastructure of the human vomeronasal organ, *J. Steroid. Biochem. Mol. Biol.*, 39, 553, 1991.

76. Segovia, S. and Guillamón, A., Effects of sex steroids on the development of the vomeronasal organ in the rat, *Dev. Brain Res.*, 5, 209, 1982.

77. Segovia, S., Paniagua, R., Nistal, M., and Guillamón, A., Effects of postpuberal gonadectomy on the neurosensorial epithelium of the vomeronasal organ in the rat, *Dev. Brain Res.*, 14, 289, 1984.

78. Imamura, K., Mori, K., Fujita, S. C., and Obata, K., Immunochemical identification of subgroups of vomeronasal nerve fibers and their segregated terminations in the accessory olfactory bulb, *Brain Res.*, 328, 362, 1985.

79. Mori, K., Imamura, K., Fujita, S. C., and Obata, K., Projections of two subclasses of vomeronasal nerve fibers to the accessory olfactory bulb in the rabbit, *Neuroscience*, 20, 259, 1987.

80. Schaeffer, C., Roos, J., and Aron, C., Accessory olfactory bulb lesions and lordosis behavior in the male rat feminized with ovarian hormones, *Horm. Behav.*, 20, 118, 1986.

81. Brennan, P. A., Hancock, D., and Keverne, E. B., The expression of the immediate-early genes c-fos, egr-1 and c-jun in the accessory olfactory bulb during the formation of an olfactory memory in mice, *Neuroscience*, 49, 277, 1992.

82. Kaba, H., Roser, A., and Keverne, E. B., Neural basis of olfactory memory in the context of pregnancy block, *Neuroscience*, 32, 657, 1989.

83. Brennan, P. A. and Keverne, E. B., Impairment of olfactory memory by local infusions of non-selective excitatory amino acid receptor antagonists into the accessory olfactory bulb, *Neuroscience*, 33, 463, 1989.

84. Beltramino, C. and Taleisnick, S., Effects of electrochemical stimulation in the olfactory bulb on the release of gonadotropin hormone in rat, *Neuroendocrinology*, 28, 320, 1979.

85. Caminero, A. A., Segovia, S., and Guillamón, A., Sexual dimorphism in accessory olfactory bulb mitral cells: a quantitative Golgi study, *Neuroscience*, 45, 663, 1991.

86. Roos, J., Roos, M., Schaeffer, C., and Aron, C., Sexual differences in the development of accessory olfactory bulbs in the rat, *J. Comp. Neurol.*, 270, 121, 1988.

87. Segovia, S., Orensanz, L. M., Valencia, A., and Guillamón, A., Effects of sex steroids on the development of the accessory olfactory bulb in the rat: a volumetric study, *Dev. Brain Res.*, 16, 312, 1984.

88. Segovia, S., Valencia, A., Calés, J. M., and Guillamón, A., Effects of sex steroids on the development of two granule cells subpopulations in the accessory olfactory bulb, *Dev. Brain Res.*, 30, 283, 1986.

89. Valencia, A., Segovia, S., and Guillamón, A., Effects of sex steroids on the development of the accessory olfactory bulb mitral cells, *Dev. Brain Res.*, 24, 287, 1986.

90. Bayer, S. A., ^{3}H-Thymidine-radiographic studies of neurogenesis in the rat olfactory bulb, *Exp. Brain Res.*, 50, 329, 1983.

91. Struble, R. G. and Walters, C. P., Light microscopic differentiation of two populations of rat olfactory bulb granule cells, *Brain Res.*, 236, 237, 1982.

92. Roselli-Austin, L., Hamilton, K. H., and Williams, J., Early postnatal development of the rat accessory olfactory bulb, *Dev. Brain Res.*, 36, 304, 1987.

93. Roos, J., Roos, M., Schaeffer, C., and Aron, C., Prepubescent hormonal control of the development of accessory olfactory bulbs in the male rat, *Dev. Brain Res.*, 47, 309, 1989.

94. Naftolin, F., Ryan, K. J., Davies, I. J., Reddy, V. V., Flores, F., Petro, Z., Khun, M., White, R. J., Takaoka, Y., and Wolin, L., The formation of estrogens by central neuroendocrine tissue, *Recent Prog. Horm. Res.*, 31, 295, 1985.

95. Naftolin, F., Ryan, K. J., and Petro, Z., Aromatization of androstenedione by the diencephalon, *J. Clin. Endocrinol. Metab.*, 33, 368, 1971.

96. Martini, L., The 5-alpha-reduction of testosterone in the neuroendocrine structures. Biochemical and physiological implications, *Endocrinol. Rev.*, 3, 1, 1982.

97. Gorski, R. A., Gordon, J. H., Shryne, J. E., and Southam, A. M., Evidence for a morphological sex difference within the medial preoptic area of the rat brain, *Brain Res.*, 148, 333, 1978.

98. Gorski, R. A., Harlan, R. E., Jacobson. C. D., Shryne, J. E., and Southam, A. M., Evidence for the existence of a sexually dimorphic nucleus in the preoptic area of the rat, *J. Comp. Neurol.*, 193, 529, 1980.

99. Gorski, R. A., Critical role for the medial preoptic area in the sexual differentiation of the brain, *Prog. Brain Res.*, 61, 129, 1984.

100. Mizukami, S., Nishizuka, M., and Arai, Y., Sexual difference in nuclear volume and its ontogeny in the rat amygdala, *Exp. Neurol.*, 79, 569, 1983.

101. Collado, P., Valencia, A., Del Abril, A., Rodriguez-Zafra, M., Pérez-Laso, C., Segovia, S., and Guillamón, A., Effects of estradiol on the development of sexual dimorphism in the bed nucleus of the accessory olfactory tract in the rat, *Dev. Brain Res.*, 75, 285, 1993.

102. Pérez-Laso, C., Segovia, S., Collado, P., Rodriguez-Zafra, M., Del Abril, A., and Guillamón, A., Effects of estradiol on the development of sex differences in the accessory olfactory bulb in the rat, submitted.

103. Valencia, A., Collado, P., Calés, J. M., Segovia, S., Pérez-Laso, C., Rodriguez-Zafra, M., and Guillamón, A., Postnatal administration of dihydrotestosterone to male rat abolishes sexual dimorphism in the accessory olfactory bulb: a volumetric study, *Dev. Brain Res.*, 68, 132, 1992.

104. Simerly, R. B., Chang, C., Muramatsu, M., and Swanson, L. W., Distribution of androgen and estrogen receptor mRNA-containing cells in the rat brain: an *in situ* hybridization study, *J. Comp. Neurol.*, 294, 76, 1990.

105. Beyer, C. and Feder, H. H., Sex steroids and afferent input: their roles in brain sexual differentiation, *Annu. Rev. Physiol.*, 49, 439, 1987.

106. McEwen, B. S., Non-genomic and genomic effects of steroids on neural activity, *TIPS*, 12, 141, 1991.

107. Schumacher, M., Rapid membrane effects of steroid hormones: an emerging concept in neuroendocrinology, *TINS*, 13, 359, 1990.

108. Segovia, S., Pérez-Laso, C., Rodriguez-Zafra, M., Calés, J. M., Del Abril, A., De Blas, M. R., Collado, P., Valencia, A., and Guillamón, A., Early postnatal diazepam exposure alters sex differences in the rat brain, *Brain Res. Bull.*, 26, 899, 1991.

109. Pérez-Laso, C., Valencia, A., Rodriguez-Zafra, M., Calés, J. M., Guillamón, A., and Segovia, S., Perinatal administration of diazepam alters sexual dimorphism in the rat olfactory bulb, *Brain Res.*, 634, 1, 1994.

110. Collado, P., Guillamón, A., Valencia, A., and Segovia, S., Sexual dimorphism in the bed nucleus of the accessory olfactory tract in the rat, *Dev. Brain Res.*, 56, 263, 1990.

111. McShane, S., Glaser, L., Greer, E. R., Houtz, J., Tong, M. F., and Diamond, M. C., Cortical asymmetry — a preliminary study: neurons-glia, female-male, *Exp. Neurol.*, 99, 353, 1988.

112. Canic, J.-A., Vaccaro, D. E., Livingston, E. M., Leeman, S. E., Ryan, K. J., and Fox, T. O., Localization of aromatase and 5-alpha- reductase to neuronal and non-neuronal cells in fetal rat hypothalamus, *Brain Res.*, 372, 277, 1986.

113. Collado, P., Segovia, S., Calés, J. M., Pérez-Laso, C., Rodriguez-Zafra, M., Guillamón, A., and Valencia, A., Female's DHT controls sex differences in the rat bed nucleus of the accessory olfactory tract, *Neurorep.*, 3, 327, 1992.

114. Stumpff, W. E. and Sar, M., The olfactory system as target organ for steroid hormones, in *Olfaction and Endocrine Regulation*, Breipohl, W., IRL Press, London, 11, 1982.

115. Roselli, C. E., Ellinwood, W. E., and Resko, J. A., Regulation of brain aromatase activity in rats, *Endocrinology*, 114, 192, 1984.

116. De Olmos, J. S., Alheid, G. F., and Beltramino, C. A., Amygdala, in *The Rat Nervous System. Forebrain and Midbrain*, Paxinos, G., Ed., Academic Press, Sydney, 223, 1985.

117. Ju, G. and Swanson, L. W., Studies on the cellular architecture of the bed nuclei of the stria terminalis in the rat. I. Cytoarchitecture, *J. Comp. Neurol.*, 280, 587, 1989.

118. Ju, G., Swanson, L. W., and Simerly, R. B., Studies on the cellular architecture of the bed nuclei of the stria terminalis in the rat. II. Chemoarchitecture, *J. Comp. Neurol.*, 280, 603, 1989.

119. Del Abril, A., Segovia, S., and Guillamón, A., The bed nucleus of the stria terminalis in the rat: regional sex differences controlled by gonadal steroids early after birth, *Dev. Brain Res.*, 32, 295, 1987.

120. Hines, M., Davis, F., Coquelin, A., Goy, R. W., and Gorski, R. A., Sexually dimorphic regions in the medial preoptic area and the bed nucleus of the stria terminalis of the guinea pig: a description and an investigation of their relationship to gonadal steroids in adulthood, *J. Neurosci.*, 5, 40, 1985.

121. Allen, L. S. and Gorski, R. A., Sex difference in the bed nucleus of the stria terminalis of the human brain, *J. Comp. Neurol.*, 302, 697, 1990.

122. Del Abril, A., Guillamón, A., and Segovia, S., El nucleo de la estria terminal de la rata: diferencias de sexo en la población neuronal de la región medial anterior, *Trab. Inst. Cajal*, 76, 230, 1987.

123. Guillamón, A., Segovia, S., and Del Abril, A., Early effects of gonadal steroids on the neuron number in the medial posterior and the lateral division of the bed nucleus of the stria terminalis in the rat, *Dev. Brain Res.*, 44, 281, 1988.
124. Staudt, J. and Dörner, G., Structural changes in the medial and central amygdala of the male rat, following neonatal castration and androgen treatment, *Endocrinology.*, 67, 296, 1976.
125. Nishizuka, M. and Arai, Y., Sexual dimorphism in synaptic organization in the amygdala its dependence on neonatal hormone environment, *Brain Res.*, 212, 31, 1981.
126. Nishizuka, M. and Arai, Y., Organizational action of estrogen on synaptic pattern in the amygdala: implications for sexual differentiation of the brain, *Brain Res.*, 213, 422, 1981.
127. Nishizuka, M. and Arai, Y., Regional difference in sexually dimorphic synaptic organization of the medial amygdala, *Exp. Brain Res.*, 49, 462, 1983.
128. Numan, M., Neural basis of maternal behavior in the rat, *Psychoneuroendocrinology*, 13, 47, 1988.
129. Dörner, G., *Hormones and Brain Differentiation*, Amsterdam, Elsevier, 1976.
130. Pfaff, D. and Keiner, M., Atlas of estradiol-concentrating cells in central nervous system of the female rat, *J. Comp. Neurol.*, 151, 121, 1973.
131. Sar, M. and Stumpf, W. E., Cellular localization of androgens in the brain and pituitary after injection of tritiated testosterone, *Experientia*, 28, 1364, 1972.
132. Dörner, G. and Staudt, J., Structural changes in the preoptic anterior hypothalamic area of the male rat, following neonatal castration and androgen substitution, *Neuroendocrinology*, 4, 136, 1968.
133. Byne, W. and Bleier, R., Medial preoptic sexual dimorphisms in the guinea pig. I. An investigation on their hormonal dependence, *J. Neurosci.*, 7, 2688, 1987.
134. Commins, D. and Yahr, P., Adult testosterone levels influence the morphology of a sexually dimorphic area in the mongolian gerbil brain, *J. Comp. Neurol.*, 244, 132, 1984.
135. Tobet, S. A., Zahniser, D. J., and Baum, M. J., Differentiation in male ferrets of a sexually dimorphic nucleus of the preoptic/anterior hypothalamic area requires prenatal estrogen, *Neuroendocrinology*, 44, 299, 1986.
136. Viglietti-Panzica, C., Panzica, G. C., Priori, M. G., Galcagni, M., Ansemitti, G. C., and Balthazart, J., A sexually dimorphic nucleus in the quail preptic area, *Neurosci. Lett.*, 64, 129, 1986.
137. Swaab, D. F. and Fliers, E., A sexually dimorphic nucleus in the human brain, *Science*, 228, 1112, 1985.
138. Hofman, M. A. and Swaab, D. F., The sexually dimorphic nucleus of the preoptic area in the human brain: a comparative morphometric study, *J. Anat.*, 164, 55, 1989.
139. Swaab, D. F. and Hofman, M. A., Sexual differentiation of the human hypothalamus: ontogeny of the sexually dimorphic nucleus of the preoptic area, *Dev. Brain Res.*, 44, 314, 1988.
140. Simerly, R. B., Swanson, L. W., and Gorski, R. A., Demonstration of a sexual dimorphism in the distribution of serotonin inmunoreactive fibers in the medial preoptic nucleus of the rat, *J. Comp. Neurol.*, 225, 151, 1984.
141. Bloch, G. J. and Gorski, R. A., Cytoarchitectonic analysis of the SDN-POA of the intact and gonadectomized rat, *J. Comp. Neurol.*, 275, 604, 1988.
142. Bloch, G. J. and Gorski, R. A., Estrogen/progesterone treatment in adulthood affects the size of several components of the medial preoptic area in the male rat, *J. Comp. Neurol.*, 275, 613, 1988.
143. Rhees, R. W., Shryne, J. E., and Gorski, R. A., Onset of the hormone-sensitive perinatal period for sexual differentiation of the sexually dimorphic nucleus of the preoptic area in female rats, *J. Neurobiol.*, 21, 781, 1990.
144. Rhees, R. W., Shryne, J. E., and Gorski, R. A., Termination of the hormone-sensitive period for differentiation of sexually dimorphic nucleus of the preoptic area in male and female rats, *Dev. Brain Res.*, 52, 17, 1990.
145. Jacobson, C. D., Csernus, V. J., Shryne, J. E., and Gorski, R. A., The influence of gonadectomy, androgen exposure or a gonadal graft in the neonatal rat on the volume of the sexually dimorphic nucleus of the preoptic area, *J. Neurosci.*, 1, 1142, 1981.
146. Döhler, K. D., Coquelin, A., Davis, F., Hines, M., Shryne, J. E., and Gorski, R. A., Differentiation of the sexually dimorphic nucleus in the preoptic area of the rat brain is determined by the perinatal hormone environment, *Neurosci. Lett.*, 33, 295, 1982.
147. Döhler. K. D., Coquelin, A., Davis, F., Hines, M., Shryne, J. E., and Gorski, R. A., Pre- and postnatal influence of testosterone propionate and diethylstilbestrol on differentiation of the sexually dimorphic nucleus of the preoptic area in male and female rats, *Brain Res.*, 302, 291, 1984.
148. Döhler, K. D., Coquelin, A., Davis, F., Hines, M., Shryne, J. E., Sickmoller, P. M., Jarzab, B., and Gorski, R. A., Pre- and postnatal influence of an estrogen antagonist and an androgen antagonist on differentiation of the sexually dimorphic nucleus of the preoptic area in male and female rats, *Neuroendocrinology*, 42, 443, 1986.
149. Döhler, K. D., Coquelin, A., Hines, M., Davis, F., Shryne, J. E., and Gorski, R. A., Hormonal influence on sexual differentiation of rat brain anatomy, in *Hormones and Behavior in Higher Vertebrates*, Balthzart, J., Pröve, E., and Gilles, R., Eds., Berlin, Springer, 1983, 194.
150. Döhler, K. D., Hines, M., Coquelin, A., Davis, F., Shryne, J. E., and Gorski, R. A., Pre- and postnatal influence of diethylstilbestrol on differentiation of sexually dimorphic nucleus in the preoptic area of the female rat brain, *Neuroendocrinol. Lett.*, 4, 361, 1982.
151. Ohe, E. Effects of aromatase inhibitor on sexual differentiation of SDN-POA in rats, *Acta Obstet. Gynaecol. JPN.*, 46, 227, 1994.

152. McCarthy, M. M., Schlenker, E. H., and Pfaff, D. W., Enduring consequences of neonatal treatment with antisense oligodeoxynucleotides to estrogen receptor messenger ribonucleic acid on sexual differentiation of rat brain, *Endocrinology*, 133, 433, 1993.

153. Döhler, K. D., Hancke, J. L., Srivastava, S. S., Hoffman, C., Shryne, J. E., and Gorski, R. A., Participation of estrogens in female sexual differentiation of the brain; neuroanatomical, neuroendocrine and behavioral evidence, *Prog. Brain Res.*, 61, 99, 1984.

154. Döhler, K. D., Ganzenmüller, C., and Veit, C., The development of sex differences and similarities in brain anatomy, physiology and behavior is under complex hormonal control, in *The Development of Sex Differences and Similarities in Behavior*, Haug, M., Whalen, R. E., Aron, C., and Olsen, K., Eds., Kluwer Academic Publishers, Dordrecht, 1993, 341.

155. Teresawa, E., Wiegand, S. J., and Bridson, W. E., A role for medial preoptic nucleus on afternoon of proestrus in female rats, *Am. J. Physiol.*, 238, 533, 1980.

156. Bleier, R., Byne, N., and Siggelkow, I., Citoarchitectonic sexual dimorphisms of the medial preoptic and anterior hypothalamic areas in guinea pig, rat, hamster, and mouse, *J. Comp. Neurol.*, 212, 118, 1982.

157. Symerly, R. B., Swanson, L. W., and Gorski, R. A., The distribution of monoaminergic cells and fibers in a periventricular preoptic nucleus involved in the control of gonadotropin release: immunohistochemical evidence for a dopaminergic sexual dimorphism, *Brain Res.*, 330, 55, 1985.

158. Symerly, R. B., Hormonal control of the development and regulation of tyrosine hydroxylase expression within a sexually dimorphic population of dopaminergic cells in the hypothalamus, *Mol. Brain Res.*, 6, 297, 1989.

159. Arai, Y., Nishizuka, M., Murakami, S., Miyakawa, M., Machida, M., Takeuchi, H., and Sumida, H., Morphological correlates of neuronal plasticity to gonadal steroids: sexual differentiation of the preoptic area, in *The Development of Sex Differences and Similarities in Behavior*, Haug, M., Whalen, R. E., Aron, C., and Olsen, K. L., Eds., Kluwer Academic Publishers, Dordrecht, 1993, 311.

160. Del Abril, A., Segovia, S., and Guillamón, A., Sexual dimorphism in the parastrial nucleus of the rat preoptic area, *Dev. Brain Res.*, 52, 11, 1990.

161. De Olmos, J. S. and Ingram, W. R., The projection field of the stria terminalis in the rat brain: an experimental study, *J. Comp. Neurol.*, 146, 303, 1972.

162. Kevetter, C. A. and Winans, S., Connections of the corticomedial amygdala in the golden hamster. I. Efferents of the "vomeronasal amygdala," *J. Comp. Neurol.*, 197, 81, 1981.

163. Krettek, J. E. and Price, J. L., Amygdaloid projections to subcortical structures within the basal forebrain and brainstem in the rat and cat, *J. Comp. Neurol.*, 178, 225, 1978.

164. Leonard, C. M. and Scott, J. V., Origin and distribution of the amygdalofugal pathways in the rat: an experimental neuroanatomical study, *J. Comp. Neurol.*, 141, 313, 1971.

165. Beltramino, C. and Taleisnick, S., Ventral premammillary nuclei mediate pheromonal-induced LH release stimuli in the rat, *Neuroendocrinology*, 41, 119, 1985.

166. Pfaff, D. W., *Estrogens and Brain Function: Neural Analysis of a Hormone-Controlled Mammalian Reproductive Behavior*, Springer, New York, 1980.

167. Cottingham, S. L. and Pfaff, D. W., Interconnectedness of steroid hormone-binding neurons: existence and implications, in *Morphology of Hypothalamus and its Connections.*, Ganten, D. and Pfaff, D. W., Eds., Springer, Berlin, 1986, 223.

168. Matsumoto, A. and Arai, Y., Sex difference in volume of the ventromedial nucleus of the hypothalamus in the rat, *Endocrinol. Jpn.*, 30, 277, 1983.

169. Dörner, G. and Staudt, J., Structural changes in the hypothalamic ventromedial nucleus of the male rat, following neonatal castration and androgen treatment, *Neuroendocrinology*, 4, 278, 1969.

170. Matsumoto, A. and Arai, Y., Male-female difference in synaptic organization of the ventromedial nucleus of the hypothalamus in the rat, *Neuroendocrinology*, 42, 232, 1986.

171. Pozzo-Miller, L. D. and Aoki, A., Stereological analysis of the hypothalamic ventromedial nucleus. II. Hormone-induced changes in synaptogenetic pattern, *Dev. Brain Res.*, 61, 189, 1991.

172. García-Segura, L. M., Chowen, J. A., Párducz, A., and Naftolin, F., Gonadal Hormones as promoters of structural synaptic plasticity: cellular mechanisms, *Prog. Neurobiol.*, 44, 279, 1994.

173. Pérez, J., Naftolin, F., and García-Segura, L. M., Sexual differentiation of synaptic connectivity and neuronal plasma membrane in the arcuate nucleus of the rat hypothalamus, *Brain Res.*, 527, 116, 1990.

174. Madeira, M. D., Sousa, N., Cadete-Leite, A., Lieberman, A. R., and Paula-Barbosa, M. M., The supraoptic nucleus of the adult rat hypothalamus displays marked sexual dimorphism which is dependent on body weight, *Neuroscience*, 52, 497, 1993.

175. Paula-Barbosa, M. M., Sousa, N., and Madeira, M. D., Ultrastructural evidence of sexual dimorphism in supraoptic neurons: a morphometric study, *J. Neurocytol.*, 22, 697, 1993.

176. Suarez, I., Bodega, G., Rubio, M., and Fernández, B., Sexual dimorphism in the distribution of glial fibrillary acidic protein in the supraoptic nucleus of the hamster, *J. Anat.*, 178, 79, 1991.

177. Hofman, M. A., Goudsmit, E., Purba, J. S., and Swaab, D. F., Morphometric analysis of the supraoptic nucleus in the human brain, *J. Anat.*, 172, 259, 1990.

178. Goudsmit E., Hofman, M. A., Fliers, E., and Swaab, D. F., The supraoptic and paraventricular nuclei of the human hypothalamus in relation to sex, age and Alzheimer's disease, *Neurobiol. Aging*, 11, 529, 1990.

179. Lin, L. P., Lee, Y., Tohyama, M., and Shiosaka, S., A sex-specific cytochrome P-450 (F-1) colocalized with various neuropeptides in the paraventricular and supraoptic nuclei of female rat, *Neuroendocrinology*, 54, 127, 1991.

180. Hagihara, K., Shiosaka, S., Lee, Y., Kato, J., Hatano, O., Takakusu, A., Emi, Y., Omura, T., and Tohyama, M., Presence of sex difference of cytochrome P-450 in the rat preoptic area and hypothalamus with reference to coexistence with oxytocin, *Brain Res.*, 515, 69, 1990.

181. Blanco, E., Carretero, J., Sánchez, F., Riesco, J. M., and Vazquez, R., Sex-specific effects of Met-enkephalin treatment on vasopressin immunoreactivity in the rat supraoptic nucleus, *Neuropeptides*, 13, 115, 1989.

182. Chung, S. K., McCabe, J. T., and Pfaff, D. W., Estrogen influences on oxytocin mRNA expression in preoptic and anterior hypothalamic regions studied by *in situ* hybridization, *J. Comp. Neurol.*, 307, 281, 1991.

183. Crespo, D., Cos, S., and Fernández-Viadero, C., Ultrastructural changes in hypothalamic supraoptic nucleus neurons of ovariectomized estrogen-deprived young rats, *Neurosci. Lett.*, 133, 253, 1991.

184. Warenbourg, W. and Poulain, P., Presence of estrogen receptor immunoreactivity in the oxytocin-containing magno-cellular neurons projecting to the neurohypophysis in the guinea-pig, *Neuroscience,* 40, 41, 1991.

185. Breedlove, S. M. and Arnold, A. P., Hormone accumulation in a sexually dimorphic motor nucleus of the rat spinal cord, *Science*, 210, 564, 1980.

186. Breedlove, S. M. and Arnold, A. P., Sexually dimorphic motor nucleus in the rat lumbar spinal cord: response to adult hormone manipulations, absence in androgen-insensitive rats, *Brain Res.*, 225, 297, 1981.

187. Breedlove, S. M. and Arnold, A. P., Hormonal control of a developing neuromuscular system. I. Complete desmas-culinization of the spinal nucleus of the bulbocavernosus in male rats using the antiandrogen flutamide, *J. Neurosci.*, 3, 417, 1983.

188. Grisham, W., Casto, J. M., Kashon, M. L., Ward, I. L., and Ward, O. B., Prenatal flutamide alters sexually dimorphic nuclei in the spinal cord of male rats, *Brain Res.*, 578, 69, 1992.

189. Breedlove, S. M. and Arnold, A. P., Hormonal control of a developing neuromuscular system. II. Sensitive periods for the androgen induced masculinization of the rat spinal nucleus of the bulbocavernosus, *J. Neurosci.*, 3, 424, 1983.

190. Sengelaub, D. R., Nordeen, E. J., Nordeen, K. W., and Arnold, A. P., Hormonal control of neuron number in sexually dimorphic spinal nuclei of the rat: III. Differential effects of the androgen dihydrotestosterone, *J. Comp. Neurol.*, 280, 637, 1989.

191. Breedlove, S. M., Jacobson, C. D., Gorski, R. A., and Arnold, A. P., Masculinization of the female rat spinal cord following a single injection of testosterone propionate but not estradiol benzoate, *Brain Res.*, 237, 173, 1982.

192. Goldstein, L. A. and Sengelaub, D. R., Hormonal control of neuron number in sexually dimorphic spinal nuclei of the rat: IV. Masculinization of the spinal nucleus of the bulbocavernosus with testosterone metabolites, *J. Neurobiol.*, 21, 719, 1990.

193. Jordan, C. L., Breedlove, M. S., and Arnold, A. P., Sexual dimorphism and the influence of neonatal androgen in the dorsolateral motor nucleus of the rat lumbar spinal cord, *Brain Res.*, 249, 309, 1982.

194. Nordeen, E. J., Nordeen, K. W., Sengelaub, D. R., and Arnold, A. P., Androgens prevent normally occurring cell death in a sexually dimorphic spinal nucleus, *Science*, 229, 671, 1985.

195. Breedlove, S. M., Jordan, C. L., and Arnold, A. P., Neurogenesis of motoneurons in the sexually dimorphic spinal nucleus of the bulbocavernosus in rats, *Dev. Brain Res.*, 9, 39, 1983.

196. Fishman, R. B. and Breedlove, S. M., Local perineal implants of anti-androgen block masculinization of the spinal nucleus of the bulbocavernosus, *Dev. Brain Res.*, 70, 283, 1992.

197. Rand, M. N. and Breedlove, S. M., Androgen locally regulates rat bulbocavernosus and levator ani size, *J. Neurobiol.*, 23, 17, 1992.

198. Forger, N. G., Lynn, L. H., Roberts, S. L., and Breedlove, S. M., Regulation of motoneuron death in the spinal nucleus of the bulbocavernosus, *J. Neurobiol.*, 23, 1192, 1992.

199. Forger, N. G. and Breedlove, S. M., Sexual dimorphism in human and canine spinal cord: role of early androgen, *Proc. Natl. Acad. Sci. U.S.A.*, 83, 7527, 1986.

200. Clemens, L. G., Wagner, C. K., and Ackerman, A. E., A sexually dimorphic motor nucleus: steroid sensitive afferents, sex differences and hormonal regulation, in *The Development of Sex Differences and Similarities in Behavior*, Haug, M., Whalen, R. E., Aron, C., and Olsen, K. L., Eds., Kluwer Academic Publishers, Dordrecht, 1993, 19.

201. Lang, R. E., Heil, J., Ganten, D., Hermann, K., Rascher, W., and Unger, T., Effects of lesions in the paraventricular nucleus of the hypothalamus on vasopressin and oxytocin contents in brainstem and spinal cord of rat, *Brain Res.*, 260, 326, 1983.

202. Melis, M. R., Argiolas, A., and Gessa, G. L., Oxytocin induced penile erection and yawning: site of action in the brain, *Brain Res.*, 398, 259, 1986.

203. Palkovits, M., Afferents onto neuroendocrine cells, in *Morphology of Hypothalamus and its Connections*, Ganten, D. and Pfaff, D., Eds., Springer, Berlin, 1986, 197.

204. Silverman, A. J., Hoffman, D. L., and Zimmerman, E. A., The descending connections of the paraventricular nucleus of the hypothalamus (PVN), *Brain Res. Bull.*, 6, 47, 1981.

205. Weiss, M. L. and Hatton, G. I., Collateral input to the paraventricular and supraoptic nuclei in rat. I. Afferent from the subfornical organ and the anteroventral third ventricle region, *Brain Res. Bull.*, 24, 231, 1990.

206. Guillamón, A., De Blas, M. R., and Segovia, S., Effects of sex steroids on the development of the locus coeruleus in the rat, *Dev. Brain Res.*, 40, 306, 1988.

207. De Blas, M. R., Segovia, S., and Guillamón, A., Effects of postpuberal gonadectomy on cell population of the locus coeruleus in the rat, *Med. Sci. Res.*, 18, 355, 1990.

208. Luque, J. M., De Blas, M. R., Segovia, S., and Guillamón, A., Sexual dimorphism of the dopamine-beta-hydroxylase-immunoreactive neurons in the rat locus coeruleus, *Dev. Brain Res.*, 67, 211, 1992.

209. Roselli, C. E., Horton, L. E., and Resko, J. A., Distribution and regulation of aromatase activity in the rat hypothalamus and limbic system, *Endocrinology*, 117, 2471, 1985.

210. Grisham, W., Kerchner, M., and Ward, I. L., Prenatal stress alters sexually dimorphic nuclei in the spinal cord of male rats, *Brain Res.*, 551, 126, 1991.

211. Toran-Alleran, C. D., Organotypic culture of the developing cerebral cortex and hypothalamus: relevance to sexual differentiation, *Psychoneuroendocrinology*, 16, 7, 1991.

212. Gladue, B. A. and Clemens, L.G., Androgenic influences on feminine sexual behavior in male and female rats: defeminization blocked by prenatal antiandrogen treatment, *Endocrinology*, 103, 1702, 1978.

213. Rodriguez-Zafra, M., De Blas, M. R., Pérez-Laso, C., Cales, J. M., Guillamón, A., and Segovia, S., Effects of perinatal diazepam exposure on the sexually dimorphic rat locus coeruleus, *Neurotoxicol. Neuroteratol.*, 15, 139, 1993.

214. Del Cerro, M. C. R., Izquierdo, M. A. P., Pérez-Laso, C., Rodriguez-Zafra, M., Guillamón, A., and Segovia, S., Early postnatal diazepam exposure facilitates maternal behavior in virgin female rats, *Brain Res. Bull.*, 38, 143, 1995.

215. Orchinik, M. and McEwen, B., Novel and classical actions of neuroactive steroids, *Neurotransmission*, 9, 1, 1993.

216. Akwa, Y., Young, J., Kabbadj, K., Zucman, D., Vourch, C., Jung-Tectas, I., Hu, Z. I., Le Goascogne, C., Jo, D. H., Corpechot, C., Simon, P., Baulieu, E. E., and Robel, P., Neurosteroids: biosynthesis, metabolism and function of pregnenolone and dehydroepiandrosterone in the brain, *J. Steroid Biochem. Mol. Biol.*, 40, 71, 1991.

217. Baulieu, E. E. and Robel, P., Neurosteroids: a new brain function?, *J. Steroid Biochem. Mol. Biol.*, 37, 395, 1990.

218. Goy, R. W. and McEwen, B. S., *Sexual Differentiation of the Brain*, Cambridge, The MIT Press, 1980.

219. Arnold, A. P. and Gorski, R. A., Gonadal steroid induction of structural sex differences in the central nervous system, *Annu. Rev. Neurosci.*, 7, 413, 1984.

220. De Vries, G. J., De Bruin, J. P. C., Uylings, H. B. M., and Corner, M. A., Eds., *Sex Differences in the Brain, Elsevier, Amsterdam*, 1984.

221. McEwen, B. S., Our changing ideas about steroid effects on an ever-changing brain, *Semin. Neurosci.*, 3, 497, 1991.

222. Pilgrim, C. and Hutchison, J. B., Developmental regulation of sex differences in the brain: can the role of gonadal steroids be redefined?, *Neuroscience*, 60, 843, 1994.

223. Matsumoto, A., Synaptogenic action of sex steroids in developing and adult neuroendocrine brain, *Psychoneuroendocrinology*, 16, 25, 1991.

224. Reisert, I. and Pilgrim, C., Sexual differentiation of monoaminergic neurons — genetic or epigenetic?, *TINS*, 14, 468, 1991.

Chapter 9

Gonadal Hormones and the Sexual Differentiation of the Nervous System: Mechanisms and Interactions

A.P. Payne

CONTENTS

I. INTRODUCTION

Adult vertebrates often exhibit sexual dimorphisms of behavior (one sex showing one pattern of behavior, the opposite sex a different pattern) or sex differences in the amount of a particular kind of behavior. This is particularly true of the social behaviors such as courtship and copulation, aggressive behavior, and parental behavior and play, although nonsocial behaviors such as exploratory behavior may also show sex differences; in addition, there are sex differences in sensorimotor and neuroendocrinological capacities.[1-4] Sex differences in behavior are not situational, since they often persist when the sexes are separate. Furthermore, while in adulthood they are often clearly linked to hormone levels (gonadectomy may abolish a behavior which can then be restored by appropriate hormone therapy; behavior may show cyclic or seasonal variation), in the higher vertebrates at least it is often difficult, or even impossible, to induce an animal to show the behavior of the opposite sex simply by administering the "wrong" hormones. Whatever the plasticity of adult social behavior patterns, it does appear that they display a degree of gender specificity that presupposes some form of "hard-wiring."

Gonadal steroids are taken up into a variety of regions of the CNS where their effects can broadly be classed as "activational" or "organizational". Activational effects are those which (1) maintain the

0-8493-7633-5/96/$0.00+$.50

structure or activity of the mature nervous system or (2) facilitate the expression of particular patterns of neural activity or behavior. By contrast, organizational effects are those which, during development, serve to determine the structure of the nervous system or the activities it is capable of carrying out. As simple examples, hormones may be necessary in adulthood to allow the animal to display courtship and copulatory behavior (thus having an activational role at that time) but they may have been necessary during development to organize the capacity to display these behaviors, i.e., to organize the neural substrate which underlies them.[5,6]

The purpose of this chapter is to examine the evidence for sexual differentiation of the nervous system and of specific regions within it which might form the neurobiological substrate for dimorphic activities and, most particularly, to consider the mechanisms by which that differentiation can occur.

II. SEXUAL DIFFERENTIATION AS A GENERAL FEATURE OF DEVELOPMENT

A. THE TIMING AND SPECIFICITY OF HORMONES INVOLVED IN SEXUAL DIFFERENTIATION OF THE NERVOUS SYSTEM

Once sex differences in the activity of the adult nervous system were accepted as relatively nonplastic phenomena, most insight has been gained from the simple assumption that the nervous system is no different from any other part of the body which can undergo sexual differentiation during development (such as the reproductive tract, the external genitalia, etc.). A wealth of experimental, manipulative studies on sexual differentiation of the reproductive system in mammals[7-11] suggested the following characteristics should be considered:

1. That sex differences in the nervous system were unlikely to be due to genetic differences.
2. Rather, they were likely to be due to the action of hormones on the developing nervous system.
3. That the male pattern was likely to be induced by androgens such as testosterone (or its metabolites), with the female pattern being (largely) a default pattern.
4. That differentiation was likely to occur during a particular "critical period" or window of developmental opportunity, the hormonal milieu during that critical period being crucial for development but, outside the critical period, having relatively little impact.

In male mammals, androgens are produced by the fetal testis (1) during a critical period for differentiation of the reproductive system and genitalia which is prenatal and (2) during a period which may be pre- or postnatal and which coincides with sexual differentiation of the nervous system.[12-15] Although there is acknowledged to be male/female overlap in blood androgen titers,[16] the general finding in the rat is of consistently higher mean androgen levels in males from day E18 to P5.[17-20] From the point of view of experimental intervention, this means that sexual differentiation of the mammalian central nervous system (CNS) can be investigated by postpartum hormone manipulations, unlike differentiation of the reproductive tract and genitalia where differentiation is prenatal and experimental intervention usually involves the less satisfactory administration of hormones to the pregnant mother. In humans, plasma androgen levels are significantly higher in male infants compared to females after birth[21] — perhaps for several months[22] — a finding which appears true for a range of primate species.[23-25] However, the timing of sexual differentiation of the human brain remains an open question; Swaab and Hoffman[26] suggest that some regions of the human brain do not begin to exhibit dimorphisms until after four years.

Although testosterone is the major product of the testis, it is believed to act directly at relatively few sites during development. Rather, it is considered a prohormone which acts at many target tissues only after conversion to more active metabolites: (1) after aromatization to 17β-estradiol and (2) after 5α-reduction to 5α-dihydrotestosterone.

In particular, there is evidence to suggest that testosterone per se is responsible for maintenance of the Wolffian duct and its derivatives, 5α-dihydrotestosterone is responsible for masculinization of the urogenital sinus and tubercle, while 17β-estradiol is responsible for masculinization of the nervous system.[8,10,27,28] Evidence for the latter comes from a variety of approaches, including the localized implantation of estrogens into the CNS[29] and the blockage of aromatase activity by androstanediones.[28]

In the mouse, males have higher hypothalamic aromatase activity than females, as well as higher numbers of aromatase-immunoreactive cells.[29,30] Beyer and co-workers have since suggested that there is no intrinsic sex difference in aromatase activity between male and female hypothalamic neurons *in vitro,* but that aromatase activity is increased by androgens; moreover, androgens may also increase the survival during development of aromatase-immunoreactive neurons.[31,32] This may help to resolve an apparently fundamental species difference, viz. that prenatal exposure to DHT has been said to alter behavior in female primates, a contrast to the findings for other laboratory species where DHT is thought to be relatively inactive.[33] However, since androgens stimulate brain aromatase activity,[34,35] the effects of DHT could well be indirect. Confirmatory evidence for this view comes from work on the developing rhesus monkey, where sex differences can be found in aromatase activity in the mediobasal hypothalamus and amygdala, but there are no sex differences in 5α-reductase activity or androgen receptor levels.[36] Recently, Compaan et al.[37] have shown that neonatal brain aromatase activity is higher in aggressive strains of mice than in nonaggressive strains; aggression is one of the social behaviors that depends upon perinatal androgens.[38-40]

Control of Fetal Steroid Levels

The testis is the source of circulating androgens in fetal blood, and there is a specific prenatal population of Leydig cells to produce steroids; these Leydig cells remain active into the postnatal period and then disappear to be replaced by an adult population derived from fibroblast-like precursors.[41,42]

It is less clear what control mechanisms operate on the early testis. In human fetuses, placental syncytiotrophoblast cells produce human chorionic gonadotrophin (hCG) under the influence of trophic factors produced by cytotrophoblast cells; this process reaches a peak by nine weeks, after which placental hCG decreases (for review see Silver-Khodr[43]). By contrast, pituitary LH/FSH is detectable starting at 9 to 11 weeks, and rises thereafter. These hormones are higher in female fetuses than males, probably because testicular androgens are having inhibitory feedback effects.[44] It would seem that gonadotrophins (first of a placental source but later, and perhaps more importantly, of a fetal pituitary source) are present during the critical period of fetal androgen production and masculinization of the reproductive tract and genitalia in humans. A detailed review of human fetal pituitary function and clinical disorders is provided by Cuttler,[45] while a consideration of possible hypothalamic controls is provided by Gilmore.[46]

It is also probable that the pituitary is responsible for controlling postnatal androgen production leading to changes in adult brain activity. Thus, chronic LHRH agonist administration during weeks 2 to 4 after birth to male rhesus monkeys retards puberty (as determined by LH and testosterone secretion) and decreases the capacity for copulatory behavior.[47]

Finally, it is likely that the activity of the fetal/postnatal pituitary is either different from (or at least less well honed than) its activity in adulthood. In the sheep, the fetal hypothalamo-pituitary system exhibits a degree of functional control during gestation, but matures over a relatively wide gestational range and may not be fully developed even at birth.[48] Much of its prenatal activity is pulsatile, and the possibility of independent pulse generators has been considered.[49]

B. LABORATORY EXPERIMENTS

A very large number of experiments have been performed in which hormones have been manipulated in pre- or postnatal pups of laboratory species and adult behavior examined. Manipulations have largely followed two routes:

1. Suppression of androgens in natural XY males, either by surgical castration, by the administration of Leydig cell cytotoxins such as ethylidene dimethane sulfonate (pharmacological castration), by the administration of anti-androgens such as cyproterone or flutamide, and by the administration of compounds that interfere with subsequent metabolism of testosterone into 5α-dihydrotestosterone or estradiol.
2. Administration of androgens/estrogens to natural XX females by injection or implantation.

A great many experiments using these types of manipulation have demonstrated quite clearly that sexually dichotomous adult brain functions, such as the capacity to display a wide range of sex-typical behaviors (including sexual behavior, aggressive behavior, parental behavior, play behavior), or sex-typical physiological controls (e.g., gonadotrophin release) can be readily and predictably altered to that of the opposite sex by appropriate hormone intervention.[9,50-52] In some cases, by the use of threshold hormone regimens, it is possible to distinguish between "masculinization" and "defeminization" of the nervous system as separate processes.[53-55] Similar effects of hormones on human gender-specific behavior and orientation have also been considered.[56,57]

C. NATURAL PHENOMENA

It is obvious that most information on sex differences in the nervous system and their genesis has come via experimental manipulations involving hormone administration or elimination, and these are referred to throughout this paper. However, there are also some naturally occurring phenomena which give insight and therefore should be considered.

Testicular feminizing syndrome — Testicular feminizing syndrome in humans is a mutation in which body tissues are insensitive to androgens so that, even though the genetic, gonadal, and hormonal sex of the individual is male, the morphotype is unmistakably female.[56,58] Similar mutations occur in laboratory rodents; some individual male King-Holtzman rats, for example, are Tfm. In many tissues there is an 85 to 90% reduction in androgen receptors[59] while in others, such as muscle, receptor numbers are normal but they fail to bind DNA.[60] Androgen levels may be higher than normal.[61] Sexually dimorphic regions of the spinal cord (such as the spinal nucleus of bulbocavernosus (SNB) and dorsolateral nucleus (DLN), see below) are considerably reduced in male testicular feminizing syndrome mutant (Tfm) rats compared to controls.[62,63]

Prenatal stress — Prenatal stress is known to adversely affect the ability of male rodents to show masculine copulatory patterns in adulthood and to increase the capacity to exhibit female patterns.[64,66,67] The effects of prenatal stress on a wide variety of social and nonsocial behaviors is reviewed by Archer and Blackman.[68] Stressing pregnant rats (restraint and temperature rise) during the third trimester of pregnancy (E14 to 20) decreases the size of the sexually dimorphic nucleus of the medial preoptic area (SDN-MPOA) by some 50% in male offspring, but is without apparent affect on female offspring.[69] There is normally assumed to be an inverse relationship between pituitary-gonadal and pituitary-adrenal activity, so that stress may suppress androgen production. Indeed, there are positive correlations between the reduced size of the sexually dimorphic nucleus of the preoptic area (SDN-POA), lowered plasma androgen titers, and reduced frequency of sexual activity.[70] However, corticosterone administration during pregnancy also significantly decreases the volume of the SDN-MPOA.[71] Moreover, a further possibility which must be borne in mind is that stress conditions trigger opioid release in adults[72,73] and that prenatal exposure to opiates also reduces the capacity to display masculine sexual behavior in adulthood[65,74] so that a nonhormonal explanation could also be enlisted.

Position in uterine horn — Individual differences in sex-typical behavior and reproductive morphology have often been noted. One possibility, at least for species that deliver multiple young, is that this is due to effects from developing siblings. The classic example is that of dizygotic twin farm animals such as cattle, where female calves whose sibling is a male develop as "freemartins," sterile females with partially masculinized reproductive systems.[75] Here, the twins are genuinely consanguineous *in utero*, and the female is influenced by androgens emanating from the male calf. Even in laboratory rodents, the position within the uterine horn can influence female fetuses. Thus, females located between two male fetuses (2M females) have higher amniotic fluid and blood levels of testosterone during development, masculinized anogenital distances, and a greater degree of aggressiveness as adults.[76,77]

III. STRUCTURAL DIMORPHISMS WITHIN THE CNS

One of the most important discoveries concerning sexual differentiation of the nervous system has been that there are many regions which exhibit actual sexual dimorphisms of structure. These dimorphisms range from the relative size of a region, the cell types it contains, the levels of transmitters/peptides occurring within it, to patterns of neurite branching and synapse termination.

Not surprisingly, many areas of the nervous system which (1) exhibit known morphological or neurochemical dimorphisms and (2) are organized by hormonal influences during development, are linked to reproductive activities. These regions would include the anterior hypothalamus,[28,78] the MPOA,[79-82] the supraoptic nucleus,[83] the suprachiasmatic nucleus,[84] the arcuate nucleus,[85] the ventromedial nucleus (VMN),[86] the amygdala,[87,88] the bed nucleus of the stria terminalis,[89] the vomeronasal organ and accessory olfactory bulb,[90,91] a variety of motor neuron groups such as the cremasteric nucleus,[92] the spinal nucleus of bulbocavernosus,[93] and the dorsolateral nucleus,[94] as well as pelvic autonomic ganglia.[95,96] However, there are others for which the association is more tenuous, including the cerebral cortex,[97] the hippocampus,[98-101] the locus coeruleus,[102,103] parts of the corpus callosum,[104,105] (but see Going and Dixson[106]), the anterior commissure,[107] and the superior cervical ganglion.[108,109]

Because the bulk of experimental evidence comes from two sexually dimorphic regions of the CNS, (1) the MPOA and (2) motor neuron groups in the lumbosacral spinal cord, these will be considered briefly at the outset of the chapter.

A. THE SEXUALLY DIMORPHIC NUCLEUS OF THE MEDIAL PREOPTIC AREA

It was originally reported (Figure 1) that complex morphological sex differences occur within the POA in terms of the pattern of termination of afferent synaptic endings from a nonamygdaloid source.[79] Subsequently, Gorski and co-workers[80,81] reported a sex difference in the size of a nucleus within the MPOA of the rat such that the sex of a brain could be reliably determined on a routinely stained section with the naked eye. This nucleus is sexually dimorphic in size, with males of many mammalian species (including guinea pig, hamster, and mouse,[78] ferret,[28] and man[82,110]) possessing a larger SDN-MPOA than females; the cell density appears to be similar in the two sexes, thus there is a genuine difference in neuron numbers.[81] The administration of androgens to female laboratory species increases the size of

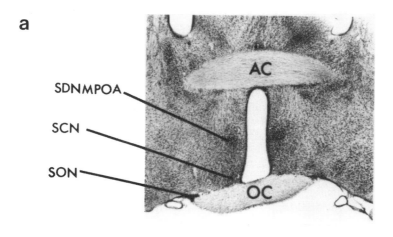

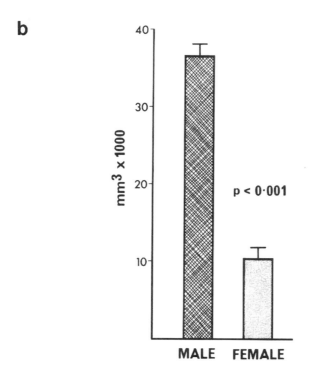

Figure 1   (a) A coronal section through the diencephalon of a male Albino Swiss rat at the level of the MPOA. AC = anterior commissure; OC = optic chiasm; SCN = suprachiasmatic nucleus; SON = supraoptic nucleus; and SDNMPOA = sexually dimorphic nucleus of the MPOA (Original magnification × 15). (b) The volume of the SDN-MPOA in adult male and female Albino Swiss rats (mm^3 × 1000).

the SDN-MPOA, while depleting androgens in males decreases its adult size (reviewed Gorski[111]). Hormone administration to females and neonatally castrated males shows that the critical period is terminated by day P6.[112]

The MPOA is involved in gonadotrophin secretion[52,113] and sexual behavior,[114-116] and it is not surprising that authors have shown deficits in gonadotrophin secretion[117] and masculine mating behavior following discrete lesions to the SDN-MPOA;[118,119] DeJonge et al.[120] have made the point that feminine mating behavior is unaffected by these lesions. The demonstration by LeVay[110] that the SDN-MPOA is smaller in homosexual men than in heterosexual ones has excited great interest, but its significance is, as yet, unclear.

B. THE SPINAL NUCLEUS OF BULBOCAVERNOSUS AND THE DORSOLATERAL NUCLEUS

There are several sexually dimorphic groups of motor neurons in the spinal cord. The SNB (sometimes called the dorsomedial (DM) group) which contains five times more neurons in males than females, and the DLN contains twice as many neurons in males than in females; both occur within the region L5-S1 in the rat (Figure 2).[94,121,122] These two groups are believed to be the counterpart of Onuf's nucleus which occurs in the sacral region of the cord in man and several other mammalian species, and supplies perineal musculature.[123] The SNB/DLN motor neurons innervate the perineal muscles bulbocavernosus, ischiocavernosus, and levator ani which involute soon after birth in the female rat[124] and which are, in the male, implicated in erection and penile reflexes associated with cervical stimulation and copulatory plug formation.[125-127] The small number of surviving motor neurons in females, together with some of the male motor neurons in the SNB/DLN pools, innervate urethral and anal sphincters.[62,93,128-132]

The pattern of innervation to the perineal muscles is quite circumscribed in the rat, with specific muscles being innervated by specific nuclear groups, but appears less so in the mouse.[133]

If androgens are administered to female rat pups at the time of birth, the number of motor neurons found in adulthood in the SNB/DLN groups is greatly increased and the muscles are retained.[121,124,134]

IV. MECHANISMS OF HORMONE ACTION UNDERLYING SEXUAL DIFFERENTIATION OF THE CNS

Among the many processes involved in the development of the nervous system are (1) cell division, (2) cell migration, (3) cell death, and (4) neurite formation, synaptogenesis and the development of appropriate connections. All or some of these processes might conceivably be hormone dependent during a critical period of sexual differentiation within the developing nervous system. For major reviews, see Arnold and Gorski[135] and Sengelaub.[136]

A. CELL DIVISION

Autoradiographic ³H-thymidine labeling experiments demonstrate that adult SNB and DLN spinal cord motor neurons (Figure 2) are derived by mid-gestation (E12), long before the critical period during which androgens might affect the development of these groups.[137] Moreover, although females given perinatal androgens are subsequently found to have more SNB/DLN motor neurons, labeling experiments again demonstrate that these were laid down by E12.[129] There is, therefore, no evidence that androgens produce a wave of cell division in this system. Further, there is no evidence that androgens cause other cell types to differentiate into motor neurons. In Tfm rats, there is a normal rise in SNB cell numbers prior to birth (showing that the rise itself is androgen independent) followed by a fall to female levels which cannot be prevented by androgen administration.[138]

In sexually dimorphic neural circuits controlling mate calling in frogs, gonadal differentiation and subsequent steroid accumulation by neurons does not occur until late stages of development, by which time neurogenesis is complete.[139,140] Again, a steroid effect on neurogenesis therefore seems unlikely.

The only system in which a steroid effect on cell division has been proposed is avian song centers; these centers are unusual because they not only exhibit a number of sexual dimorphisms but also seasonal and other variations in adulthood.[141,142] Most particularly, there is evidence:

1. That the number of neurons in song centers such as the hyperstriatum ventrale pars caudale (Hvc) increases during the first three months post hatching, when birds such as the male zebra finch are learning songs.[143,144]
2. That this increase is due to the generation of new neurons in areas such as the Hvc and Area X, and not to the migration of neurons from other sites.[145]

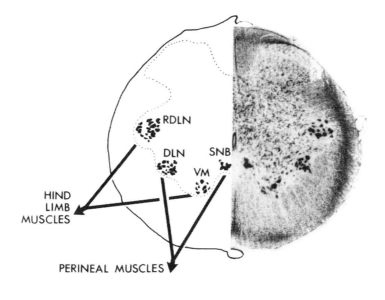

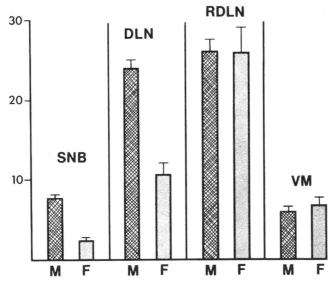

Figure 2 (Top) A transverse section through the spinal cord of a male Albino Swiss rat (L5) showing four groups of motor neurons. SNB = spinal nucleus of bulbocavernosus; VM = ventromedial nucleus; DLN = dorso-lateral nucleus; and RDLN = retrodorsolateral nucleus (Original magnification × 25). (Bottom) The number of motor neurons in these four groups in adult male and female Albino Swiss rats. There are significant sex differences in the number of SNB and DLN neurons (which supply perineal muscles and sphincters) but not in VM and RDLN neurons which supply the hind limb (neurons/50 μm section).

3. That the bulk of these newly formed neurons are androgen dependent.[144]
4. That new neurons and glial cells can be generated within the adult bird Hvc and, combining ^{3}H-thymidine and HRP studies, that these neurons are incorporated into the appropriate neural circuits.[146,147] They therefore have very different characteristics from, e.g., mammalian dimorphic systems.

B. CELL MIGRATION

Labeling of developing cells in the SDN-MPOA with ^{3}H-thymidine revealed (1) that cells destined for the SDN-MPOA are still being formed during day E18, whereas cells for surrounding parts of the MPOA are formed entirely by E16[148] and (2) a complex vee-shaped pattern of cell migration from the walls of the third ventricle. Initially, cells traveled downward to a position just dorsal to the suprachiasmatic nucleus; from there, the migration route turned upward and laterally to the definitive position of the nucleus.[149] The whole process occurred over a period of several days. Gorski[111] speculated (1) that neurons which had not arrived in that time period would not make apppropriate synaptic contacts and would not survive and (2) that androgens might increase the duration of the time period, permitting more cells to arrive and thus resulting in a physically larger nucleus.

Evidence for neuronal migration in other sexually dimorphic systems (whether or not androgen controlled) is less compelling at present. Gorlick and Kelley[140] have considered that vocalization centers in the frog hindbrain may become dimorphic due to migration. A medial pool of laryngeal motor neurons is lost during development, but the surviving lateral motor pool (with more neurons in males than females) consists of neurons with medially directed dendrites (longer in males than in females). These authors have postulated that this might be due to androgen-dependent growth of dendrites leading to some shift of medial laryngeal neurons into the lateral pool. In sexually dimorphic groups of motor neurons in the L5-S1 region of the rat spinal cord, Sengelaub and Arnold[63] have suggested that SNB and DLN neurons initially form a common population from which SNB neurons migrate to their adult position. Evidence includes small numbers of elongated motor neurons which can be seen within the adult cord midway between the SNB and DLN nuclei; it is suggested that these are cells which have halted while still on migration. Moreover, there are said to be glial formations linking the two sites along which motor neurons could migrate.[150]

C. (PROGRAMMED) CELL DEATH

Neuronal cell death is an important feature of the development of the CNS, since neurons are produced to excess and then reduced in number. In the case of the rat SNB, neuron numbers rise from E18 to E22 (the day before birth in most strains), probably as a result of migration from lateral neuron pools in the L5-S1 region. Thereafter, numbers decline until they reach adult levels by postnatal day 10. However, the decline is greater in females than in males, leading to the characteristc sex difference in neuron numbers.[63,151] The fall is much smaller than normal in females given androgens, but much larger than normal in male testicular feminizing syndrome mutant (Tfm) rats.[62,128,138] A similar sex difference occurs in the number of cells in the DLN of the rat; again, this is due to sex differences in the rate of cell loss which can be demonstrated to be androgen dependent.[138,152]

D. GROWTH AND SYNAPTOGENESIS

Not only does the male rat SNB contain far more motor neurons than the female SNB,but those neurons are twice as large on average.[62] Size as a characteristic can vary according to adult hormone levels (see below), but large differences in cell body size are determined during sexual differentiation, and the critical period during which androgens affect cell size in the SNB terminates by day P12.[153] The development of SNB dendrites appears to be biphasic with (1) a period of exuberant dendritic outgrowth until four weeks after birth followed by (2) a period of retraction until seven weeks. Androgens are necessary for phase one, but the rise in androgens at puberty also brings phase two to a halt, establishing the adult dendritic length.[154] Even so, SNB neurons have very long dendrites which decussate and establish contact with contralateral SNB neurons.[136]

The earliest modern experimental study of structural sexual dimorphisms within the CNS dealt with sex differences in the pattern of synaptic termination of axons from nonamygdaloid neurons in the rat MPOA. Raisman and Field[79,155] showed that there was an adult sex difference in the proportion of surviving synapses ending on (1) synaptic shafts and (2) synaptic spines, following lesioning of the stria terminalis. They went on to show that perinatal hormone manipulations resulted in predictable changes in the patterns of synaptic termination. Sex differences have also been found in synaptic arrangements within the medial amygdala[87] and the VMN of the hypothalamus;[156] again, these are responsive to perinatal hormone manipulation.

A rather different effect of early androgens on synapses occurs at neuromuscular junctions. Here Jordan et al.[157,158] reported that the elimination of multiple synapses on muscles (which normally occurs some two weeks after birth[159,160]) is delayed in perineal muscles which are both sexually dimorphic and possess sexually dimorphic innervations, such as levator ani. They described a rapid phase of synapse elimination between 2 to 4 weeks after birth, followed by a slow phase beginning in the third month. Furthermore, they subsequently found that androgen administration delayed synapse elimination further, thus preventing the loss of multiple innervation.[157,158]

An important diencephalic area implicated in the display of sexual behavior is the VMN of the hypothalamus, a major site for eliciting sexual behavior in the female rat.[161] There are sex differences in the levels of cytosolic progesterone receptors in the VMN, and appropriate perinatal hormone manipulation can alter levels toward that of the opposite sex (e.g., if males receive the aromatase inhibitor ATD, or if females receive estradiol). In particular, there are sex differences in the number of dendritic spine and shaft synapses in the ventrolateral (but not the ventromedial) part of the VMN, with male rats possessing more of both than females; perinatal castration of males decreases and perinatal androgen treatment to females increases synapse frequency in adulthood.[86]

The branching pattern of dendrites has also been found to show sex differences in several parts of the brain, such as the preoptic region of the rhesus monkey[162] and the hippocampus. In the latter, male CA3 neurons have more distal branching of dendrites, while females have more proximal branching.[99,100] The granule cell layer of the dentate gyrus is also thicker in males than females, but is increased in females as a result of postnatal androgen administration.[101]

V. PATTERNS OF SEXUAL DIFFERENTIATION WITHIN THE CNS

In 1991, an influential commentary by Reisert and Pilgrim[163] postulated two quite separate mechanisms for the sexual differentiation of monoaminergic neural systems — an *epigenetic* mechanism which depended upon the hormonal milieu during development (akin to the classical "androgen-dependent" mechanism) and another, possibly *genetic,* mechanism. It is almost certainly the case that the factors influencing sexual differentiation within the CNS include nonhormonal ones; it is less clear (1) how nonhormonal factors interact with hormonal ones, (2) how they come to influence differentiation, and (3) whether they themselves are hormonally controlled.

Because what is differentiated within the CNS is populations of neurons (the number of neurons within the population, their morphology, connectivity, or other characteristics), and because there is a wealth of data demonstrating that perinatal hormone manipulation can influence the differentiation of each dimorphic system that has been reported, it is not sensible to imagine a pattern of sexual differentiation which does not include a hormonal component. However, it is possible to imagine at least four simple schemes, depicted in Figure 3 type A-D. A *type A* pattern envisages an effect of hormones on a neuronal population during a critical period of embryonic or early postnatal development. A *type B* pattern envisages hormone influences on neurones being supplemented by additional controls from surrounding components of the developing CNS (e.g., glial cells, extracellular matrix, trophic factors, transmitters, or peptides) or specifically from cells — whether other neurons or muscles — which are the normal contacts or targets of the sexually differentiating neurons. Two refinements would be if hormones acted only upon the surrounding tissues/target cells so that hormone effects upon the sexually differentiating neuron population were indirect (a *type C* pattern) or if hormones acted on *both* the differentiating neuron population *and* the surrounding tissues/target cells, and the latter also influenced the development of the neurone population (a *type D* pattern). These four basic patterns could be made more complex if (1) some processes leading to differentiation (e.g., differential neuron cell death) were hormonally controlled and others (e.g., neuron migration) were controlled by surrounding tissues or (2) hybrids occurred between the four basic types.

VI. INTERACTIONS DURING SEXUAL DIFFERENTIATION

Several kinds of evidence exist for thinking that, although androgens clearly act directly on developing neurons, there is considerable scope for more complex mechanisms. This section seeks to analyze two quite different regions of the mammalian CNS and the development of sexual dimorphisms within them.

A. INTERACTIONS BETWEEN NEURONS, THEIR TARGETS, ANDROGENS AND AMINES: SEXUALLY DIMORPHIC MOTOR NEURON/TARGET MUSCLE MODELS

Three known sexual dimorphisms involve groups of motor neurons within the CNS and their muscles; these are

The SNB/DLN nuclei — These motor neuron pools in L5-S1 of the rat cord which innervate perineal muscles and sphincters have been discussed in detail above (see Figure 2).

162

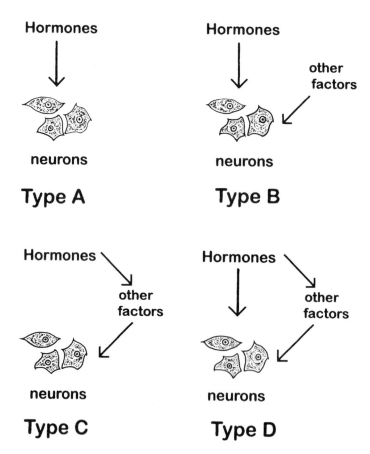

Figure 3 Possible patterns of hormone action in inducing morphological sex differences in the CNS during a critical period of sexual differentiation.

The cremasteric nucleus — The cremasteric nucleus is a group of motor neurons which, in the rat, occurs in segments L1-L2. It contains about 300 neurons per side in males, but only some 90 in females.[91,164] The latter authors have suggested that, as the cremaster muscle is responsible for raising and lowering the testes in response to temperature changes, the cremasteric nucleus may function as a spinal temperature control center. They have commented that the nucleus'considerable serotonergic input may relate to thermoregulation, while its substance P innervation may relate to primary afferent nociceptive and thermal information from the testes, scrotum, and cremaster. Castration of males on the day of birth leads to a considerable reduction in the number of cremasteric nucleus motor neurons.[164]

Medullary motor neuron pools which form part of cranial nerves IX and X and which innervate the laryngeal apparatus of amphibia — Amphibians such as *Xenopus*[165,166] and *Rana pipiens*[167] possess a complex vocalization circuitry in which a variety of telencephalic and diencephalic regions project to a pretrigeminal nucleus within the dorsal tegmental area of the medulla (DTAM) which, in turn, controls the activity of motor neurons in parts of cranial nerve nuclei IX and X. In *Xenopus*, males have some 1200 large laryngeal motor neurons while females possess only 750 smaller ones.[168] Several connections of DTAM are much larger in males than females.[169] The bipennate muscles in the larynx of *Xenopus* contain equivalent numbers of fibers in both sexes (some 4000) before metamorphosis; thereafter, muscle fibers in males increase some eightfold in number, a rise which is stimulated by androgen administration to immature males and prevented by the anti-androgen flutamide.[170]

In each of the three cases, the targets of these motor neurons are muscle groups which are themselves dimorphic. In some cases, it can be demonstrated that these muscle groups become dimorphic as a response to the hormonal environment during development, as well as requiring androgens for size maintenance during adulthood.[121,124,134,171,172]

Developing muscles are widely believed to exert trophic effects on the motor neurons which innervate them[173,174] and contact with target muscles converts a motor neuron from an "elongating neuron" to a "transmitting neuron" with alterations to its phenotype which also allow it to receive increased synaptic input.[175] The assumption is also made that, in the absence of such contacts with target muscles, the developing motor neuron does not become successfully integrated into the neuronal circuitry of the developing spinal cord and dies.

Fishman and Breedlove[176] have shown that while (1) neonatal female rats given subcutaneous testosterone alone have increases in SNB motor neuron numbers as expected, (2) this effect could be reversed in females by local implantation of an anti-androgen into the perineal muscles. Because systemically administered anti-androgen was much less effective, this was interpreted as androgens acting directly to spare the target muscles while sparing SNB neurons indirectly and secondarily through a muscle-nerve trophic effect. However, there remains the possibility that the localized intramuscular anti-androgen could be retrogradely transported along the axons of SNB nerves to affect receptor sites on the cell bodies. A further difficulty is presented by studies in which DLN motor neuron numbers can be increased by treatments that do not also result in the sparing of perineal muscles (e.g., prenatal dihydrotestosterone administration[134]). Moreover, Hauser and Toran-Allerand[177] reported that androgens increased the survival of lumbosacral motor neurons in cultured spinal cord in the *presence* or *absence* of muscle; they commented that a direct neuronotrophic effect of androgens must be considered in parallel to the well-documented role of the target muscle in motor neuron survival.

There are a wide variety of sex differences in neurotransmitter levels in various parts of the adult nervous system.[178] In development, serotonergic and noradrenergic fibers are among the first descending pathways to make contact with α-motor neurons,[179,180] arriving at their destinations well in advance of nonaminergic ones. Serotonergic fibers have a sexually dimorphic pattern of termination in parts of the ventral horn, including the cremasteric nucleus[164] and the L5-S1 region.[181] Serotonin has trophic effects during development,[182-184] and it is thus quite possible that descending serotonergic fibers affect developing motor neurons. In particular, administration of *p*-chlorophenylalanine (*p*-CPA), which blocks serotonin synthesis, results in an increase in SNB neuron numbers in both sexes.[185] Interactively, *p*-CPA appears to increase the duration of the critical period during which androgens can act to maintain SNB and DLN neuron numbers, but treatment is without effect on the nondimorphic VMN located within the same lumbosacral segments.[186] This type of differentiation, where androgens act on motor neuron numbers and target muscles, while there are also transmitter effects on neuron survival, would appear to be a mixture of type B and D mechanisms. However, if it could be demonstrated that transmitter levels within the cord were themselves hormonally controlled it would be a true type D pattern. The observation that sex differences occur in the pattern of serotonin terminals suggests this could be true, but direct experimental studies are lacking at present.

B. INTERACTIONS BETWEEN NEURONS, HORMONES, AND AMINES: THE SDN-MPOA

Because this region does not contain motor neurons, the complication of hormone effects on developing peripheral target tissues does not arise as it does in the SNB/DLN model discussed above. However, there are transient sex differences in hypothalamic 5HT levels during postnatal development which can be influenced by prior hormone manipulation[187-189] as well as a permanent sex difference in adulthood in the density of 5HT-immunoreactive terminals surrounding the SDN-MPOA.[190] In this example, the administration of *p*-CPA can enlarge (i.e., masculinize) the nucleus, at least within carefully defined observation periods.[191] It is also of interest to note that 5HT suppression with *p*-CPA during the second week after birth can alter the capacity to display masculine sexual behavior in adulthood.[189,192,193] If serotonin does have trophic effects (albeit negative ones) on the development of this region, and if serotonin levels are suppressed by androgens, this could well be argued as a pure type D differentiation pattern.

C. THE SIGNIFICANCE OF INTERACTIONS BETWEEN HORMONES AND OTHER NEUROMODULATORS

The importance of type D differentiation patterns is likely to be that they not only ensure the appropriate development of sexually dimorphic regions by mechanisms such as the regulation of cell death, but they also ensure that the spared neurons are in the correct place and making appropriate connections by linking neuron survival to a further factor such as transmitter availability from a wholly different neuron. It is important to note

1. That it is not necessary for cell survival and connectivity to occur at the same time; indeed, there is likely to be a gap between the two phases. That said, these models of sexual dimorphism really

presuppose two stages of cell death involving sexually dimorphic neuron pools: one which is dependent upon the hormonal milieu during the classical "critical period" and the other some time (days?) later which relates to appropriate connectivity.

2. That it is not essential for the aminergic effect to be hormonally controlled, and that Reisert and Pilgrim's[163] suggestions of epigenetic and genetic effects are perfectly tenable. That the pattern of aminergic terminals (or the levels of amines during transient periods of development) do actually appear to be hormonally controlled[187,190,194] is an added refinement. Many other transmitters or peptides may prove to have sexually dimorphic distributions within the CNS; for example, there is a dimorphic pattern of Met-enkephalin fibers within the periventricular region of the MPOA (denser in females) which can be modified by neonatal and adult hormone manipulations.[195] The innervation of rat cremasteric nucleus motor neurons (L1) is also sexually dimorphic, with the male nucleus receiving far more 5HT and substance P terminals than the female nucleus; the former is from descending brainstem pathways, the latter from local intraspinal sources in laminae IV and V.[164,196,197]

It is interesting in the examples of the SDN-MPOA and the SDN/DLN motor neuron groups cited above, that serotonin suppression (as opposed to augmentation) has a trophic effect. Sikich et al.[198] demonstrated that stimulation of 5HT-1A receptors by 8-hydroxy-dipropylamino-tetralin (8-OH-DPAT) decreases the branching of neurites by 70% and reduces total neurite length by 50% in cultures of fetal frontal cortex cells. Similar findings have been reported for terminal outgrowth of raphe neurons in rat after treatment with 5-methoxytryptamine[199] and growth cone responses to serotonin in invertebrates.[200,201]

Other pharmacological treatments which interfere with peptides or transmitters may do so in ways in which a hormonal component could also be inferred. For example, while prenatal exposure to methadone reduces striatal acetylcholine in both sexes of rat neonates, it reduces hindbrain acetylcholine levels in male pups only.[202] Female rats given testosterone propionate as neonates show anovulatory sterility; this effect is enhanced by the simultaneous administration of tropolone, a catechol-*O*-methyltransferase (COMT) inhibitor.[203]

Gonadal hormones may also interact with other types of hormone. For example, Toran-Allerand et al.[204] reported that estrogen could have synergistic interactions with insulin and growth hormone in regulating neurite outgrowth *in vitro*.

D. GLIAL CELLS AND THEIR INVOLVEMENT IN DEVELOPMENT

Very little is known at present about the possible involvement of glial cells in the development of sexual dimorphisms. There are certainly adult sex differences in the distribution of glial fibrillary acid protein (GFAP) in, for example, the supraoptic nucleus[205] and similar differences might also occur during development. Glial cells may not only guide neurons to their definitive position,[206,207] but glial cells and glycoconjugates may form temporary boundaries ("cordones") within the developing nervous system to act either as landmarks or boundaries leading to morphological compartmentalization.[208] The possibility that an early population of glial cells helps the migration of SNB/DLN neurons[176] has already been referred to above, and similar features have been described in relation to adult avian song centers.[209] Glial cells are themselves known to have steroid receptors[210] and Melcangi et al.[211] reported that 5α-reductase activity is high in oligodendrocytes in many areas of the brain during early postnatal development and before the myelination process begins, falling thereafter. Because this enzyme converts testosterone to the relatively inactive 5α-dihydrotestosterone (and thus prevents any possible conversion to estradiol), glial cells could provide a means by which neurons are shielded from the masculinizing effects of estradiol. This could be an important local mechanism in the sexual differentiation process which has yet to be fully recognized.

VII. ASYMMETRIES AND PATHOLOGIES

The possible role of androgens in the development of brain asymmetries has become a focus of attention in recent years. In particular, in both the avian and mammalian forebrain, centers associated with vocal behavior are often asymmetric, the activity associated with them is hormone dependent in adulthood, and the centers themselves are sexually dimorphic and organized by hormones during early development.[141,142,212-215]

Interestingly, this has led to considerations of asymmetries in primates, including man, and a variety of pathologies of brain function associated with them. Sholl and Kim[216] reported asymmetries in androgen receptor densities in male fetal rhesus monkey brains, with androgen receptor density higher in the right frontal lobe than in the left, but lower in the right temporal lobe than in the left; these differences did not occur in female fetuses. It was postulated that this might result in sex differences in development of these regions. In humans, there are not only asymmetries in brain function, notably in language, but also the assumption that these asymmetries are more pronounced in men.[217,218] Indeed, Geschwind and colleagues[219,220] proposed that unusually high androgen levels during development might cause atypical asymmetric growth and suppressed thymus function leading to learning and language dysfunctions (dyslexias, stammering) and immune disorders, particularly in left-handed males. In particular, androgen exposure may result in pathologies of brain development, such as cortical ectopias[221,222] which could, in turn, underlie functional deficits.

VIII. DO SEXUALLY DIMORPHIC REGIONS OF THE CNS RETAIN HORMONE-DEPENDENT RESPONSIVENESS/PLASTICITY IN ADULTHOOD?

This is an important question, since a negative reply would suggest that sexual dimorphisms arise as a result of factors to which the neurons are unresponsive at any time before or after. This is unlikely to be the case. We have already considered that hormones have an *organizational effect* on differentiation of the CNS and its capacities during a perinatal "critical period." Nevertheless, they also act in adulthood to allow behaviors to occur — an *activational effect*. Thus, gonadectomizing adult male or female mammals decreases their ability to display many behaviors such as courtship and copulatory behavior and aggression, but these can then be restored by appropriate hormone treatment.

Activational effects presuppose that the neurobiological substrate underlying sex-typical behavior is being maintained by hormones. This is indeed the case.[223,224] Thus, hormone availability in adulthood controls cell soma size and the dendritic field size of SNB neurons (an activational action), but not the number of neurons (which are determined by hormones during the critical perinatal period — an organizational effect).[62,225,226] In the white-footed mouse *(Peromyscus leucopus)*, Forger and Breedlove[227] reported that the perikarya of SNB cells altered seasonally with day length, presumably through an appropriate hormonal mechanism. One exception may be the spinal nucleus of bulbocavernosus in some mouse strains, where adult castration in adulthood is said to affect the number of SNB cells in C57 strain, the cell size in DBA strain, but both characteristics in their F1 hybrid offspring.[228]

In the adult female rat VMN, there is induction of synapse formation by estradiol over the estrous cycle, but estrogens cannot influence synapse formation in the male VMN.[229,230] Gap junction mRNA expression between SNB and DLN motor neurons is androgen dependent in adulthood, as is β-actin and β-tubulin expression.[231,232] The number of synapses in the hippocampus reaches a peak before proestrus and then declines;[233] antagonists of intracellular progestin receptors (e.g., RU486) block the decrease, implying that hormone effects on synapse elimination act through genomic receptors.[234] A more complex proposal is that hormones may affect spatial memory through action on the hippocampus. Thus, stress leads to CA3 pyramidal cell dendrite atrophy and adrenalectomy results in dentate gyrus granule cell loss (both impair spatial memory), while estrogen administration leads to increases in the spine density of CA1 pyramidal cells and memory enhancement.[98,99,235,236]

Nonspecific changes in volume of regions of the sexually dimorphic MPOA could be achieved by castration of the male rat[237] or by estrogen/progesterone treatment,[238] or by various hormone treatments in adult gerbils.[119] However, whether the effects were due to cell number, size, or dendritic field is unclear.

Similar adult remodeling of synapses by hormones occurs in the arcuate nucleus of the hypothalamus,[239,240] where anti-estrogen antibodies block the loss of axosomatic synapses,[241] and the midbrain central grey.[242]

Gap junctions occur between dendrites of many types of neurons[243] implying electrical coupling. In the rat SNB, castration decreases (and subsequent androgen administration increases) the number and size of these gap junctions,[244] although the same effects could not be demonstrated in the DLN.[245]

Castration increases the proportion of SNB motor neurons which are positive for CGRP from about 50 to 80%; androgen replacement decreases this again.[246] It is unclear why this should occur, but CGRP may have some trophic role since it is involved in nicotinic acetylcholine receptor synthesis.[247]

IX. ARE SEXUALLY DIMORPHIC REGIONS OF THE CNS UNIQUE?

This is an important question because a positive reply suggests that certain regions of the CNS develop according to quite different rules. Several factors must be borne in mind. First, sex differences in the CNS may be far more widespread than is currently appreciated; indeed, it has been suggested that we should consider all regions dimorphic until proven otherwise. Secondly, it is likely that neurons in sexually dimorphic regions use developmental cues which have become especially magnified, i.e., that they represent one end of a spectrum of responsiveness. All striated muscles can bind and respond trophically to hormones such as testosterone,[248] thus any group of α-motor neurons could theoretically depend upon hormone-dependent muscles for their survival. The fact that neurons at L1 (cremasteric) and L5-S1 (SNB/DLN) do so crucially could therefore be an exaggeration of a widespread phenomenon. Indeed, some authors suggest that all α-motor neurons may possess receptors for androgens at some stages of development.[249,250] Hauser and Toran-Allerand[177] found that androgens significantly increased the number of motor neurons in mouse lumbosacral cord cultures; because the neurons had been labeled with ^{3}H thymidine on E10, it was possible to conclude that androgens must have prevented naturally occurring cell loss. However, the authors were at pains to point out that increased survival was found at all cord levels, whether or not these included sexually dimorphic regions.

Yu[251,252] has shown that lesioning nondimorphic motor cranial nerves (e.g., XII) can result in up to 30% death of cell bodies within the brainstem nucleus. However, if androgens are administered from the time of lesioning, cell body loss is halved. Similar data have been obtained and reviewed by Jones.[253] This affords the interesting possibility that studies of sexually dimorphic regions might lead to greater insight into more general principles of plasticity and survival.

X. CONCLUSIONS AND FUTURE DIRECTIONS

The 1960s and early 1970s witnessed a plethora of studies on sex differences in adult behavior, sensory and learning capacities, and neurophysiological controls, and how these differences might depend upon the hormonal milieu during development. That period of study, now largely over, established important principles.

1. Sex differences were not innate (genetically determined), but could be modified by quite simple hormone manipulations, provided these were carried out during a "critical period" of CNS development.
2. The capacity to show male patterns of behavior or other CNS activity depended on the presence of hormones during the critical period and that the feminized brain was the default pattern.
3. Estrogens were as potent as androgens in causing masculinization of the CNS.

Aside from matters such as the timing of the critical period and hormone specificity, these findings demonstrated that the brain and its capacities were little different from other parts of the body which underwent sexual differentiation during development.

From the late 1970s up to the present, researchers have concentrated on demonstrating structural sex differences in the CNS, some of which are remarkable for their simplicity and robustness. These sex differences range from the wholly ultrastructural to differences which can be seen with the naked eye. For most of these dimorphisms, sufficient experimental work has been carried out to demonstrate that they obey the same laws as behavior and other CNS activity; namely that they are not genetically determined but depend on hormones during the same critical period of development. It is important to recognize that the discovery of sex differences in CNS structure has not simply demonstrated a potential underpinning substrate for sex-typical behaviors, it has also legitimized that whole field of research in the eyes of many for whom "behavior" remains a far less tangible biological characteristic. Also, more work is needed to link specific structural sex differences in the CNS to specific sex differences in behavior or other CNS activity. The findings that humans show sex differences in CNS structure (and that differences even occur in individuals of the same sex but different sexual orientation) begs a number of questions, not least on cause and effect.

What should the next decade bring? There are a number of basic biological problems. In some cases they have been addressed in other arenas, but remain cinderella topics in sexual differentiation of the CNS. Where sex dimorphisms depend on differential cell death (and many seem to) what mechanism — necrosis or apoptosis — is involved? We have begun to show that hormones may not act alone to produce dimorphisms, but act with other factors such as neurotransmitters (behaving as trophic agents), but this area remains relatively unexplored. Furthermore, the action of hormones and other factors may not simply be on the neurons themselves; the role of glial cells and target tissues (especially muscle) will remain a fruitful area of investigation.

Finally, there remain several areas where thinking on sexual differentiation of the nervous system has produced "special pleading" which has perhaps hindered research. First, the widespread assumption has been that dimorphisms are established during the critical period and are then impervious to change.

While this may be true in the broad sense, the finding that hormones can act during adulthood to modify features such as cell body size, dendrite length, and synaptic contacts should not be too surprising. The analogy is again with other regions of the body which undergo sexual differentiation (such as the reproductive tract), where adult hormone manipulations can produce quite dramatic changes in organ size and activity. Secondly, the realization that sexually dimorphic regions, while undeniably special, are not unique may have considerable repercussions, for example, in the fields of injury and repair. If neurons in dimorphic regions require androgens/estrogens for survival, then common mechanisms may be operating with neurons whose survival rate after injury can be increased by androgen treatment; in particular, other factors which interact with hormones in the establishment of dimorphic regions should be examined in injury. Because differentiation is a ubiquitous developmental process by which normal structure is arrived at, the tendency has been to imagine abnormalities during sexual differentiation of the nervous system as being of the "too little" or "too much" variety (for example, under- or overmasculinization of behavior). The possible link between androgen levels during development and various pathological conditions involving cortical ectopias, language difficulties, and immune deficiencies suggests a complex level of interaction which clearly warrants examination.

REFERENCES

1. Harris, G.W. and Levine, S., Sexual differentiation of the brain and its experimental control, *J. Physiol. (London),* 181, 379–400, 1965.
2. Whalen, R.E. and Edwards, D.A., Hormonal determinants of the development of masculine and feminine behavior in male and female rats, *Anat. Rec.,* 157, 173–180, 1967.
3. Goy, R.W. and McEwen, B.S., *Sexual Differentiation of the Brain,* Cambridge, MIT Press, 1980.
4. MacLusky, N.J. and Naftolin, F., Sexual differentiation of the central nervous system, *Science,* 211, 1294–1303, 1981.
5. Arnold, A.P. and Breedlove, S.M., Organizational and activational effects of sex steroids on brain and behavior: a reanalysis, *Horm. Behav.,* 19, 469–498, 1985.
6. Baum, M.J., Frank Beach's research on the sexual differentiation of behavior and his struggle with the "organizational" hypothesis, *Neurosci. Biobehav. Rev.,* 14, 201–206, 1990.
7. Clarke, I.J., Scaramuzzi, R.J., and Short, R.V., Effects of testosterone implants in pregnant ewes on their female offspring, *J. Embryol. Exp. Morphol.,* 36, 87–99, 1976.
8. Imperato-McGinley, J., Binienda, Z., Arthur, A., Mininberg, D.T., Vaughan, E.D., and Quimby, F.W., The development of a male pseudohermaphrodite rat using an inhibitor of the enzyme 5α-reductase, *Endocrinology,* 116, 807–812, 1985.
9. Neumann, F., Steinbeck, H., and Hahn, J.D., Hormones and brain differentiation, in *The Hypothalamus,* Martini, L., Motta, M., and Fraschini, F., Eds., Academic Press, New York, 1970, 1–35.
10. Wilson, J.D., George, F.W., and Griffin, J.E., The hormonal control of sexual development, *Science,* 211, 1278–1284, 1981.
11. Wilson, J.D., Sexual differentiation of the gonads and of the reproductive tract, *Biol. Neonate,* 55, 322–330, 1989.
12. De Moor, P., Verhoeven, G., and Heyns, W., Permanent effects of foetal and neonatal testosterone secretion on steroid metabolism and binding, *Differentiation,* 1, 241–253, 1973.
13. Corbier, P., Kerdelhue, B., Picon, R., and Roffi, J., Changes in testicular weight and serum gonadotropin and testosterone levels before, during and after birth in the perinatal rat, *Endocrinology,* 103, 1985–1991, 1978.
14. Resko, J.A., Gonadal hormones during sexual differentiation in vertebrates, in *Handbook of Behavioral Neurobiology,* Vol. 7, Adler, N., Pfaff, D., and Goy, R.W., Eds., Plenum Press, New York, 1985, 21–42.
15. Gondos, B., Testicular development, in *Fetal and Neonatal Physiology,* Polin, R.A. and Fox, W.M., Eds., W.B. Saunders, Philadelphia, 1992, 1864–1870.
16. Turkelson, C.M., Dunlap, J.L., MacPhee, A.A., and Gerall, A.A., Assay of perinatal testosterone and influence of anti-progesterone and theophylline on induction of sterility, *Life Sci.,* 21, 1149–1158, 1978.
17. Pang, S.F., Caggiula, A.R., Gay, V.L., Goodman, R.L., and Pang, C.S.F., Serum concentrations of testosterone, estrogens, luteinizing hormone and follicle-stimulating hormone in male and female rats during the critical period of neural sexual differentiation, *J. Endocrinol.,* 80, 103–110, 1979.
18. Slob, A.K., Ooms, M.P., and Vreeburg, J.T.M., Prenatal and early postnatal sex differences in plasma and gonadal testosterone and plasma luteinizing hormone in female and male rats, *J. Endocrinol.,* 87, 81–87, 1980.
19. Weisz, J. and Ward, I.L., Plasma testosterone and progesterone titers of pregnant rats, their male and female fetuses, and neonatal offspring, *Endocrinology,* 106, 306–316, 1980.
20. Roffi, J., Chami, F., Corbier, P., and Edwards, D.A., Testicular hormones during the first few hours after birth augment the tendency of adult male rats to mount receptive females, *Physiol. Behav.,* 39, 625–628, 1987.
21. Mizuno, M., Lobotsky, J., Lloyd, C.W., Kobayashi, T., and Murasawa, Y., Plasma androstenedione and testosterone during pregnancy and in the newborn, *J. Clin. Endocrinol. Metab.,* 28, 1133–1142, 1968.
22. Forest, M.G., Sizonenko, P.C., Cathiard, A.M., and Bertrand, J., Hypohyso-gonadal function in humans during the first year of life. I. Evidence for testicular activity in early infancy, *J. Clin. Invest.,* 53, 819–828, 1974.

168

23. Winter, J.S.D., Faiman, C., Hobson, W.C., and Reyes, F.I., The endocrine basis of sexual development in the chimpanzee, *J. Reprod. Fertil.,* Suppl. 28, 131–138, 1980.
24. Plant, T.M., A striking diurnal variation in plasma testosterone in infant male rhesus monkeys *(Macaca mulatta), Neuroendocrinology,* 35, 370–373, 1982.
25. Dixson, A.F., Plasma testosterone concentrations during postnatal development in the male common marmoset, *Folia Primatol.,* 47, 166–170, 1986.
26. Swaab, D.F. and Hoffman, M.A., Sexual differentiation of the human hypothalamus: ontogeny of the sexually dimorphic nucleus of the preoptic area, *Dev. Brain Res.,* 44, 314–318, 1988.
27. Döhler, K.-D., Coquelin, A., Davis, F., Hines, M., Shryne, J.E., and Gorski, R.A., Pre- and postnatal influence of testosterone propionate and diethylstilbestrol on differentiation of the sexually dimorphic nucleus of the preoptic area in male and female rats, *Brain Res.,* 302, 291–295, 1984.
28. Tobet, S.A., Zahniser, D.J., and Baum, M.J., Differentiation in male ferrets of a sexually dimorphic nucleus of the preoptic/anterior hypothalamic area requires prenatal estrogen, *Neuroendocrinology,* 44, 299–308, 1986.
29. Christensen, L.W. and Gorski, R.A., Independent masculinization of neuroendocrine systems by intracerebral implants of testosterone or estradiol in the neonatal female rat, *Brain Res.,* 146, 325–340, 1978.
30. Beyer, C., Wozniak, A., and Hutchison, J.B., Sex-specific aromatization of testosterone in mouse hypothalamic neurons, *Neuroendocrinology,* 58, 673–681, 1993.
31. Beyer, C., Green, S.J., Barker, P.J., and Hutchison, J.B., Aromatase-immunoreactivity is localized specifically in neurones in the developing mouse hypothalamus and cortex, *Brain Res.,* 638, 203–210, 1994a.
32. Beyer, C., Green, S.J., and Hutchison, J.B., Androgens influence sexual differentiation of embryonic mouse hypo-thalamus aromatase neurons *in vitro, Endocrinology,* 135, 1220–1226, 1994b.
33. Goy, R.W., Bercovitch, F.B., and McBrair, M.C., Behavioral masculinization is independent of genital masculinization in prenatally androgenised female rhesus macaques, *Horm. Behav.,* 22, 552–571, 1988.
34. Steimer, T. and Hutchinson, J.B., Androgen increases formation of behaviorally effective estrogen in dove brain, *Nature,* 292, 345–347, 1981.
35. Roselli, C.E., Ellinwood, W.E., and Resko, J.A., Regulation of brain aromatase activity in rats, *Endocrinology,* 114, 192–200, 1984.
36. Sholl, S.A. and Kim, K.L., Aromatase, 5–alpha-reductase, and androgen receptor levels in the fetal monkey brain during early development, *Neuroendocrinology,* 52, 94–98, 1990a.
37. Compaan, J.J., van Wattum, G., de Reiter, A.J., van Oortmessen, G.A., Vrosthaas, J.M., and Bohus, B., Genetic differences in female house mice in aggressive response to sex steroid hormone treatment, *Physiol. Behav.,* 54, 899–902, 1993.
38. Bronson, F.H. and Desjardins, C., Neonatal androgen administration and adult aggressiveness in female mice, *Gen. Comp. Endocrinol.,* 15, 320–325, 1978.
39. Payne, A.P., Neonatal androgen administration and aggression in the female golden hamster during interactions with males, *J. Endocrinol.,* 63, 497–506, 1974.
40. Payne, A.P., and Swanson, H.H., Neonatal androgenization and aggression in the male golden hamster, *Nature (London),* 239, 282–283, 1972.
41. Lording, D.W. and de Kretser, D.M., Comparative ultrastructural histochemical studies of the interstitial cells of the rat testis during fetal and postnatal development, *J. Reprod. Fertil.,* 29, 261–269, 1972.
42. Mendis-Handagama, S.M.L.C., Risbridger, G.P., and de Kretser, D.M., Morphometric analysis of the components of the neonatal and the adult rat testis interstitium, *Int. J. Androl.,* 10, 525–534, 1987.
43. Siler-Khodr, T.M., Endocrine and paracrine function of the human placenta, in *Fetal and Neonatal Physiology,* Polin, R.A. and Fox, W.M., Eds., W.B. Saunders, Philadelphia, 1992, 74–85.
44. Kaplan, S.L. and Grumbach, M.M., The ontogenesis of human foetal hormones. II. Luteinizing hormone (LH) and follicle stimulating hormone (FSH), *Acta Endocrinol.,* 81, 808–829, 1976.
45. Cuttler, L., LH and FSH secretion in the fetus and newborn, in *Fetal and Neonatal Physiology,* Polin, R.A. and Fox, W.M., Eds., W.B. Saunders, Philadelphia, 1992, 1797–1806.
46. Gilmore, D.P., Neurosecretions and neuro-transmitters of the fetal hypothalamus, in *Fetal and Neonatal Physiology,* Polin, R.A. and Fox, W.M., Eds., W.B. Saunders, Philadelphia, 1992, 1779–1785.
47. Mann, D.R., Gould, K.G., Collins, D.C., and Wallen, K., Blockade of neonatal activation of the pituitary-testicular axis: effect on peripubertal luteinizing hormone and testosterone secretion and on testicular development in male monkeys, *J. Clin. Endocrinol. Metab.,* 68, 600–607, 1989.
48. Gluckman, P.D., Maturation of hypothalamic-pituitary function in the ovine fetus and neonate, in *The Fetus and Independent Life,* Ciba Foundation Symposium 86, 1991, 5–42.
49. Gluckman, P.D., The onset and organization of hypothalamic control in the fetus, in *The Physiological Development of the Fetus and Newborn,* Jones, C.T. and Nathanielsz, P.W., Eds., Academic Press, London, 1985, 103–111.
50. Pfaff, D.W. and Zigmond, R.E., Neonatal androgen effects on sexual and non-sexual behavior of adult rats tested under various hormone regimes, *Neuroendocrinology,* 7, 129–145, 1971.
51. Södersten, P., Sexual differentiation: do males differ from females in behavioral sensitivity to gonadal hormones?, *Prog. Brain Res.,* 61, 257–270, 1984.
52. Dyer, R.G., Sexual differentiation of the forebrain-relationship to gonadotrophin secretion, *Prog. Brain Res.,* 61, 223–236, 1984.

53. Payne, A.P., Neonatal androgen administration and sexual behavior: behavioral responses and hormonal responsiveness of female golden hamster, *Anim. Behav.,* 27, 242–250, 1979.

54. Vreeburg, J.T.M., van der Vaart, P.D.M., and van der Schoot, P., Prevention of central defeminization but not masculinization in male rats by inhibition neonatally of estrogen biosynthesis, *J. Endocrinol.,* 74, 375–382, 1977.

55. Perakis, A. and Stylianopoulou, F., Effects of a prenatal androgen peak on rat brain sexual differentiation, *J. Endocrinol.,* 108, 281–285, 1986.

56. Money, J, and Ehrhardt, A.A., *Man and Woman, Boy and Girl,* Johns Hopkins Press, Baltimore, MD, 1972.

57. Reinisch, J.M. and Sanders, S.A., Prenatal gonadal steroidal influences on gender-related behavior, *Prog. Brain Res.,* 61, 407–416, 1984.

58. Griffin, J.E. and Wilson, J.D., The androgen resistance syndromes: 5α-reductase deficiency, testicular feminization, and related syndromes, in *The Metabolic Basis of Inherited Disease,* 6th ed, Scriver, C.R., Beaudet, A.L., Sly, W.S., and Valle, D., Eds., McGraw-Hill, New York, 1989, 1919–1944.

59. Naess, O., Haug, E., Ahramandel, A., Aakvaag, A., Hansson, V., and French, F., Androgen receptors in the anterior pituitary and central nervous system of the androgen "insensitive" (Tfm) rat: correlation between receptor binding and effects of androgens on gonadotropin secretion, *Endocrinology,* 99, 1295–1303, 1976.

60. Max, S.R., Cytosolic androgen receptor in skeletal muscle from normal and testicular feminization mutant (Tfm) rats, *Biochem. Biophys. Res. Commun.,* 101, 792–799, 1981.

61. Wilson, J.D., Sexual differentiation, *Annu. Rev. Physiol.,* 40, 279–306, 1978.

62. Breedlove, S.M. and Arnold, A.P., Sexually dimorphic motor nucleus in the rat spinal cord: response to adult hormone manipulation, absence in androgen insensitive rats, *Brain Res.,* 255, 297–307, 1981a.

63. Sengelaub, D.R. and Arnold, A.P., Development and loss of early projections in a sexually dimorphic rat spinal nucleus, *J. Neurosci.,* 6, 1613–1620, 1986.

64. Ward, I.L., Effects of maternal stress on the sexual behavior of male offspring, *Monogr. Neural Sci.,* 9, 169–175, 1983.

65. Ward, O.B., Orth, J.M., and Weisz, J.A., A possible role of opiates in modifying sexual differentiation, *Monogr. Neural Sci.,* 9, 194–200, 1983.

66. McEwen, B.A., Gonadal steroids and brain development, *Biol. Reprod.,* 22, 43–48, 1980.

67. Rhees, R.W. and Fleming, D.E., Effects of malnutrition, maternal stress, or ACTH injections during pregnancy on sexual behavior of male offspring, *Physiol. Behav.,* 27, 879–882, 1980.

68. Archer, J.E. and Blackman, D.E., Prenatal psychological stress and offspring behavior in rats and mice, *Dev. Psychobiol.,* 4, 193–248, 1971.

69. Anderson, D.K., Rhees, R.W., and Fleming, D.E., Effects of prenatal stress on differentiation of the sexually dimorphic nucleus of the preoptic area (SDN-POA) of the rat brain, *Brain Res.,* 332, 113–118, 1985.

70. Anderson, D.K., Fleming, D.E., Rhees, R.W., and Kinghorn, E., Relationships between sexual activity, plasma testosterone, and the volume of the sexually dimorphic nucleus of the preoptic area in prenatally stressed and non-stressed rats, *Brain Res.,* 370, 1–10, 1986.

71. Karaviti, L., Schoonmaker, J., Shryne, J.E., and Gorski, R.A., Corticoids and sexual differentiation of brain structure, *Physiol. Abstr.,* 25, 315, 1982.

72. Scallet, A.C., Effects of conditioned fear and environmental novelty on plasma β-endorphin in the rat, *Peptides,* 3, 203–206, 1982.

73. Young, E.A., Lewis, J., and Akil, H., The preferential release of beta-endorphin from the anterior pituitary lobe by corticotropin releasing factor (CRF), *Peptides,* 7, 603–607, 1986.

74. Johnston, H.M., Payne, A.P., and Gilmore, D.P., Perinatal exposure to morphine affects adult sexual behavior of the male golden hamster, *Pharmacol. Biochem. Behav.,* 42, 41–44, 1992.

75. Hunter, R.H.F., *Sex Determination, Differentiation and Intersexuality in Placental Mammals,* Cambridge University Press, 1995.

76. vom Saal, F.S. and Bronson, F.H., The variation in length of the estrous cycle in mice due to former intrauterine proximity to male fetuses, *Biol. Reprod.,* 22, 777–780, 1980.

77. vom Saal, F.S., The intrauterine position phenomenon in mice, in *The Biology of Aggression,* Brain, P.F. and Benton, D., aan den Rijn, Alphen, Eds., and Sijthof & Noordhoff, 1981, 231–236.

78. Bleier, R., Byne, W., and Siggelkow, I., Cytoarchitectonic sexual dimorphisms of the medial preoptic and anterior hypothalamic areas in guinea pig, rat, hamster, and mouse, *J. Comp. Neurol.,* 212, 118–130, 1982.

79. Raisman, G. and Field, P.M., Sexual dimorphism in the neuropil of the preoptic area of the rat and its dependence on neonatal androgen, *Brain Res.,* 54, 1–29, 1973.

80. Gorski, R.A., Gordon, J.H., Shryne, J.E., and Southam, A.M., Evidence for a morphological sex difference within the medial preoptic area of the rat brain, *Brain Res.,* 148, 333–346, 1978.

81. Gorski, R.A., Harlan, R.E., Jacobson, C.D., Shryne, J.E., and Southam, A.M., Evidence for the existence of a sexually dimorphic nucleus in the preoptic area of the rat, *J. Comp. Neurol.,* 193, 529–539, 1980.

82. Swaab, D.F. and Fliers, E., A sexually dimorphic nucleus in the human brain, *Science,* 228, 1112–1115, 1985.

83. Hofman, M.A., Goudsmit, E., Purba, J.S., and Swaab, D.F., Morphometric analysis of the supraoptic nucleus in the human brain, *J. Anat.,* 172, 259–270, 1990.

84. Le Blond, C.B., Morris, S., Karakiulakis, G. Powell, R., and Thomas, P.J., Development of sexual dimorphism in the suprachiasmatic nucleus of the rat, *J. Endocrinol.,* 95, 137–145, 1982.

85. Pérez, J., Naftolin, F., and Garciá-Segura, L.M., Sexual differentiation of synaptic connectivity and neuronal plasma membrane in the arcuate nucleus of the rat hypothalamus, *Brain Res.*, 527, 116–122, 1990.
86. Matsumoto, A. and Arai, Y., Male-female differences in synaptic organization of the ventromedial nucleus of the hypothalamus of the rat, *Neuroendocrinology*, 42, 232–236, 1986.
87. Nishizuka, M. and Arai, Y., Sexual dimorphisms in synaptic organization in the amygdala and its dependence on neonatal hormone environment, *Brain Res.*, 212, 31–38, 1981.
88. Mitzukami, S., Nishizuka, M., and Arai, Y., Sexual difference in nuclear volume and its ontogeny in the rat amygdala, *Exp. Neurol.*, 79, 569–581, 1983.
89. Allen, L.S. and Gorski, R.A., Sex differences in the bed nucleus of the stria terminalis of the human brain, *J. Comp. Neurol.*, 302, 697–706, 1990.
90. Segovia, S. and Guillamón, A., Effects of sex steroids on the development of the vomeronasal organ in the rat, *Dev. Brain Res.*, 5, 209–212, 1982.
91. Nagy, J.I. and Senba, E., Neural relations of cremaster motoneurons, spinal cord systems and the genitofemoral nerve in the rat, *Brain Res. Bull.*, 15, 609–627, 1985.
92. Segovia, S., Orensanz, L.M., Valencia, A., and Guillamón, A., Effects of sex steroids on the development of the accessory olfactory bulb in the rat: a volumetric study, *Dev. Brain Res.*, 16, 312–314, 1984.
93. Breedlove, S.M. and Arnold, A.P., Hormone accumulation in a sexually dimorphic motor nucleus of the rat spinal cord, *Science*, 210, 564–566, 1980.
94. Jordan, C.L., Breedlove, S.M., and Arnold, A.P., Sexual dimorphism and the influence of neonatal androgen in the dorsolateral motor nucleus of the rat lumbar spinal cord, *Brain Res.*, 259, 309–314, 1982.
95. Greenwood, D., Coggleshall, R.E., and Hulsebosch, C.E., Sexual dimorphism in numbers of neurons in the pelvic ganglia of adult rats, *Brain Res.*, 340, 160–162, 1985.
96. Melvin, J.E., McNeill. T.H., Hervonen, A., and Hamill, R.W., Organizational role of testosterone on the biochemical and morphological development of the hypogastric ganglion, *Brain Res.*, 485, 1–10, 1989.
97. Muñoz-Cueto, J.A., García-Segura, L.M., and Ruiz-Marcos, A., Developmental sex differences and effect of ovariectomy on the number of cortical pyramidal cell dendritic spines, *Brain Res.*, 519, 64–68, 1990.
98. Gould, E., Westlind-Danielsson, A., Frankfurt, M., and McEwen, B.S., Sex differences and thyroid hormone sensitivity of hippocampal pyramidal neurons, *J. Neurosci.*, 10, 996–1003, 1990.
99. Gould, E., Woolley, C.S., Frankfurt, M., and McEwen, B.S., Gonadal steroids regulate dendritic spine density in hippocampal pyramidal cells in adulthood, *J. Neurosci.*, 10, 1286–1291, 1990.
100. Juraska, J., Sex differences in "cognitive" regions of the rat brain, *Psychoneuroendocrinology*, 16, 1–3, 1991.
101. Roof, R., The dentate gyrus is sexually dimorphic in prepubescent rats: testosterone plays a significant role, *Brain Res.*, 610, 148–157, 1993.
102. Guillamón, A., de Blas, M.R., and Segovia, S., Effects of sex steroids on the development of the locus coeruleus in the rat, *Dev. Brain Res.*, 40, 306–310, 1988.
103. Luque, J.M., de Blas, M.R., Segovia, S., and Guillamón, A., Sexual dimorphisms of the dopamine-beta-hydroxylase-immunoreactive neurons in the rat locus coeruleus, *Dev. Brain Res.*, 67, 211–215, 1992.
104. de Lacoste-Utamsing, C. and Holloway, R.L., Sexual dimorphisms in the human corpus callosum, *Science*, 216, 1431–1432, 1986.
105. Yoshi, I, Barker, W., Apicella, A., Chang, J., Sheldon, J., and Duara, R., Measurements of the corpus callosum (cc) on magnetic resonance (mr) scans: effects of age, sex, handedness and disease, *Neurology*, 36, Suppl. 1, 133, 1986.
106. Going, J.J. and Dixson, A., Morphometry of the adult human corpus callosum: lack of sexual dimorphism, *J. Anat.*, 174, 163–167, 1990.
107. Allen, L.S. and Gorski, R.A., Sexual dimorphism of the anterior commissure and massa intermedia of the human brain, *J. Comp. Neurol.*, 312, 97–104, 1991.
108. Wright, L.L. and Smolen, A.J., Effects of neonatal castration or treatment with dihydrotestosterone on numbers of neurons in the rat superior cervical sympathetic ganglion, *Dev. Brain Res.*, 20, 314–316, 1985.
109. Beaston-Wimmer, P. and Smolen, A.J., Gender differences in neurotransmitter expression in the rat superior cervical ganglion, *Dev. Brain Res.*, 58, 123–128, 1991.
110. LeVay, S.A., A difference in hypothalamic structure between heterosexual and homosexual men, *Science*, 253, 1034–1037, 1991.
111. Gorski, R.A., Sexual differentiation of the brain: mechanisms and implications for neuroscience, in *From Message to Mind*, Easter, S.S., Barald, K.F., and Carlson, B.M., Eds., Sinauer Assoc., Sunderland, MA, 1988, 256–271.
112. Rhees, R.W., Shryne, J.E., and Gorski, R.A., Termination of the hormone-sensitive period for differentiation of the sexually dimorphic nucleus of the preoptic area in male and female rats, *Dev. Brain Res.*, 52, 17–23, 1990.
113. Everett, J.W., Central neural control of reproductive functions of the adenohypophysis, *Physiol. Rev.*, 44, 373–431, 1964.
114. Hart, B.L., Medial preoptic-anterior hypothalamic area and the socio-sexual behavior of male dogs: a comparative neuropsychological analysis, *J. Comp. Physiol. Psychol.*, 86, 328–349, 1974.
115. Van de Poll, N.E. and Van Dis, H., The effect of medial preoptic-anterior hypothalamic lesions on bisexual behavior of the male rat, *Brain Res. Bull.*, 4, 505–511, 1979.

116. Bermond, B., Effects of medial preoptic hypothalamus anterior lesions on three kinds of behavior in the rat: intermale aggressive, male-sexual and mouse-killing behavior, *Aggress. Behav.,* 8, 335–354, 1982.

117. Preslock, J.P. and McCann, S.M., Lesions of the sexually dimorphic nucleus of the preoptic area: effects upon LH, FSH and prolactin in rats, *Brain Res. Bull.,* 18, 127–134, 1987.

118. Arendash, G.W. and Gorski, R.A., Effects of discrete lesions of the sexually dimorphic nucleus of the preoptic area or other medial preoptic regions on the sexual behavior of male rats, *Brain Res. Bull.,* 10, 147–154, 1983.

119. Commins, D. and Yahr, P., Adult testosterone levels influence the morphology of a sexually dimorphic area in the mongolian gerbil brain, *J. Comp. Neurol.,* 224, 132–140, 1984.

120. DeJonge, F.H., Swaab, D.F., Ooms, M.P., Endert, E., and Van de Poll, N.E., Developmental and functional aspects of the human and rat sexually dimorphic nucleus of the preoptic area, in *Hormones, Brain and Behaviour in Vertebrates. 1. Sexual Differentiation, Neuroanatomical Aspects, Neurotransmitters and Neuropeptides,* Balthazart, J., Ed., Karger, Basel, 1990, 121–136.

121. Tobin, A.M. and Payne, A.P., Perinatal androgen administration and the maintenance of sexually dimorphic and non-dimorphic lumbosacral motor neurone groups in female Albino Swiss rats, *J. Anat.,* 177, 47–53, 1991.

122. Grisham. W., Casto, J.M., Kashon, M.L., Ward, I.L., and Ward, O.B., Prenatal flutamide alters sexually dimorphic nuclei in the spinal cord of male rats, *Brain Res.,* 578, 69–74, 1992.

123. Forger, N.G. and Breedlove, S.M., Sexual dimorphism in human and canine spinal cord: role of early androgen, *Proc. Natl. Acad. Sci. U.S.A.,* 83, 7527–7531, 1986.

124. Cihák, R., Gutmann, E., and Hanzlíková, V., Involution and hormone-induced persistence of the M. sphincter (levator) ani in female rats, *J. Anat.,* 106, 93–110, 1970.

125. Hart, B.L., Testosterone regulation of penile reflexes in male rats, *Science,* 155, 1283–1284, 1967.

126. Hart, B.L. and Melese-d'Hospital, P.Y., Penile mechanisms and the role of the striated penile muscles in penile reflexes, *Physiol. Behav.,* 31, 807–813, 1983.

127. Sachs, B.C., Role of penile muscles in penile reflexes, copulation and induction of pregnancy in the rat, *J. Reprod. Fertil.,* 66, 433–443, 1982.

128. Breedlove, S.M. and Arnold, A.P., Hormonal control of a developing neuromuscular system. II. Sensitive periods for the androgen induced masculinization of the rat spinal nucleus of the bulbocavernosus, *J. Neurosci.,* 3, 424–432, 1981b.

129. Breedlove, S.M., Steroid influences on the development and function of a neuromuscular system, *Prog. Brain Res.,* 61, 147–170, 1984.

130. Schrøder, H.D., Organization of the motoneurons innervating the pelvic muscles of the male rat, *J. Comp. Neurol.,* 192, 567–587, 1980.

131. McKenna, K.E. and Nadelhaft, I., The organization of the pudendal nerve in the male and female rat, *J. Comp. Neurol.,* 248, 532–549, 1986.

132. Ueyama, T., Arakawa, H., and Mizuno, N., Central distribution of efferent and afferent components of the pudendal nerve in the rat, *Anat. Embryol.,* 177, 37–49, 1987.

133. Wagner, C.K. and Clemens, L.G., Anatomical organization of the sexually dimorphic perineal muscular system in the house mouse, *Brain Res.,* 499, 93–100, 1989.

134. Davidson, T., Tobin, A.M., and Payne, A.P., Maintenance of the spinal nucleus of bulbocavernosus and perineal muscles in female Albino Swiss rats treated with perinatal dihydrotestosterone, *J. Reprod. Fertil.,* 90, 619–623, 1990.

135. Arnold, A.P. and Gorski, R.A., Gonadal steroid induction of structural sex differences in the central nervous system, *Annu. Rev. Neurosci.,* 7, 413–442, 1984.

136. Sengelaub, D.R., Cell generation, migration, death and growth in neural systems mediating social behavior, in *Advances in Comparative and Environmental Physiology. 3. Molecular and Cellular Basis of Social Behavior in Vertebrates,* Balthazart, J., Ed., Springer-Verlag, Berlin, 1989, 239–267.

137. Breedlove, S.M., Jordan, C.L., and Arnold, A.P., Neurogenesis of motoneurons in the sexually dimorphic spinal nucleus of the bulbocavernosus in rats, *Dev. Brain Res.,* 9, 39–43, 1983.

138. Sengelaub, D.R. and Arnold, A.P., II. Development of the spinal nucleus of the bulbocavernosus in androgen-insensitive *(Tfm)* rats, *J. Comp. Neurol.,* 280, 630–636, 1989b.

139. Gorlick, D.L. and Kelley, D.B., The ontogeny of androgen receptors in the CNS of *Xenopus laevis, Dev. Brain Res.,* 26, 193–201, 1986.

140. Gorlick, D.L. and Kelley, D.B., Neurogenesis in the vocalization pathway of *Xenopus laevis, J. Comp. Neurol.,* 257, 614–627, 1987.

141. De Voogd, T.J., The avian song system: relating sex differences in behavior to dimorphism in the central nervous system, *Prog. Brain Res.,* 61, 171–184, 1984.

142. Nordeen, E.J. and Nordeen, K.W., Hormones and the structuring of neural pathways controlling avian song, in *Hormones, Brain and Behavior in Vertebrates. 1. Sexual Differentiation. Neuroanatomical Aspects, Neurotransmitters and Neuropeptides, Comp. Physiol.,* Vol. 8, Karger, Basel, 1990, 92–103.

143. Bottjer, S.W., Gleassner, S.L., and Arnold, A.P., Ontogeny of brain nuclei controlling song learning and behavior in zebra finches, *J. Neurosci.,* 5, 1556–1562, 1985.

144. Bottjer, S.W., Miesner, E.A., and Arnold, A.P., Changes in neuronal number, density and size account for increases in volume of song-control nuclei during song development in zebra finches, *Neurosci. Lett.,* 67, 263–268, 1986.

145. Nordeen K.W. and Nordeen, E.J., Sex differences in the songbird brain involve neurons born during adolescence, *Soc. Neurosci. Abstr.,* 12, 1214, 1986.

146. Goldman, S.A. and Nottebohm, F., Neuronal production, migration and differentiation in a vocal control nucleus of the adult female canary brain, *Proc. Natl. Acad. Sci. U.S.A.,* 60, 2390–2394, 1983.

147. Paton, J.A. and Nottebohm, F., Neurons generated in the adult brain are recruited into functional circuits, *Science,* 225, 1046–1048, 1984.

148. Jacobson, C.D. and Gorski, R.A., Neurogenesis of the sexually dimorphic nucleus of the medial preoptic area in the rat, *J. Comp. Neurol.,* 196, 519–529, 1981.

149. Jacobson, C.D., Davis, F.C., and Gorski, R.A., Formation of the sexually dimorphic nucleus of the preoptic area: neuronal growth, migration and changes in cell number, *Dev. Brain Res.,* 21, 7–18, 1985.

150. Fishman, R.B. and Breedlove, S.M., Radial glia in the fetal rat spinal cord may guide the migration of sexually dimorphic motoneurons, *Soc. Neurosci. Abstr.,* 12, 1219, 1986.

151. Nordeen E.J., Nordeen, K.W., Sengelaub, D.R., and Arnold, A.P., Androgens prevent normally occurring cell death in a sexually dimorphic spinal nucleus, *Science,* 229, 671–673, 1985.

152. Sengelaub, D.R. and Arnold, A.P., Hormonal control of neuron number in the sexually dimorphic spinal nuclei of the rat. I. Testosterone-regulated death in the dorsolateral nucleus, *J. Comp. Neurol.,* 280, 622–629, 1989a.

153. Lee, J.H., Jordan, C.L., and Arnold, A.P., Critical period for androgenic regulation of soma size of sexually dimorphic motoneurons in rat lumbar spinal cord, *Neurosci. Lett.,* 98, 79–84, 1989.

154. Goldstein, L.A., Kurz, E.M., and Sengelaub, D.R., Androgen regulation of dendritic growth and retraction in the development of a sexually dimorphic spinal nucleus, *J. Neurosci.,* 10, 935–946, 1990.

155. Raisman, G. and Field, P.M., Sexual dimorphism in the preoptic area of the rat, *Science,* 173, 731–734, 1971.

156. Pozzo Miller, L.D. and Aoki, A., Stereological analysis of the hypothalamic ventromedial nucleus. II. Hormone-induced changes in the synaptogenic pattern, *Dev. Brain Res.,* 61, 189–196, 1991.

157. Jordan, C.L., Letinsky, M.S., and Arnold, A.P., The role of gonadal hormones in neuromuscular synapse elimination in rats. I. Androgen delays the loss of multiple innervation in the levator ani muscle, *J. Neurosci.,* 9, 229–238, 1989a.

158. Jordan, C.L., Letinsky, M.S., and Arnold, A.P., II. Multiple innervation persists in the adult levator ani muscle after juvenile androgen treatment, *J. Neurosci.,* 9, 239–247, 1989b.

159. Riley, D.A., Spontaneous elimination of nerve terminals from the endplates of developing skeletal myofibers, *Brain Res.,* 134, 279–285, 1977.

160. Thompson, W.J., Changes in the innervation of mammalian skeletal muscle fibers during postnatal development, *TINS,* 9, 25–28, 1986.

161. Pfaff, D.W., *Estrogen and Brain Function,* Springer-Verlag, New York, 1990.

162. Ayoub, D.M. and Greenough, W.T., and Juraska, J.M., Sex differences in dendritic structure in the preoptic area of the juvenile macaque monkey brain, *Science,* 219, 197–198, 1983.

163. Reisert I. and Pilgrim, C., Sexual differentiation of monoaminergic neurons — genetic or epigenetic?, *TINS,* 14, 468–473, 1991.

164. Newton, B.W. and Hamill, R.W., Target regulation of the serotonin and substance P innervation of the sexually dimorphic cremaster nucleus, *Brain Res.,* 485, 149–156, 1989.

165. Tobias, M. and Kelley, D.B., Physiological characterization of the sexually dimorphic larynx in *Xenopus laevis, Soc. Neurosci. Abstr.,* 11, 496, 1985.

166. Kelley, D.B., Neuroeffectors for vocalization in *Xenopus laevis:* hormonal regulation of sexual dimorphism, *J. Neurobiol.,* 17, 231–248, 1986.

167. Schmidt, R., Neural correlates of frog calling. Masculinization by androgens, *Horm. Behav.,* 17, 94–102, 1983.

168. Simpson, H., Tobias, M., and Kelley, D.B., Origin and identification of fibers in the cranial nerve IX-X complex of *Xenopus laevis:* Lucifer yellow backfills *in vitro, J. Comp. Neurol.,* 244, 430–444, 1986.

169. Wetzel, D., Kelley, D.B., and Hearter, U., A proposed efferent pathway for mate calling in South African clawed frogs *Xenopus laevis, J. Comp. Physiol.,* 157, 749–761, 1985.

170. Sassoon, D and Kelley, D.B., The sexually dimorphic larynx of *Xenopus laevis:* development and androgen regulation, *Am. J. Anat.,* 177, 457–472, 1986.

171. Wainman, P. and Shipounoff, G.C., The effects of castration and testosterone propionate on the striated perineal musculature of the rat, *Endocrinology,* 29, 975–978, 1941.

172. Venable, J.H., Constant cell populations in normal, testosterone-deprived and testosterone-stimulated levator ani muscles, *Am. J. Anat.,* 119, 263–270, 1966.

173. McManaman, J.L., Oppenheim, R.W., Prevette, D., and Marchetti, D., Rescue of motoneurons from cell death by a purified skeletal muscle polypeptide: effects of the ChAT development factor, CDF, *Neuron,* 4, 891–898, 1990.

174. Oppenheim, R.W., Cell death during development of the nervous system, *Annu. Rev. Neurosci.,* 14, 453–501, 1991.

175. Lowrie, M.B. and Vrbová, G., Dependence of postnatal motoneurones on their targets: review and hypothesis, *TINS,* 15, 80–84, 1992.

176. Fishman, R.B. and Breedlove, S.M., Local perineal implants of anti-androgen block masculinization of the spinal nucleus of the bulbocavernosus, *Dev. Brain Res.,* 70, 283–286, 1992.

177. Hauser, K.F. and Toran-Allerand, C.D., Androgen increases the number of cells in fetal mouse spinal cord cultures: implications for motoneuron survival, *Brain Res.,* 485, 157–164, 1989.

178. De Vries, G.J., Sex differences in neurotransmitter systems, *J. Neuroendocrinol.*, 2, 1–13, 1990.

179. Micevych, P.E., Coquelin, A., and Arnold, A.P., Immunohistochemical distribution of substance P, serotonin, and methionine enkephalin in sexually dimorphic nuclei of the rat lumbar spinal cord, *J. Comp. Neurol.*, 248, 235–244, 1986.

180. Rajaofetra, N., Sandillon, F., Geffard, M., and Privat, A., Pre- and post-natal ontogeny of serotonergic projections to the rat spinal cord, *J. Neurosci. Res.*, 22, 305–321, 1989.

181. Kojima, M. and Sano, Y., Sexual differences in the topographical distribution of serotoninergic fibers in the anterior column of the rat lumbar spinal cord, *Anat. Embryol.*, 170, 117–121, 1984.

182. Lauder, J.M., Wallace, J.A., and Krebs, H., Roles for serotonin in neuroembryogenesis, *Adv. Exp. Med. Biol.*, 133, 477–506, 1981.

183. Lauder, J.M., Hormonal and humoral influences on brain development, *Psychoneuroendocrinology*, 8, 121–155, 1983.

184. Mattson, M.P., Neurotransmitters in the regulation of neuronal cytoarchitecture, *Brain Res. Rev.*, 13, 179–212, 1988.

185. Cowburn, P.J. and Payne, A.P., The effects of serotonin manipulation during the postnatal period on the development of sexually dimorphic and non-dimorphic lumbosacral motor neuron groups in the Albino Swiss rat, *Dev. Brain Res.*, 66, 59–62, 1992.

186. Cowburn, P.J. and Payne, A.P., Androgens and indoleamines interact to control sexual dimorphisms in the rat spinal cord, *Neurosci. Lett.*, 169, 101–104, 1994.

187. Ladosky, W. and Gaziri, L.C.J., Brain serotonin and sexual differentiation of the nervous system, *Neuroendocrinology*, 6, 168–174, 1970.

188. Hardin, C.M., Sex differences and the effects of testosterone injections on biogenic amine levels of neonatal rat brain, *Brain Res.*, 62, 286–290, 1973.

189. Johnston, H.M, Payne, A.P., Gilmore, D.P., and Wilson, C.A., Neonatal serotonin reduction alters the adult feminine sexual behavior of golden hamsters, *Pharmacol. Biochem. Behav.*, 35, 571–575, 1990.

190. Simerly, R.B., Swanson, L.W., and Gorski, R.A., Demonstration of a sexual dimorphism in the distribution of serotonin-immunoreactive fibers in the medial preoptic nucleus of the rat, *J. Comp. Neurol.*, 225, 151–166, 1984.

191. Handa, R.J., Hines, M., Schoomaker, J.N., Shryne, J.E., and Gorski, R.A., Evidence that serotonin is involved in the sexually dimorphic development of the preoptic area in the rat brain, *Dev. Brain Res.*, 30, 278–282, 1986.

192. Gladue, B.A., Humphreys, R.R., DeBold, J.F., and Clemens, L.G., Ontogeny of biogenic amine systems and modifications in the rat, *Pharmacol. Biochem. Behav.*, 7, 253–258, 1977.

193. Wilson, C.A., Pearson, J.R., Hunter, A.J., Tuohy, P.A., and Payne, A.P., The effects of neonatal manipulation of hypothalamic serotonin levels upon adult sexual activity in the adult rat, *Pharmacol. Biochem. Behav.*, 24, 1175–1183, 1986.

194. Siddiqui, A., Gilmore, D.P., and Clark, J., Regional differences in the indoleamine content of the rat brain: effects of neonatal castration and androgenization, *Biogenic Amines*, 6, 105–114, 1989.

195. Watson, R.E., Hoffman, G.E., and Wiegand, S.J., Sexually dimorphic opioid distribution in the preoptic area: manipulation by gonadal steroids, *Brain Res.*, 398, 157–163, 1986.

196. Gibson, S.J., Bloom, S.R., and Polak, J.M., A novel substance P pathway linking the dorsal and ventral horn in the upper lumbar segments of the rat spinal cord, *Brain Res.*, 301, 243–251, 1984.

197. Uda, K., Okamura, H., Kawakami, F., and Ibata, Y., Sexual differences in the distribution of substance P immunoreactive fibers in the ventral horn of the rat spinal cord, *Neurosci. Lett.*, 64, 157–162, 1986.

198. Sikich, L., Hickok, J.M., and Todd, R.D., 5-HT1A receptors control neurite branching during development, *Dev. Brain Res.*, 56, 269–274, 1990.

199. Shemer, A.V., Azmitia, E.C., and Whitaker-Azmitia, P.M., Dose-related effects of prenatal 5-methoxy-tryptamine (5-MT) on development of serotonin terminal density and behavior, *Dev. Brain Res.*, 59, 59–63, 1991.

200. Haydon, P.G., McCobb, D.P., and Kater, S.B.J., Serotonin selectively inhibits growth cone motility and synaptogenesis of specific identified neurons, *Science*, 226, 561–564, 1984.

201. Haydon, P., McCobb, D.P., and Kater, S.B.J., Dopamine and serotonin inhibition of neurite elongation of different identified neurons, *J. Neurobiol.*, 18, 197–215, 1986.

202. Guo, H., Enters, E.K., McDowell, K.P., and Robinson, S.E., The effect of prenatal exposure to methadone on neurotransmitters in neonatal rats, *Dev. Brain Res.*, 57, 296–298, 1990.

203. Reznikov, A.G., Nosenko, N.D., and Tarasenko, L.V., Augmentation of the sterilizing effect of neonatal androgenization with tropolone, a catechol-O-methyltransferase inhibitor, in female rats, *Neuroendocrinology*, 52, 455–459, 1990.

204. Toran-Allerand, D., Leland, E., and Pfeninger, K.H., Estrogen and insulin synergism in neurite outgrowth enhancement *in vitro*: mediation of steroid effects by interaction with growth factors, *Dev. Brain Res.*, 41, 87–100, 1988.

205. Suárez, I., Bodega, G, Rubio, M., and Fernandez, B., Sexual dimorphisms in the distribution of glial fibrillary acidic protein in the supraoptic nucleus of the hamster, *J. Anat.*, 178, 79–82, 1991.

206. Hatten, M.E., Fishell, F., Stitt, T.N., and Mason, C.A., Astroglia as a scaffold for development of the CNS, *Semin. Neurosci.*, 2, 455–465, 1990.

207. Rakic, P., Principles of neural cell migration, *Experientia*, 46, 882–891, 1990.

208. Steindler, D.A., Glial boundaries in the developing nervous system, *Annu. Rev. Neurosci.*, 16, 445–470, 1993.

209. Nottebohm, F., Buskirk, D.R., Burd, G.D., and O'Loughlin, B., Glial fibers in adult canary brain are thought to act as pathways for putative migrating neuroblasts, *Soc. Neurosci. Abstr.*, 11, 964, 1985.

210. Ovaida, H., Vlodavsky, I., Abramsky, O., and Weidenfeld, J., Binding of hormonal steroids to isolated oligodendroglia and astroglia grown *"in vitro"* on a naturally produced extracellular matrix, *Clin. Neuropharmacol.,* 7, 307–311, 1984.

211. Melcangi, R.C., Celotti, F., Ballabio, M., Castano, P., Poletti, A., Milani, S., and Martini, L., Ontogenetic development of the 5α-reductase in the rat brain: cerebral cortex, hypothalamus, purified myelin and isolated oligodendrocytes, *Dev. Brain Res.,* 44, 181–188, 1988.

212. Nottebohm, F., Neural lateralization of vocal control in a passerine bird. I. Song, *J. Exp. Zool.,* 177, 229–261, 1971.

213. Nottebohm, F. and Nottebohm, M., Left hypoglossal dominance in the control of canary and white-crowned sparrow song, *J. Comp. Physiol.,* A 108, 171–192, 1976.

214. Holman, S.D. and Hutchison, J.B., Lateralized action of androgen on development of behavior and brain sex differences, *Brain Res. Bull.,* 27, 261–265, 1991.

215. Holman, S.D. and Hutchison, J.B., Lateralization of a steroid-sensitive brain area associated with vocal behavior, *Behav. Neurosci.,* 107, 186–193, 1993.

216. Sholl, S.A. and Kim, K.L., Androgen receptors are differentially distributed between right and left cerebral hemispheres of the fetal male rhesus monkey, *Brain Res.,* 516, 122–126, 1990b.

217. Galaburda, A.M., Rosen, G.D., and Sherman, G.F., Individual variability in cortical organization: its relationship to brain laterality and implications to function, *Neuropsychologia,* 28, 529–546, 1990.

218. Witelson, S.F., Neural sexual mosaicism: sexual differentiation of the human temporo-parietal region for functional asymmetry, *Psychoneuroendocrinology,* 16, 131–153, 1991.

219. Geschwind, N. and Behan, P., Left-handedness: association with immune disease, migraine and developmental learning disorder, *Proc. Natl. Acad. Sci. U.S.A.,* 79, 5097–5100, 1982.

220. Geschwind, N. and Galaburda, A.M., Cerebral lateralisation. Biological mechanisms, association and pathologies: I. A hypothesis and a program for research, *Arch. Neurol.,* 42, 428–459, 1985.

221. Galaburda, A.M., The pathogenesis of childhood dyslexia, *Res. Publ. Assoc. Res. Nerv. Ment. Dis.,* 66, 127–137, 1988.

222. Galaburda, A.M. and Kemper, T.L., Cytoarchitectonic abnormalities in developmental dyslexia: a case study, *Ann. Neurol.,* 6, 94–100, 1979.

223. De Voogd, T.J., Androgens can affect the morphology of mammalian CNS neurons in adulthood, *TINS,* 10, 341–342, 1987.

224. Arnold, A.P., Hormonally induced synaptic reorganization in the adult brain, in *Hormones, Brain and Behavior in Vertebrates. 1. Sexual Differentiation, Neuroanatomical Aspects, Neurotransmitters and Neuropeptides,* Balthazart, J., Ed., Karger, Basel, 1990, 82–91.

225. Kurz, E.M., Sengelaub, D.R., and Arnold, A.P., Androgens regulate dendritic length of sexually dimorphic mammalian motoneurons in adulthood, *Science,* 232, 395–398, 1986.

226. Sasaki, M. and Arnold, A.P., Androgenic regulation of dendritic trees of motoneurons in the spinal nucleus of the bulbocavernosus: reconstruction after intracellular iontophoresis of horseradish peroxidase, *J. Comp. Neurol.,* 308, 11–27, 1991.

227. Forger, N.G. and Breedlove, S.M., Seasonal variation in mammalian striated muscle mass and motoneuron morphology, *J. Neurobiol.,* 18, 155–165, 1987.

228. Wee, B.E.F. and Clemens, L.G., Characteristics of the spinal nucleus of the bulbocavernosus are influenced by genotype in the house mouse, *Brain Res.,* 424, 305–310, 1987.

229. McEwen, B.S., Coirini, H., Westlind-Danielsson, A., Frankfurt, M., Gould, E., Schumacher, M., and Woolley, C., Steroid hormones as mediators of neural plasticity, *J. Steroid Biochem. Molec. Biol.,* 40, 1–14, 1991.

230. Frankfurt, M., Gonadal steroids and neuronal plasticity. Studies in the adult rat hypothalamus, *Ann. N.Y. Acad. Sci.,* 743, 45–60, 1994.

231. Matsumoto, A., Hormonally induced synaptic plasticity in the adult neuroendocrine brain, *Zool. Sci.,* 9, 679–695, 1992.

232. Matsumoto, A., Arai, Y., Urano, A., and Hyodo, S., Effect of androgen on the expression of gap junction and β-actin mRNAs in adult rat motoneurons, *Neurosci. Res.,* 14, 133–144, 1992.

233. Woolley, C.S. and McEwen, B.S., Estradiol mediates fluctuations in hippocampal synapse density during the estrous cycle in the adult rat, *J. Neurosci.,* 12, 2549–2554, 1992.

234. Woolley, C.S. and McEwen, B.S., Roles of estradiol and progesterone in regulation of hippocampal dendritic spine density during the estrous cycle in the rat, *J. Comp. Neurol.,* 336, 293–306, 1993.

235. Gould, E., Woolley, C.S., and McEwen, B.S., The hippocampal formation: morphological changes induced by thyroid, gonadal and adrenal hormones, *Psychoneuroendocrinology,* 16, 67–84, 1991.

236. Luine, V.N., Steroid hormone influences on spatial memory, *Ann. N.Y. Acad. Sci.,* 743, 201–211, 1994.

237. Bloch, G.J. and Gorski, R.A., Cytoarchitectonic analysis of the SDN-POA of the intact and gonadectomized rat, *J. Comp. Neurol.,* 275, 604–612, 1988a.

238. Bloch, G.J. and Gorski, R.A., Estrogen/progesterone treatment in adulthood affects the size of several components of the medial preoptic area in the male rat, *J. Comp. Neurol.,* 275, 613–622, 1988b.

239. Garcia-Segura, L.M., Baetons, D., and Naftolin, F., Synaptic remodeling in arcuate nucleus after injection of estradiol valerate in adult female rats, *Brain Res.,* 366, 131–136, 1986.

240. Garcia-Segura, L.M., Hertnandez, P., Olmos, G., Tranque, P.A., and Naftolin, F., Neuronal membrane remodeling during the oestrus cycle: a freeze fracture study in the arcuate nucleus of the rat hypothalamus, *J. Neurocytol.*, 17, 377–383, 1988.

241. Leedom, L., Lewis, C., Garcia-Segura, L.M., and Naftolin, F., Regulation of arcuate nucleus synaptology by estrogen, *Ann. N.Y. Acad. Sci.*, 743, 61–71, 1994.

242. Chung, S.K., Pfaff, D.W., and Cohen, R.S., Estrogen-induced alterations in synaptic morphology in the midbrain central grey, *Exp. Brain Res.*, 69, 522–530, 1988.

243. Dudek, F.E., Andrew, R.D., MacVicar, B.A., Snow, R.W., and Taylor, C.P., Recent evidence for, and possible significance of, gap junctions and electrotonic synapses in the mammalian brain, in *Basic Mechanisms of Neuronal Hyperexcitability*, Jasper, H.H. and van Gelder, N.M., Eds., Alan R. Liss, New York, 1983, 31–73.

244. Matsumoto, A., Arnold, A.P, Zampighi, G.A., and Micevych, P.E., Androgenic regulation of gap junctions between motoneurons in rat spinal cord, *J. Neurosci.*, 8, 4177–4183, 1988.

245. Matsumoto, A., Arnold, A.P., and Micevych, P.E., Gap junctions between lateral spinal neurons in the rat, *Brain Res.*, 495, 362–366, 1989.

246. Popper, P. and Micevych, P.E., The effect of castration on calcitonin gene-related peptide in spinal motor neurons, *Neuroendocrinology*, 50, 338–343, 1989.

247. New, H.V. and Mudge, A.W., Calcitonin gene-related peptide regulates muscle acetylcholine receptor synthesis, *Nature (London)*, 323, 809–811, 1986.

248. Dubé, J.Y., Lesage, R., and Tremblay, R.R., Androgen and estrogen binding in rat skeletal and perineal muscle, *Can. J. Biochem.*, 54, 50–55, 1976.

249. Sar, M. and Stumpf, W.E., Androgen concentration in motor neurons of cranial nerves and spinal cord, *Science*, 197, 77–80, 1977.

250. Simerly, R.B., Chang, C., Muramatsu, M., and Swanson, L.W., Distribution of androgen and estrogen receptor mRNA-containing cells in the rat brain: an *in situ* hybridization study, *J. Comp. Neurol.*, 294, 76–95, 1990.

251. Yu, W.-H.A., Responsiveness of hypoglossal neurons to testosterone in pre-pubertal rats, *Brain Res. Bull.*, 13, 667–672, 1984.

252. Yu, W.-H.A., Administration of testosterone attenuates neuronal loss following axotomy in the brainstem motor nuclei of female rats, *J. Neurosci.*, 9, 3908–3914, 1989.

253. Jones, K.J., Androgenic enhancement of motor neuron regeneration, *Ann. N.Y. Acad. Sci.*, 743, 141–164, 1994.

Chapter 10

Steroid Control of Hypothalamic Function in Reproduction

Virendra B. Mahesh and Darrell W. Brann

CONTENTS

0-8493-7633-5/96/$0.00+$.50
© 1996 by CRC Press, Inc.

I. INTRODUCTION

Steroid secretion from the ovaries, testes, and the adrenals is regulated by tropic hormones secreted by the anterior pituitary gland. The anterior pituitary peptide hormone secretion is regulated by central nervous system (CNS) neurotransmitters. CNS neurotransmitters and anterior pituitary hormones are regulated in turn by negative feedback by circulating steroids and short-loop and ultra short-loop feedback effects within the CNS-pituitary complex. Added to this is an additional level of complexity in which estrogens and progesterone can exert a positive feedback effect on gonadotropin secretion in the ovulatory cycle.

In the human menstrual cycle, the falling levels of estradiol and progesterone from day 23 to 28 of the previous cycle result in an increase of FSH and LH due to escape from the negative feedback effects of these steroids (Figure 1). During the first five days, the elevated levels of FSH stimulate follicular growth and estradiol production. The increasing amounts of estradiol produced by the ovary reduce circulating FSH levels. Nevertheless, the dominant follicle continues to grow because of the higher levels of FSH receptors induced by the action of estradiol and FSH which result in increased sensitivity of the granulosa cells in the dominant follicle to circulating FSH levels. The sustained estradiol levels and the accompanying progesterone increases between days 12 to 14 are responsible for inducing the preovulatory gonadotropin surge leading to ovulation.

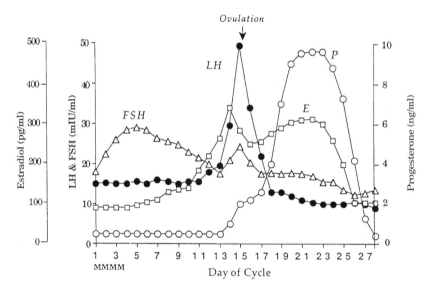

Figure 1 Changes in serum estradiol (□), progesterone (○), LH (●), and FSH (△) during the human menstrual cycle. (From Brann, D. W., Mills, T. M., and Mahesh, V. B., *Female Reproduction: The Ovulatory Cycle In Reproductive Toxicology,* Witorch, R. J., Ed., Raven Press, New York, 1995, 23-44. With permission.).

The regulation of gonadotropin secretion during the ovulatory cycle has been a subject of extensive investigation and several excellent reviews have appeared in the literature on the subject.[1-6] This chapter will focus on the regulation of gonadotropin secretion during the ovulatory cycle by ovarian and adrenal steroids in describing the effect of steroids on modulating hypothalamic-pituitary function.

II. ESTROGEN REGULATION OF HYPOTHALAMIC-PITUITARY FUNCTION

A. POSITIVE AND NEGATIVE FEEDBACK EFFECT OF ESTRADIOL

It is well established that ovariectomy results in an elevation of serum LH and FSH levels which can be suppressed by estrogen replacement. This negative feedback system is maintained primarily by the small amounts of estradiol in circulation in a variety of animal species including humans because ovariectomy results in a decrease in uterine weight. The amount of estradiol necessary to restore serum luteinizing hormone (LH) and follicle stimulating hormone (FSH) levels to basal levels found in the intact immature rat also restores the uterine weight to that found in intact rats of similar age.[7] Higher levels of estradiol result in an estrogen-induced surge of LH demonstrating a positive feedback effect of estrogens on gonadotropin secretion, whereas very high levels of estrogens inhibit gonadotropin secretion.[7]

B. EFFECT OF ESTROGENS ON THE INDUCTION OF PUBERTY AND THE PREOVULATORY GONADOTROPIN SURGE

Detailed studies in the intact animal have shown that the administration of estrogens invariably results in the initial lowering of gonadotropins followed by the manifestation of a positive feedback effect several hours later.[8] The positive feedback effects of estrogens have important physiological effects on the onset of puberty and the regulation of the preovulatory surge of gonadotropins. It has been demonstrated by a number of investigators that the administration of estrogens and aromatizable androgens advanced the onset of puberty in the rat.[9-11] Furthermore, changes in uterine weight and stimulation of endometrial stroma, endometrial luminal epithelium, and myometrium occurred prior to the first change in serum LH around the time of puberty.[12] A linear increase in the percentage of large follicles also occurred from 22 to 40 days of age with day 32 showing the first evidence of increased estrogen secretion, day 34 showing the first increase in LH, and the preovulatory surge of gonadotropins occurring on day 38.[12] The increase in estrogen production on day 32 in this study was attributed to increased percentage of large follicles.[12] Increased levels of estradiol, progesterone, and testosterone in serum along with increased sensitivity of the pituitary to GnRH was also reported to occur in another study prior to vaginal opening and the first ovulation.[13] In the human, aromatizable androgens such as dehydroepiandrosterone and testosterone also show increases before the rise of serum LH and FSH during puberty, thereby emphasizing the role of adrenal and ovarian steroids as an important component of pubertal changes.[14]

Estradiol has been considered to be an important ovarian signal responsible for triggering the preovulatory surge of gonadotropins resulting in ovulation. Ovariectomy in rats, hamsters, and sheep abolishes the preovulatory surge of gonadotropins, and the administration of estradiol or estradiol benzoate reinstates at least a partial surge.[7,15-22] Furthermore, the administration of antibodies to estradiol or estrogen antagonists abolishes the gonadotropin surge.[23-25] The ability of estradiol to induce surges of LH similar to the preovulatory LH surge has been extensively demonstrated in monkeys as well as women.[26-29]

C. CHANGES IN SENSITIVITY OF THE PITUITARY TO GnRH DURING THE OVULATORY CYCLE

There appears to be a remarkable change in sensitivity of the pituitary to GnRH during the ovulatory cycle.[27,30,31] Because the classical action of steroid hormones involves the interaction with DNA of the activated steroid-bound intracellular receptor, a detailed study of the depletion of cytoplasmic estrogen receptors of the anterior pituitary and hypothalamus at various times in the rat ovulatory cycle and changes in sensitivity to GnRH in the release of LH was undertaken by Greeley et al.[30] Figure 2 shows that there was a remarkable correlation between the depletion of anterior pituitary cytoplasmic estrogen receptors due to nuclear binding and the increased sensitivity to GnRH in the release of LH as well as in the induction the endogenous LH surge.[32] A similar depletion of cytoplasmic estrogen receptors also occurred in the hypothalamus during the ovulatory cycle.[30] Such a relationship between estrogen receptor translocation and the gonadotropin surge was also observed in the natural onset of puberty and the pregnant mare's serum gonadotropin (PMSG) induced ovulation in the immature rat.[13,33]

The work of several investigators, confirmed by a detailed study by Ashiru and Blake,[34] has shown that the continued presence of estradiol is not needed on the day of proestrus for the preovulatory gonadotropin surge. Ovariectomy on the day of proestrus does not abolish the gonadotropin surge. Furthermore, it has been demonstrated that there is a fall in circulating estradiol during this surge. If this fall is prevented by the implantation of Silastic capsules of estradiol, the magnitude of the gonadotropin surge is decreased.[35,36]

D. HYPOTHALAMIC AND PITUITARY SITES OF ESTROGEN ACTION

The action of estradiol on the regulation of gonadotropin secretion could occur at the level of the hypothalamus as well as the pituitary. The uptake of ^{3}H-estradiol in various regions of the hypothalamus, various cell types of the pituitary, and the presence of estrogen receptors have been reported by several

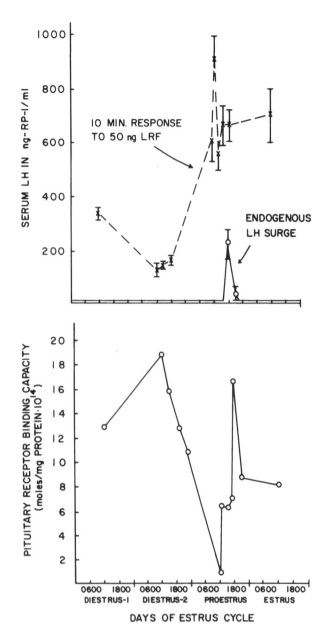

Figure 2 Pituitary cytosol estradiol receptor binding and release of LH in response to LHRH at various times in the four-day cyclic rat. There appeared to be a good correlation between the increased sensitivity of the pituitary to LHRH and decrease in cytosol estradiol receptor binding (due to nuclear translocation) with maximum sensitivity to LHRH around noon of proestrus. Also shown in the upper graph is the endogenous LH surge.[30] (From Mahesh et al., *J. Steroid Biochem.*, 6, 1025, 1975. With permission.)

investigators.[37-42] The principal hypothalamic effect of estradiol appears to be the induction of GnRH release. This is supported by studies *in vivo* in the rat[1,43], sheep[44,45] and monkey[46,47] and in hypothalamic superfusion studies *in vitro*.[48-50]

The possibility of a direct action of estradiol on the anterior pituitary was suggested by variations in responsiveness to GnRH in the release of LH at different phases of the estrous cycle and after estrogen administration.[27,30,31,51,52] However, in these experiments hypothalamic effects of estradiol could be responsible for at least a part of the overall effect. The ability of estradiol to potentiate the LH-releasing effects of GnRH directly at the level of the anterior pituitary was demonstrated by Greeley et al.[53] in 1975 in ovariectomized rats in which the hypothalamo-hypophyseal connections were interrupted by stalk-section. The direct effects of estradiol on enhancing the pituitary sensitivity to GnRH were confirmed by *in vitro* dispersed pituitary cell culture studies.[54-56] Experiments with *in vitro* pituitary cell cultures also demonstrated the negative feedback effect at short duration of estrogen treatment and a positive feedback effect after a longer duration of estrogen treatment.[57,58]

E. EFFECT OF ESTROGENS ON PITUITARY GnRH RECEPTORS

The self-priming effect of GnRH on the sensitivity of the pituitary to GnRH in the release of LH is well documented.[59-62] The increased sensitivity to GnRH in the release of LH appears to be due to an increase in GnRH receptors induced by the self-priming effect of GnRH as indicated by several *in vivo* studies.[63-65] Increases in GnRH receptor mRNA levels in superfused dispersed pituitary cells have also been demonstrated after the administration of GnRH.[66] Based on the cyclic changes in GnRH receptors during the rat ovulatory cycle and their correlation with circulating levels of estradiol, estradiol was considered to be an important regulator of GnRH receptors.[67,68] In the cycling rat anterior pituitary GnRH receptor mRNA levels were increased from 1200 to 2100 h on proestrus as compared to other times in the ovulatory cycle.[69] The direct effects of estradiol on GnRH receptor numbers have been shown to occur in ovariectomized estrogen-treated hypothalamic-pituitary disconnected sheep[70] and in cultured ovine pituitary cells *in vitro*.[71]

The negative feedback effect of estrogens on the secretion of LH in dispersed pituitary cell cultures after a short period of treatment followed by positive-feedback effects after prolonged exposure mentioned earlier[57,58] can also be explained by the reduction of GnRH receptors during the period of 0.5 to 4 h and an increase at 24 h.[72] However, no differences in GnRH receptor numbers were found during the period when estrogens sensitized the pituitary to GnRH beyond the increase observed during early proestrus.[63,73,74] Thus, an influence of estradiol on postreceptor events in the secretion of gonadotropins is postulated.

III. EVIDENCE QUESTIONING THE ROLE OF ESTRADIOL AS THE SOLE REGULATOR OF THE PREOVULATORY GONADOTROPIN SURGE

Evidence has accumulated in the literature that adrenal steroids may play an important role in the induction of puberty and the preovulatory gonadotropin surge. Adrenalectomy before 25 days of age delays puberty in the female rat and the effect of adrenalectomy can be reversed by corticosterone treatment.[75,76] Adrenalectomy in the adult cycling rat results in the attenuation of the preovulatory gonadotropin surge,[77] inhibition of follicular development and number of ovulations,[78] and increased irregular cyclicity.[79,80] The decrease in cyclicity and ovulation may be due to a suppression of FSH caused by adrenalectomy.[80]

The work of Hilliard and co-workers[81] showed that in the ovariectomized rabbit in which estrous behavior was induced by estradiol administration, copulation induced only a partial LH release, and the administration of the progesterone metabolite 20α-hydroxy-pregn-4-en-3-one was required for the full gonadotropin surge. Progesterone was also needed for the manifestation of pituitary sensitivity to GnRH comparable to that found in the afternoon of proestrus in estrogen-primed ovariectomized rats.[19,82,83] Finally, if the other source of progesterone in the ovariectomized rat, namely the adrenal, was removed by adrenalectomy, estradiol no longer could induce the preovulatory type LH release.[19] These studies suggested a prominent role for progesterone from adrenal and/or ovarian origin in inducing the preovulatory type gonadotropin surge.

IV. EVIDENCE SUPPORTING THE ROLE OF PROGESTERONE IN INDUCING THE PREOVULATORY TYPE GONADOTROPIN SURGE

The facilitative effects of progesterone on gonadotropin secretion have been demonstrated by a number of studies.[84-90] Equally well recognized are the inhibitory effects of progesterone on gonadotropin secretion.[84,88-93] The facilitative or inhibitory effect of progesterone is dependent upon the time of its administration in relationship to the anticipated gonadotropin surge.[84] Attention to the ability of progesterone to induce a gonadotropin surge in women was drawn by Odell and Swerdloff[94] in 1968 who demonstrated that when progesterone was administered to an estrogen-treated menopausal woman, it brought about a preovulatory type surge of LH and FSH. This observation was subsequently confirmed by others.[95,96]

The study of the effect of progesterone in the cycling rat is difficult because the stimulatory and inhibitory effects are dependent upon the time of administration in relationship to the preovulatory surge of gonadotropins.[84] Estrogen priming of the ovariectomized rat is essential for progesterone to exert its action on gonadotropin secretion.[85,89,90,97,98] This is due to the fact that estrogens are required for the induction of hypothalamic and anterior pituitary progesterone receptors.[99] The dose of estrogen used has to be considered very carefully because if the dose is large enough to induce a surge of gonadotropins, it imposes a time limit on the administration of progesterone for stimulatory or inhibitory effects similar to those in the cycling rat. Furthermore, it is difficult to differentiate between the effects of estrogens from those of progesterone. Therefore, Mahesh and co-workers established an immature ovariectomized rat model in which low doses of estradiol were used every 12 h that were enough to induce progesterone receptors[99] to ensure progesterone sensitivity but were not sufficient for an estrogen-induced gonadotropin surge.[89,90,100] The serum LH and FSH levels were reduced to about 40% of that found in the ovariectomized controls so that stimulation and inhibition of gonadotropins by progesterone could be readily observed. In this experimental model 0.1 mg/kg body weight of estradiol was injected per day for 4 days in immature rats ovariectomized at 26 days of age. Progesterone was able to induce either a preovulatory type gonadotropin surge or inhibit gonadotropin secretion, depending upon the dose of the steroid used.[89,90,100] Injection of a single stimulatory dose of progesterone in this model resulted in a well-defined LH surge and a prolonged FSH surge. The pattern of LH and FSH secretion was similar to that observed on the day of proestrus in the cycling rat.[90,101] Of particular interest was the fact that although the stimulatory dose of progesterone increased the pituitary content of both LH and FSH, it was only FSH that was selectively secreted in increased quantity during the night of the surge, thus, reproducing events that occur at proestrus in the cycling rat.[90,101]

Even though the above-described, low dose estrogen-primed ovariectomized rat model provided results of great interest, a disadvantage was the critical control of the dose of estradiol that was needed. The dose of 0.08 μg/kg body weight was too small and 0.2 μg/kg body weight dose of estradiol turned out to be too large.[100] This narrow dose range administered in oil in conjunction with animal variability made it difficult to achieve ideal estrogen priming that met the criteria of progesterone receptor induction, fall of serum gonadotropin levels by about 40%, and the absence of an estrogen-induced surge in every experiment. A new model using 26-day-old intact or ovariectomized rats injected with 2 μg of estradiol in ethanol-saline at 1700 h for 2 days was subsequently set up, validated, and extensively used.[102-106] A typical example of the induction of the LH and FSH surge induced by a single injection of 2 μg/kg body weight of progesterone in estrogen-primed intact immature rat is shown in Figure 3. Also shown are the changes in gonadotropin α subunit mRNA and LH-β and FSH-β subunit mRNA levels. It is noteworthy that estrogen priming for 2 days brought about only minimal changes in serum LH at 1500 and 1600 h and FSH at 1600 h. The injection of progesterone induced a surge of LH and FSH that was similar in magnitude and duration to the preovulatory surge. In addition progesterone also caused increases in LH-β subunit mRNA levels at 1000 h and FSH-β subunit mRNA levels at 1000, 1200, and 1300 h[106] which are similar to those observed in cycling rats.[107,108]

V. PHYSIOLOGICAL ROLE OF PROGESTERONE IN THE REGULATION OF THE PREOVULATORY GONADOTROPIN SURGE

Although the induction of the preovulatory type surge by progesterone in estrogen-primed animals including humans is well established, the physiological role of progesterone in the ovulatory cycle was unclear. Therefore, the progesterone receptor antagonist RU486 was used. When RU486 was administered to cycling rats on proestrus or PMSG-primed immature rats on the day of the gonadotropin surge at 1000 h, the endogenous LH and FSH surge in the afternoon was attenuated (Figure 4A) and this was accompanied by a decrease in the number of ova per ovulating rat.[109] Confirmation of the physiological role of progesterone in regulating the preovulatory surge of gonadotropins was also obtained by the use of the 3β-hydroxy steroid dehydrogenase inhibitor trilostane, which blocks progesterone synthesis. Administration of trilostane at 1130 h during proestrus significantly attenuated the preovulatory LH surge (Figure 4B).[5] A similar observation was made by DePaolo[110] in 1988 using the 3β-hydroxy steroid dehydrogenase inhibitor epostane. These data clearly indicate an important physiological role of progesterone in regulating the preovulatory surge of gonadotropins.

In the presence of adequate estrogen priming, the source of progesterone could be the ovary as well as the adrenal. In the cycling rat, the ovarian feedback for the gonadotropin surge is completed by early hours of proestrus and ovariectomy on the morning of proestrus does not interfere with the preovulatory

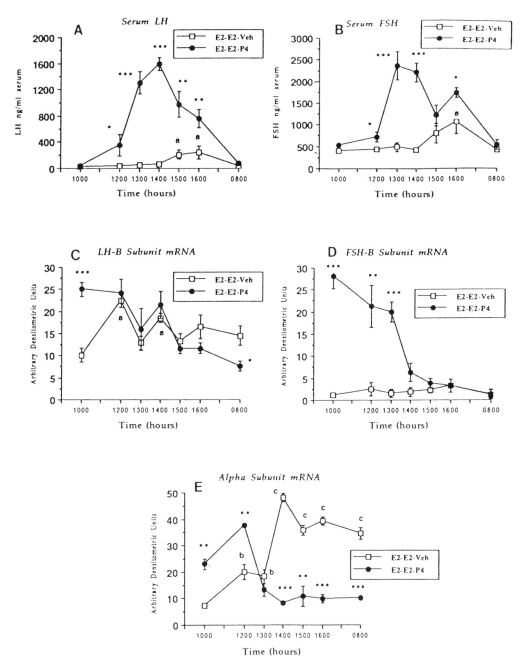

Figure 3 (A–E) Effects of progesterone on serum LH and FSH levels and LH-β, FSH-β, and α-subunit mRNA levels in the anterior pituitary in intact estrogen-primed immature rats. Immature female rats received 2 μg of estradiol at 1700 h on 27 and 28 days of age. On day 29, the animals received either vehicle (E$_2$-E$_2$-Veh) or progesterone (E$_2$-E$_2$-P$_4$); (2 mg/kg) at 0900 and were killed at various time points for serum LH and FSH measurements and mRNA measurements; n = 6 rats per group. *p <0.05, **p <0.01, and ***p <0.001 vs. E$_2$-E$_2$-Veh; a and b = p <0.05 vs. 1000 h E$_2$-E$_2$-Veh; c = p <0.01 vs. 1000, 1200, and 1300 h E$_2$-E$_2$-Veh. (From Brann et al., *J. Steroid Biochem. Mol. Biol.,* 46, 427, 1993. With permission.)

184

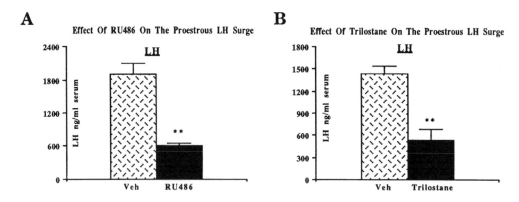

Figure 4 (A,B) Effect of progesterone receptor antagonist, RU486, or the progesterone synthesis inhibitor, trilostane, on the proestrus LH surge in cycling adult rats. Cycling adult female rats were administered either vehicle (Veh) or RU 486 (200 mg/rat, i.p.) or trilostane (Sterling-Winthrop Laboratories, Rensselaer, NY); (200 mg/kg BW i.p.) on the morning of proestrus (RU 486 at 1000 h; trilostane, 1130 h). The animals were killed at 1800 h for the measurement of serum LH. **p <0.01 vs. vehicle (Veh) controls. (From Mahesh, V. B. and Brann, D. W., *J. Steroid Biochem. Mol. Biol.*, 41, 495, 1992. With permission.)

gonadotropin surge.[111] This is also true in the PMSG-primed rat in which ovariectomy at 0800 h on the day of the preovulatory gonadotropin surge did not alter the LH surge and brought about small increases in FSH at 1800 h (Figure 5).[4] The administration of the antiprogestin RU486 at 1000 h attenuated the preovulatory gonadotropin surge. Because progesterone has a short half-life in blood, the only source of progesterone antagonized by RU486 in the ovariectomized rat could be the adrenal. These studies indicate that adrenal progesterone may have a prominent role in regulating the preovulatory surge of gonadotropins and explain the occurrence of irregular cyclicity and attenuation of the preovulatory gonadotropin surge and number of ovulations after adrenalectomy.[77-79]

VI. POSSIBLE SITES OF ACTION OF PROGESTERONE IN INDUCING THE PREOVULATORY GONADOTROPIN SURGE

The possible sites of action of progesterone in inducing the preovulatory gonadotropin surge could be the anterior pituitary in which progesterone could alter the response of the pituitary to GnRH and/or the hypothalamus where progesterone could influence GnRH release. In addition to this progesterone may also act by altering the actions of estradiol on the hypothalamic-pituitary axis.

A. MODULATION OF THE PITUITARY RESPONSIVENESS TO GnRH BY PROGESTERONE

The direct actions of progesterone on the responsiveness of the anterior pituitary to GnRH has been demonstrated by several investigators using dispersed pituitary cell cultures[54-56,112-114] and pituitaries of cycling rats superfused *in vitro*.[115] The effect of progesterone is biphasic in nature. Acute treatment with progesterone enhances and chronic progesterone treatment suppresses the pituitary responsiveness to GnRH.[56,112-114] The ability of progesterone to modulate GnRH receptors has also been studied extensively and the results have been inconclusive. Witcher et al.[116] showed a decrease in GnRH receptors of the pituitary after progesterone action whereas Attardi and Happe[117] showed no change. A more recent study shows a biphasic effect of progesterone with short duration treatment increasing and prolonged treatment decreasing GnRH receptors.[118] Krey and Kamel[119] have reported that progesterone enhances the GnRH response to secretagogues which act distal to the GnRH receptor. Thus, increasing the sensitivity of second messenger systems involved in GnRH action could lead to a response distal to the GnRH receptor.

B. RAPID RELEASE OF GnRH BY PROGESTERONE AND CHANGES IN GnRH DEGRADATION INDUCED BY PROGESTERONE

The rapid release of GnRH by progesterone *in vivo* in perfused hypothalami and by push-pull canulae experiments has been demonstrated extensively.[48-50,100,120,121] Estrogen priming is essential for this action of progesterone. The action of progesterone in bringing about this change may not be exclusively exerted through the classical intracellular progesterone receptors, because the release of GnRH occurs even when

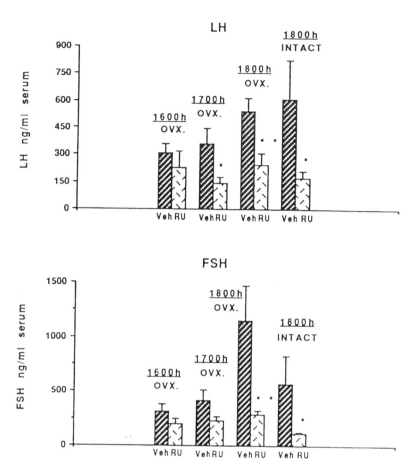

Figure 5 Effect of the antiprogestin RU486 upon preovulatory LH and FSH secretion in PMSG-primed intact immature rats and in rats ovariectomized on the morning of the PMSG-induced preovulatory LH and FSH surge. Rats that were 28 days old were administered PMSG (8 IU/0.2 ml saline s.c.) at 0800 h. Six of the eight groups of animals were ovariectomized 48 h later (0800 h, 30 days of age) under ether anesthesia. The animals received either vehicle (Veh) or RU486 (RU); (200 mg/rat in 0.1 ml ethylene glycol i.p.) 50 h later (1000 h, 30 days of age). The animals were killed at various times throughout the afternoon for LH and FSH measurements; n = at least 6 animals per group. *$p <0.05$ vs. vehicle controls; **$p <0.01$ vs. vehicle controls. (From Brann, D.W. and Mahesh, V.B., *Front. Neuroendocrinol.,* 12, 165, 1991. With permission.)

progesterone is covalently bound to bovine serum albumin and unable to enter the cell.[122] The progesterone metabolite 3α-hydroxy-5α-pregnan-20-one has also been shown to bring about gonadotropin release, presumably mediated through its effects on the GABA$_A$ receptor system to increase GnRH release.[123]

In addition to increasing the release of GnRH, progesterone may also increase the amount of GnRH available to the pituitary by suppressing GnRH degradation. Fluctuations of GnRH degrading activity have been observed during the rat ovulatory cycle with a decrease in hypothalamic activity at the time of the preovulatory gonadotropin surge.[124,125] In the ovariectomized estrogen-primed rat, a decrease in hypothalamic and plasma GnRH degrading activity occurs in response to progesterone administration.[126]

C. REGULATION OF ESTROGEN RECEPTOR DYNAMICS BY PROGESTERONE

Because estrogens have important negative and positive feedback effects on gonadotropin secretion with the acute effects of estradiol *in vivo* or *in vitro* being inhibitory to gonadotropin secretion, one possible mechanism of action of progesterone in promoting gonadotropin release could be by altering the estrogen receptor concentrations. The reduction of estrogen receptor levels by progesterone have indeed been observed in breast and uterine tissue.[127,128] The experiments of Smanik et al.[99] showed that progesterone administered 1 h before an estradiol injection in estrogen-primed ovariectomized rats reduced anterior pituitary but not hypothalamic occupied nuclear estradiol receptors. These results indicated that in addition to a rapid release of GnRH and suppression of GnRH degradation, progesterone could also overcome the acute suppressive effect of estrogens on gonadotropin secretion by decreasing occupied nuclear estradiol receptors of the pituitary.

Although our results demonstrating that progesterone decreased anterior pituitary occupied nuclear estrogen receptors were quite conclusive, other laboratories using different animal models and protocols reported that progesterone has no effects on anterior pituitary estrogen receptors.[129-132] The issue was resolved by a detailed study of the time course of progesterone action on estrogen receptors by Calderon et al.[102] Estradiol administered to the ovariectomized rat brought about the maximal induction of progesterone receptors in approximately 12 h. A single injection of progesterone resulted in nuclear accumulation of progesterone receptors for up to 2 h but not 4 h in the pituitary and for up to 4 h in the hypothalamus (Figure 6A,B). The injection of estradiol in the animal during the period of nuclear occupancy of the progesterone receptor (1 and 2 h after progesterone but not 3 h) brought about a depletion of occupied nuclear estradiol receptors of the anterior pituitary but not the hypothalamus (Figure 6C,D). These studies on the action of progesterone on occupied nuclear estradiol receptors in the anterior pituitary only during the period of the nuclear accumulation of progesterone were able to explain the results of other investigators who had apparently studied progesterone effects at other time points.

The physiological relevance of progesterone-induced reduction of occupied estrogen nuclear receptors of the pituitary had to be established next. This was done by Calderon et al.[102] by demonstrating that injection of progesterone 1 h before the final estradiol injection was able to block the effect of the estrogen on progesterone receptor synthesis, whereas injection 4 h before the final estrogen injection (in the absence of nuclear occupancy of the progesterone receptor complex) had no effect (Figure 6F). The physiological relevance of the progesterone effect on estrogen receptors was confirmed by Brann et al.[133] using a similar experimental design but with estrogen-induced prolactin secretion as an endpoint (Figure 6E). These studies clearly indicated that progesterone-induced occupied nuclear estrogen receptor depletion exerted a physiological effect comparable to reducing or interfering with the effects of estrogen in circulation. The mechanism involved in the reduction of occupied nuclear estrogen receptors appeared to be the stimulation of anterior pituitary 17β-hydroxy steroid dehydrogenase activity by progesterone.[134] This observation was further confirmed by demonstrating that if 17α-ethynyl estradiol, a synthetic estrogen that cannot be oxidized by the action of 17β-hydroxy steroid dehydrogenase, was used instead of estradiol, progesterone was unable to decrease the occupied anterior pituitary nuclear estrogen receptors.[135]

The effects of progesterone on gonadotropin secretion are also dependent upon the dose of progesterone used. Injections of low and high doses of progesterone to estrogen-primed animals stimulate gonadotropin release, whereas the intermediate doses are either ineffective or suppressive.[89,90] This finding is of considerable interest because this effect is opposite to that observed with estradiol where low and high doses suppress the post-castration rise of gonadotropins and the intermediate dose is stimulatory.[7] Furthermore, in the human menstrual cycle, serum progesterone rises at midcycle in two distinct waves. The first wave is associated with increased sensitivity to GnRH and the second wave is associated with the termination of the gonadotropin surge.[136] The modulation of estrogen receptor dynamics appears to be an important mode of action of the dose-related effect of progesterone, because the low and high dose of progesterone reduced occupied nuclear estrogen receptors in the pituitary[137] and estrogen-induced prolactin secretion,[138] whereas the intermediate dose was ineffective.

D. A MODEL EXPLAINING THE STIMULATORY AND INHIBITORY EFFECTS OF PROGESTERONE AND ITS VALIDATION

The mechanism of action of the stimulatory and inhibitory effects of progesterone depending upon its time of administration in the cycling or estrogen-primed animal in relationship to the expected preovulatory type gonadotropin surge has been difficult to explain. In 1987, Mahesh and Muldoon[139] proposed the first model to explain this effect of progesterone (Figure 7). The model was based on two assumptions. The first assumption made was that in the presence of adequate estrogen priming there would be adequate levels of GnRH receptors in the anterior pituitary and the pituitary would be adequately sensitized to GnRH in the release of gonadotropins. In addition, there would be an adequate number of progesterone receptors in the pituitary for progesterone sensitization to GnRH for the release of gonadotropins and

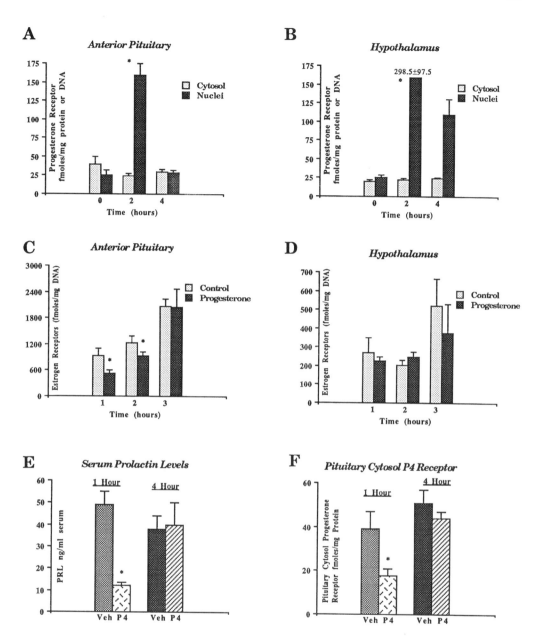

Figure 6 (A–F) Effect of progesterone (P_4) on nuclear progesterone and estrogen receptors in the anterior pituitary and the hypothalamus and on estrogen-induced prolactin release and progesterone receptor synthesis. Cytosol and nuclear progesterone receptors in the anterior pituitary gland are shown in Panel A and the hypothalamus in Panel B in 26-day-old ovariectomized rats treated with 2 μg of estradiol in ethanol saline and 0.8 mg/kg BW of progesterone 12 h after the estradiol injection, at 0, 2, and 4 h after the progesterone injection. Panels C and D show anterior pituitary and hypothalamic nuclear estrogen receptors in animals ovariectomized and treated with estradiol and progesterone as described under A and B with an additional injection of 2 μg estradiol 1, 2, and 3 h after the progesterone injection. The animals were killed 1 h after the last estradiol injection for estrogen receptor measurements. Panel E shows the effect of an injection of 0.8 mg/kg BW progesterone or prolactin release after 12 h of estrogen priming (2 μg) in ovariectomized 26-day-old female rats, which was followed by a second estradiol injection (2 μg) 1 or 4 h after progesterone. Serum prolactin was measured 12 h after the last estradiol injection. Panel F shows the effect of progesterone on estrogen-induced progesterone receptor synthesis in the anterior pituitary using the paradigm described under E. *$p < 0.05$ vs. control. (From Mahesh, V.B. and Brann, D.W., *J. Steroid Biochem. Mol. Biol.*, 41, 495, 1992. With permission.).

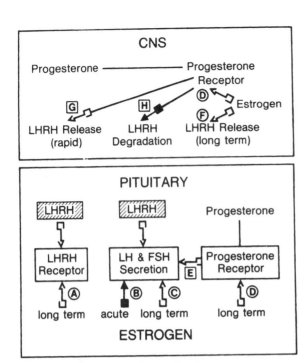

CNS

Progesterone ——————— Progesterone
Receptor

Ⓓ Estrogen

Ⓕ

[G] LHRH Release (rapid) [H] LHRH Degradation LHRH Release (long term)

PITUITARY

////LHRH////   ////LHRH////   Progesterone

LHRH Receptor | LH & FSH Secretion [E] | Progesterone Receptor

Ⓐ long term   Ⓑ acute   Ⓒ long term   Ⓓ long term

ESTROGEN

Progesterone Induced Surge Conditions:
ⒶⒸⒹⒻ not altered short term.
Ⓑ[G][H] and possibly [E] favor surge.

Inhibitory Conditions:
ⒶⒸⒹⒻ are inhibited.
Ⓑ[G][H] and possibly [E] limited by progesterone receptor availability.

←—❑ Stimulation   ////// LHRH effect
←—■ Inhibition   ❑ Progesterone effect
—— Interaction   ○ Estrogen effect

Figure 7 A model to explain stimulatory and inhibitory effects of progesterone on gonadotropin secretion. This first model does not consider the dose-dependent effects of progesterone and its effect on the pulsatile discharge of LHRH. (From Mahesh, V.B. and Muldoon, T.G., *J. Steroid Biochem.*, 27, 665, 1987. With permission.)

in the CNS for rapid release of GnRH and suppression of its degradation. In the presence of these factors which were induced by adequate estrogen priming, a single injection of progesterone would bring about: a rapid release of GnRH, suppression of GnRH degradation, antagonism of the suppressive effects of estradiol on gonadotropin secretion by reducing the occupied nuclear estrogen receptors (second assumption), and increased sensitivity of the pituitary to GnRH in the release of gonadotropins. These are conditions that would lead to a gonadotropin surge. In the absence of adequate estrogen priming, the pituitary sensitivity to GnRH would be reduced and the rapid effect of progesterone on GnRH release and inhibition of GnRH degradation would be limited due to the decreased number of progesterone receptors. Furthermore, progesterone antagonism of occupied estrogen nuclear receptors to a full extent occurs only through progesterone receptors in the presence of adequate estrogen priming. The estrogen-priming effects at the time of progesterone administration would be further slowed down by the progesterone antagonism of estrogen action. These factors would cause an inhibition of gonadotropin secretion by progesterone. This assumption was validated in subsequent experiments in which progesterone administration after 16 h of estrogen priming (as contrasted to 29 h of estrogen effects in the rat ovulatory cycle) suppressed gonadotropin secretion, whereas progesterone administration after 40 h of estrogen priming produced a preovulatory type gonadotropin-surge.[104] The second assumption that progesterone administration 1 h before a gonadotropin suppressing effect of estradiol in estrogen-primed rats would overcome the inhibitory effects of estradiol was also validated experimentally.[104]

VII. MECHANISMS INVOLVED IN THE REGULATION OF GnRH RELEASE BY PROGESTERONE

The modes of action of progesterone in inducing the preovulatory gonadotropin surge include the regulation of the hypothalamic secretion of GnRH and increasing the pituitary's sensitivity to GnRH in the release of LH and FSH. The action of progesterone in inducing the rapid release of GnRH by the hypothalamus is extensively documented.[48-50,100,120,121] These findings are supported by the observations of Brann and Mahesh[140] that the administration of a potent GnRH antagonist abolishes the ability of progesterone to induce gonadotropin surges in estrogen-primed ovariectomized rats.

In the cycling rat it has been clearly demonstrated that the preovulatory gonadotropin surge is preceded by an initial increase in GnRH which may contribute to the self-priming effect of GnRH on the pituitary followed by a second increase 2 to 3 h later which initiates the surge.[141] Such changes in hypothalamic GnRH levels and GnRH release appear only after progesterone is administered to estrogen-primed rats and not by estrogen administration, even though it may produce a small surge.[1,142] The hypothalamic changes after the administration of progesterone in the estrogen-primed rat include an increase in GnRH mRNA levels.[143,144] Furthermore, studies using c-fos protein as a marker for neuronal activation of GnRH neurons have shown that c-fos induction in GnRH neurons only occurs after the administration of progesterone to estrogen-primed rats and not by estrogen alone.[145,146] These studies confirm the important role of progesterone in inducing the preovulatory gonadotropin surge. The inhibition of peptidases that degrade GnRH has already been cited.[124-126]

In spite of the strong evidence for the regulation of GnRH secretion by progesterone in the presence of estrogens, it is unlikely that the action of progesterone is exerted directly at the GnRH-containing neuron since steroid receptors have not been found in GnRH neurons.[147-150] Thus, it is suggested that steroids act to modulate other neurotransmitters that in turn regulate the GnRH neuron. A multitude of studies have emerged which demonstrate that various neurotransmitters can control GnRH secretion by exerting either an inhibitory or a stimulatory effect. Among those exerting an inhibitory effect are the opioids, gamma amino butyric acid (GABA), and the tachykinins, although GABA and tachykinins can be stimulatory as well under certain circumstances. The catecholamines, neuropeptide Y, galanin, and excitatory amino acids have a stimulatory role on GnRH secretion. The above list is not comprehensive but contains the most well-documented transmitters in the control of GnRH secretion. This subject has been reviewed extensively.[1-6,151,152]

A. OPIOIDS

The work of several investigators has shown that the administration of the opioid antagonist naloxone leads to an increase in serum LH at every time point in the ovulatory cycle with the exception of the time of the preovulatory gonadotropin surge on proestrus.[151-155] This transient decrease in the tonic opioid tone or "the inhibitory brake" is suggested to facilitate the preovulatory gonadotropin surge on proestrus. This hypothesis is supported by an observed reduction of hypothalamic μ-opioid receptors during the preovulatory and progesterone-induced gonadotropin surges[156] as well as the documented changes in the hypothalamic opioid levels described below.[157]

Studies from our laboratory have shown that immature-ovariectomized rats treated with estradiol for two days showed a decrease in hypothalamic β-endorphin levels in the absence of a gonadotropin surge.[157] The administration of progesterone to estrogen-primed ovariectomized rats did not bring about further reduction in β-endorphin levels although it induced a gonadotropin surge. Thus, both the reduction of the opioid tone and the induction of an excitatory signal appears to be required for the manifestation of the preovulatory type gonadotropin surge. Progesterone administration, however, restored the hypothalamic β-endorphin levels to the pretreatment control levels by the next morning, whereas β-endorphin was still suppressed in the estrogen-treated controls at this time. This may explain how progesterone limits the preovulatory LH surge expression to a single day.[91] A suppression of POMC mRNA levels after two days of estrogen administration and an increase in these levels after progesterone treatment has also been reported.[158] Thus, the regulatory effects of progesterone on the opioid system is of considerable importance and unlike the GnRH neuron, β-endorphin neurons contain estrogen and progesterone receptors.[147,159]

B. GABA

There is growing evidence that the neurotransmitter GABA is an important CNS modulator of gonadotropin secretion. GABAergic neurons contain estrogen and progesterone receptors,[148-150] and glutamic acid decarboxylase containing neurons synapse with GnRH neurons in the rat medial preoptic area.[160] GABA is the major hypothalamic inhibitor of gonadotropin secretion,[161-166] and this effect is manifested largely through the $GABA_B$ receptor system.[161-163,165] Several reports also show an inhibitory role of $GABA_A$ receptors on gonadotropin secretion,[161-166] while others show a stimulatory effect of GABA on gonadotropin secretion primarily through the $GABA_A$ receptor system.[163,166-169] $GABA_A$ receptors mediated actions of GABA appear to stimulate the release of GnRH from the arcuate nuclei and the median eminence.[163] $GABA_A$ receptor activation also appears to stimulate gonadotropin release at the level of the pituitary.[166,170,171] The stimulatory and inhibitory effect of $GABA_A$ receptors appears to be critically dependent upon the steroid milieu of the experimental animal. $GABA_A$ receptor activation enhanced LH and FSH secretion in 16-day-old rats and suppressed gonadotropin secretion in 30-day-old rats.[172] $GABA_A$ activation is also involved in the action of 3α-hydroxy-5-α-pregnan-20-one induction of gonadotropin secretion in the estrogen-primed ovariectomized rat.[123] Overall, the effect of GABA is inhibitory on gonadotropin secretion and a decrease in hypothalamic GABA concentrations and in GABA pulses precedes the estrogen-induced LH pulse.[173-175]

C. TACHYKININS

The tachykinin family of peptides with the most prominent members being substance P, neurokinin A, and neuropeptide K, are found in high concentrations in the hypothalamus of the rat and the primate.[176-179] Immunoreactive substance P neurons in the arcuate nucleus project to the median eminence in the rat.[177] The rat hypothalamus contains tachykinin receptors.[180] Tachykinin neurons have estrogen receptors in the medial basal hypothalamus[181] and gonadal steroids regulate hypothalamic tachykinin levels.[182] Tachykinins and preprotachykinin mRNA have been found in the anterior pituitary as well.[183] Stimulatory[183-185] effects of substance P as well as inhibitory[186] effects of substance P on LH secretion have been reported. Similarly neuropeptide K inhibited serum LH levels in ovariectomized rats,[186,187] and the progesterone-induced surge in estrogen-primed rats[186,187] while in male rats it stimulated LH release.[188] Thus, the effects appeared to be dependent upon the steroid environment present. Age-dependent decrease in the LH releasing activity of substance P were also found in dispersed anterior pituitary cell culture experiments.[189]

D. CATECHOLAMINES

Estrogen[190] and progestin[147,191] receptors are reported to be present in catecholaminergic neurons both in and projecting to the hypothalamus, and catecholaminergic neurons have been shown to synapse on GnRH neurons in the hypothalamus.[192] Furthermore, catecholamines have been shown by numerous investigators to be potent regulators of both GnRH and gonadotropin secretion *in vivo* and *in vitro* (for review).[193] Evidence for norepinephrine involvement in progesterone-induced gonadotropin release has arisen from studies demonstrating that norepinephrine levels in the hypothalamus[194] and in plasma[195] are significantly elevated after progesterone treatment, and a correlation with LH secretion has been observed. Brann and Mahesh,[140] through the use of specific α- and β-adrenergic receptor antagonists, have demonstrated that catecholamine neurotransmission through $α_1$ and $α_2$ but not β-receptor sites, is required for the expression of progesterone-induced surges of LH and FSH in estrogen-primed ovariectomized rats. This finding suggests that a commonality exists in the pathway used by steroids to stimulate gonadotropin secretion — the involvement and mediation by the catecholamine systems. Administration of catecholamine synthesis inhibitors or α-adrenergic antagonists on proestrus also leads to abolishment of the proestrous LH surge;[196] demonstrating that catecholamine neurotransmission is an integral component in the neurotransmission line mediating not only steroid-induced gonadotropin surges but also the preovulatory gonadotropin surge.

E. NEUROPEPTIDE Y

Recent evidence suggests that neuropeptide Y (NPY) is a potent regulator of GnRH and gonadotropin secretion (for review).[197,198] Furthermore, it is found in the hypothalamus in significant concentrations which are regulated by estrogens and progesterone. Sahu et al.[199] have demonstrated that the median eminence concentrations of NPY and GnRH change in concert before the proestrus LH surge in the cycling rat. Similar parallelism is reported in hypophysial-portal plasma concentrations of NPY and GnRH.[200] Similar changes in hypothalamic NPY and GnRH occur during the progesterone-induced LH surge in estrogen-primed ovariectomized rats.[201] In pituitary cell culture studies, NPY potentiated the pituitary response to GnRH in estrogen-primed rats and the administration of progesterone did not have an additional NPY potentiation.[202] Such a potentiation of GnRH activity in the release of LH may be mediated by increased binding of GnRH to its anterior pituitary receptors.[203] A major effect of progesterone on NPY appeared to be at the level of the hypothalamus as progesterone administration to estrogen-primed rats resulted in an increase in hypothalamic NPY mRNA levels and the pituitary accumulation of NPY.[144] The administration of antibodies to NPY by either systemic or intracerebroventricular route completely abolished the progesterone-induced gonadotropin surge.[200,204]

F. GALANIN
Galanin, a 29 amino acid peptide is extensively distributed in the hypothalamus and the CNS. It is found to be co-localized in the catecholaminergic neurons[205] and within GnRH neurons[206] and other neurons innervating GnRH neurons.[206] It is found to bring about GnRH release from hypothalamic fragments *in vitro*[207] and LH release after intracerebroventricular injections *in vivo*.[208] It is cosecreted with GnRH in the hypophyseal-portal vein plasma in a pulsatile manner[209] and galanin antibodies attenuate the preovulatory LH surge.[210] Galanin also has been shown to bring about LH release from dispersed pituitary cells and enhance the action of GnRH on LH release.[209] In the PMSG-primed rat and the ovariectomized estrogen-primed rat administered progesterone, galanin mRNA levels increase, correlated tightly with the effects of progesterone in inducing the preovulatory type LH but not the FSH surge.[211] Thus, galanin appears to be an important neuropeptide which is involved in the progesterone regulation of the gonadotropin surge at the level of the anterior pituitary and the hypothalamus.

G. EXCITATORY AMINO ACIDS
1. Endogenous Excitatory Amino Acids and their Receptors
Glutamate and aspartate are the primary excitatory amino acid neurotransmitters in the CNS with glutamate being the most abundant amino acid found in the brain. Strong immunoreactivity for glutamate is found in suprachiasmatic, ventromedial, arcuate, and parvocellular and magnocellular paraventricular nuclei in the rat hypothalamus.[212,213] Immunoreactive glutamate axons are in synaptic contact with dendrites and cell bodies in medial basal hypothalamus, supraoptic, arcuate, suprachiasmatic, and paraventricular nuclei.[212,213]

Excitatory amino acid (EAA) receptors can be divided into two major subgroups: the ionotropic receptors that regulate cation-specific ion channels and the metabotropic receptors that are coupled to G proteins and modulate the production of second messengers. Thus far, only a limited amount of work has been done on metabotropic EAA receptors. The ionotropic receptors can be further subdivided into two classes: the N-methyl-D-aspartate (NMDA) receptors and the non-NMDA receptors consisting of kainate and DL-α-amino-3-hydroxy-5-methyl-4-isoxazole propionic acid (AMPA) receptors. These classifications have been made based on their selective agonists. Each of these major receptor classes are further divided into subclasses (see Reference 6 for review).

NMDA R1 receptors are found in the organum vasculosa of the lamina terminalis, preoptic area, median eminence, and arcuate, supraoptic, suprachiasmatic, and paraventricular nuclei in the hypothalamus. Kainate receptors are found in higher concentrations in the median eminence and arcuate and suprachiasmatic nuclei while other areas of the hypothalamus have lower concentrations. The distribution of AMPA receptors is similar to that of the NMDA R1 receptor (See Reference 6 for review).

2. Regulation of Gonadotropin Secretion by EAAs
The administration of either NMDA, kainate, or AMPA in male and female animals causes a rapid release of LH within 10 to 15 min in a variety of animal species. This topic has been reviewed by Brann and Mahesh.[6] NMDA has been shown to be able to release LH with every pulse administered whereas kainate releases LH only after the first injection.[214-216] The major site of action of EAAs appears to be the release of GnRH in the hypothalamus. This is based on the stimulation of GnRH release by NMDA and kainate in hypothalamic fragments *in vitro*.[217,218] The typical pattern of GnRH, LH, and FSH release after an injection of NMDA to estrogen-primed ovariectomized immature female rats is shown in Figure 8.[219] An increase in GnRH release *in vitro* was observed 3 and 5 min after NMDA administration from medial basal hypothalamus/preoptic area fragments with a rise in serum LH at 5 and 7.5 min and FSH at 7.5 min. The NMDA effect on GnRH release can be blocked by a specific NMDA receptor antagonist AP-5. The NMDA agonists are more potent in OVLT/POA release of GnRH as compared to

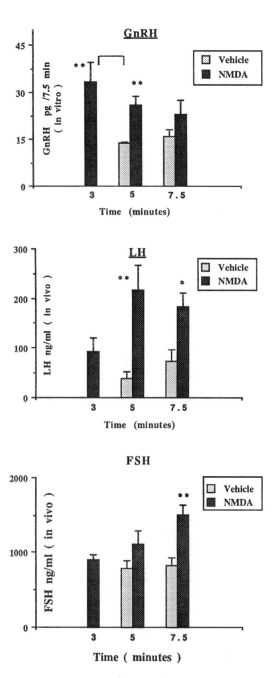

Figure 8 Serum LH and FSH concentrations (lower panels) and levels of GnRH release from individual MBH/POA *in vitro* (top panel) in relation to time after s.c. injection of NMDA (30 mg/kg BW) or saline vehicle in estrogen-primed, ovariectomized immature rats. Female rats were bilaterally ovariectomized at 26 days of age and received a 2 μg per rat injection of estradiol at 27 and 28 days of age (1700 h). On day 29 the animals received either vehicle or NMDA at 1200 h and were killed via decapitation 3, 5, and 7.5 min later. **$p <0.01$; *$p <0.05$; n = 6 animals per group. (From Brann, D.W. and Mahesh, V.B., *Endocrinology*, 128, 1541, 1991. With permission.)

non-NMDA agonists, which appear to be more potent in the ARC-ME region.[220,221] This conclusion is further supported by the observation that GnRH release from ARC/ME fragments induced by glutamate *in vitro* is blocked by the AMPA/kainate receptor antagonist, DNQX, but not by the NMDA receptor antagonist AP-7.[220]

3. Physiological Role of EAAs in the Preovulatory and Steroid-Induced Gonadotropin Surge

The physiological role of EAAs in the regulation of the preovulatory surge of gonadotropins in the cycling adult rat was first demonstrated by Brann and Mahesh[219] who showed that the administration of the NMDA antagonist MK801 completely blocked the proestrus LH surge and lowered but not blocked mean serum FSH levels (Figure 9). In the PMSG-primed immature rat in which PMSG was used to induce the first preovulatory surge of gonadotropins, the NMDA antagonist MK801 attenuated both the LH and the FSH surge.[219] The third ventricle injection of kainate/AMPA receptor antagonist DNQX in PMSG-primed immature rats also attenuated the LH and prolactin surge with no effects on the FSH surge (Figure 10).[222] Thus both NMDA and non-NMDA neurotransmission is important for the preovulatory surge of LH with NMDA neurotransmission having a role in FSH secretion as well.

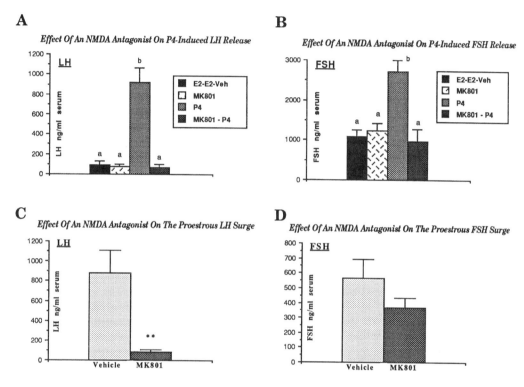

Figure 9 (A–D) Effect of an *N*-methyl-D-aspartate receptor antagonist (MK801) on the progesterone-induced LH and FSH surge in estrogen-primed, ovariectomized immature rats and upon the proestrus LH and FSH surges in cycling adult rats. (A) and (B) illustrate the effect of MK801 (0.2 mg/kg BW, s.c.) on progesterone-induced LH and FSH surges when administered 1 h prior to progesterone (1 mg/kg BW, s.c. 0900) in estrogen-primed, ovariectomized immature rats. All rats were killed at 1400 h for serum LH and FSH measurements. (C) and (D) illustrate the effect of MK801 administration on proestrus (0.2 mg/kg, s.c. at 1100 and 1500 h) on the proestrus surge of LH and FSH in cycling adult rats. b = p <0.01 vs. all other groups. **p <0.01 vs. vehicle proestrus controls.[219,223] (From Mahesh, V.B. and Brann, D.W., *J. Steroid Biochem. Mol. Biol.*, 41, 495, 1992. With permission.)

Because progesterone exerts a pivotal role in the induction of the preovulatory type gonadotropin surge in the ovariectomized animal primed with estrogens in a manner that estrogens by themselves do not induce the surge, the role of NMDA and non-NMDA neurotransmission in the progesterone-induced surge was examined in detail. The NMDA antagonist MK801 administered 1 h before the administration of progesterone completely blocked the LH and FSH surge (Figure 9),[223] whereas, the non-NMDA antagonist DNQX only blocked the LH but not the FSH surge.[222] The estrogen-induced LH surge (perhaps with participation of progesterone from the adrenal) can also be blocked in immature and adult ovariectomized rats by NMDA and non-NMDA antagonists.[224,225] The progesterone-induced GnRH mRNA is also attenuated by MK801.[226]

A

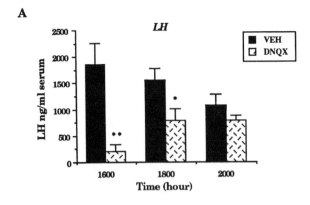

B

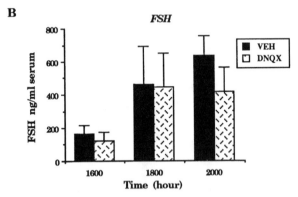

C

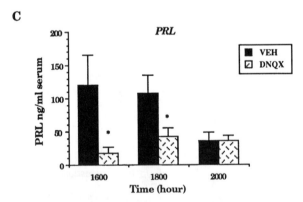

Figure 10 (A–C) Effects of intracere-broventricular administration of the non-NMDA antagonist DNQX (15 nmol × 2) on the preovulatory LH (A), FSH (B), and PRL (C) surges in the PMSG-primed immature female rat; n = 6 rats per group. Veh = PMSG + Veh; DNQX = PMSG + DNQX. **p <0.01 vs. Veh; *p <0.05 vs. Veh. (From Brann et al., *Mol. Cell. Neurosci.*, 4, 292, 1993. With permission.)

4. Regulation of EAA Receptors and Ligand Concentrations by Estrogens and Progesterone

The steroid milieu appears to be very important for the LH-releasing ability of NMDA as it has either no effect or is inhibitory to LH secretion in the ovariectomized animal not treated with estrogens.[227-229] In appropriately estrogen-primed animals, progesterone appears to significantly enhance the effects of NMDA on stimulating LH release.[230,231] These observations raise the question whether steroid treatment results in an increase in EAA receptors or the ligand itself, or a combination of the two.

NMDA receptor binding and NMDA R1 mRNA levels were not altered in male or female rats after castration and after castration and testosterone replacement in the male rat and estrogen replacement with or without progesterone in the female rat.[232] NMDA and kainate receptor binding also did not change in the hypothalamus during the onset of puberty.[233] These findings are supported by the work of Kus et al.[234] who found no effects of castration or dihydrotestosterone treatment on NMDA R1 mRNA levels in the arcuate and preoptic regions of the hypothalamus in the adult male rat. Weiland reported an increase in ^{3}H-glutamate binding in the preoptic area of ovariectomized rats treated with estrogens only when they were administered progesterone.[235] The increase in ^{3}H-glutamate binding was not displaced by NMDA and hence represented an increase in non-NMDA binding sites. Immunohistochemical studies from our laboratory suggest that this increase may be due to an increase in GluR1 subunit immunoreactivity representing AMPA receptor binding sites.[6]

In the absence of estrogen- and progesterone-induced changes in NMDA and kainate receptors, the possibility that progesterone increased glutamate and aspartate levels in the preoptic area resulting in progesterone-induced activation of EAA neurotransmission was next considered. Microdialysis studies by Ping et al.[236] in the estrogen-primed ovariectomized rat treated with progesterone showed that the release rates of glutamate and aspartate were significantly increased immediately preceding the progesterone-induced LH surge (Figure 11). Similar results were obtained by Jarry et al.[237] during the estrogen-induced LH surge while Goroll et al.[238] reported that the release rates of glutamate and aspartate are increased during puberty in the preoptic area in female rats. Thus, estrogen- and progesterone-induced EAA neurotransmission in regulating the gonadotropin surge appears to be mediated by increased EAA levels in the preoptic area as well as an increase in AMPA receptors.

5. Role of EAA in Pulsatile LH and FSH Secretion

In ovariectomized female rats both NMDA receptor antagonist AP-5 and the non-NMDA receptor antagonist DNQX significantly suppressed the LH pulse frequency, LH pulse amplitude, and mean and trough LH levels.[239] AP-5 suppressed LH pulse amplitude and mean and trough LH levels more effectively than DNQX (Figures 12 and 13). The FSH pulse amplitude and mean and trough FSH levels were suppressed by AP-5, whereas FSH pulse frequency was not altered. On the other hand, DNQX did not alter any parameter of FSH secretion. Single injections of AP-5 administered to the castrated male rat in doses and manner similar to the female rat resulted in suppression of pulse amplitude but not frequency and mean and trough levels of LH and FSH.[240] DNQX did not alter any parameter of LH secretion and only the mean levels of FSH were slightly reduced. Prolonged administration of DNQX reduced LH pulse amplitude and mean and trough levels of LH similar to AP-5. Similar results were reported in the male rat after systemic administration of AP-5.[241] Thus, EAAs appear to drive the GnRH pulse generator or modulate its activity with females showing greater sensitivity than males. Furthermore, the pulsatile discharge of LH and FSH is more sensitive to NMDA neurotransmission as compared to non-NMDA neurotransmission.

VIII. ROLE OF STRESS ON GONADOTROPIN SECRETION

Stress is an important modulator of gonadotropin secretion and its effects can be stimulatory or inhibitory depending upon the type and length of stress. This topic has been reviewed recently.[242]

Acute stress has been shown to increase LH secretion in estrogen-primed female rats[243-245] and increase fertility in aged, noncycling mice.[246] The possible role of adrenal progesterone in the facilitation of the preovulatory surge has already been mentioned.[4] The corticotrophin-releasing hormone (CRH) mRNA levels are elevated on the afternoon of proestrus.[247] Furthermore, serum ACTH, progesterone, and corticosterone are elevated on the day of proestrus and the rise in adrenal venous blood levels of progesterone are much higher than those in ovarian venous blood at early proestrus afternoon.[248-250]

Work from our laboratories has shown that the acute administration of ACTH to intact (Figure 14A) or ovariectomized (Figure 14B), estrogen-primed immature rats resulted in a dramatic increase in serum LH levels.[104,251] A similar increase was observed in serum FSH levels as well.[251] However, if intact or ovariectomized rats were not pretreated with estrogens, they did not show any changes in serum LH or FSH after ACTH administration.[251] Furthermore, adrenalectomy (Figure 14C) or pretreatment with the progesterone-glucocorticoid receptor antagonist RU486 (Figure 14E) abolished the gonadotropin-stimulating effect of ACTH. These results clearly indicated that adrenal steroids stimulated by ACTH were responsible for modulating the gonadotropin release. The administration of ACTH did indeed bring about large increases in progesterone and corticosterone secretion (Figure 14D).[104] Of these, progesterone but not corticosterone was able to stimulate LH release in estrogen-primed rats on the day of its administration (Figure 14F).[104] These results clearly point out the important role of adrenal progesterone in modulating the preovulatory gonadotropin surge.

Other mechanisms of the acute effects of ACTH on the adrenal in the stimulation of gonadotropin secretion may be mediated by increases in steroids such as deoxycorticosterone which interacts with the

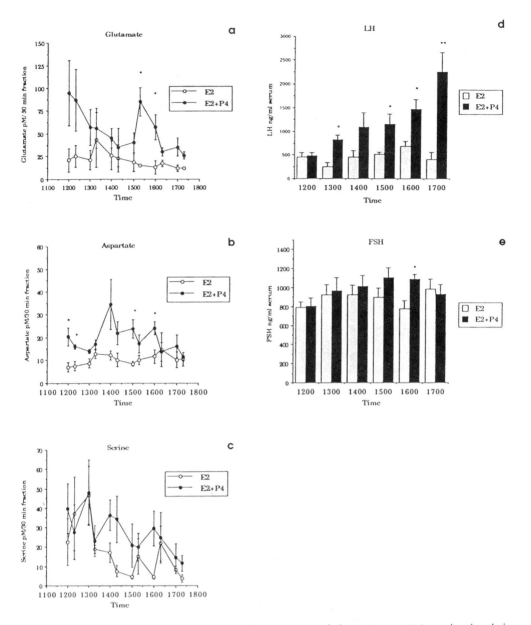

Figure 11 (a–e) Effect of progesterone upon preoptic release rats of glutamate, aspartate, and serine during the time progesterone induces an LH and FSH surge in the estrogen-primed, ovariectomized rat. Adult female rats were ovariectomized for 14 days. They were then injected with 5 µg estradiol benzoate in oil for 2 days at 1700 h. On the third day, they received either progesterone (1 mg/rat) or vehicle at 0900 h. Preoptic area perfusate samples were collected between 1200 to 1730 h. E_2 = estradiol-treated rats; $E_2 + P_4$ = estradiol plus progesterone-treated rats. *p <0.05 vs. E_2. (From Ping et al., *Neuroendocrinology*, 59, 318, 1994. With permission.)

progesterone receptor and causes a stimulation of LH and FSH secretion[103] and ovulation[252] similar to progesterone. Such facilitative actions, however, occur only in the presence of adequate estrogen priming. Corticosteroids have also been shown to bring about selective release of FSH.[104,105,252-255] Furthermore, corticosteroids may have direct ovarian effects such as stimulation of tissue plasminogen activator activity in granulosa cells that plays an important role in the process of ovulation.[256]

The inhibitory actions of stress of prolonged duration on gonadotropin secretion may be mediated by a prolonged increase in ACTH and progesterone, the chronic effects of which are suppressive on

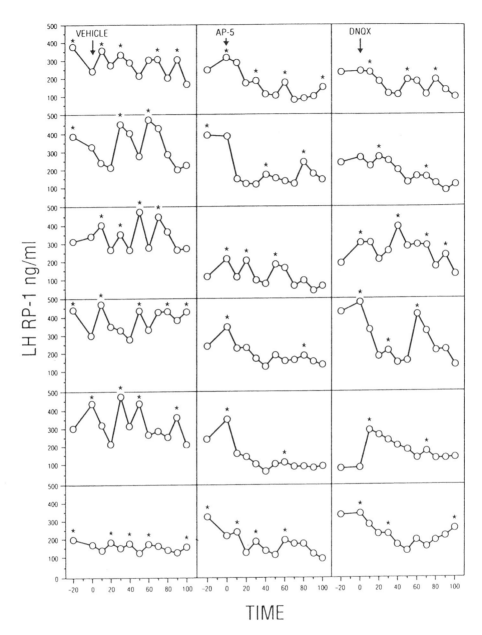

Figure 12 Pattern of LH secretion profile for six ovariectomized adult rats in each of the treatment groups: (1) vehicle, (2) AP-5, and (3) DNQX. LH concentrations are plotted as a function of time after initiation of treatment at 0 min. The asterisks above the open circles denote LH pulses, which were defined and identified using the computer algorithm Pulsar program. Arrows indicate the time of vehicle, AP-5, or DNQX injections. (From Ping et al., *Endocrinology,* 135, 113, 1994. With permission.)

gonadotropin secretion.[1-6] Furthermore, hyperprolactinemia is induced by stress and this has an inhibitory role on gonadotropin secretion (for review).[257] CRH is increased in stress and CRH has been shown to decrease GnRH release in the hypophysial-portal circulation.[258] Finally, prolonged elevation of glucocorticoids due to stress or the administration of glucocorticoids for a prolonged period could have direct suppressive effects on gonadotropin secretion. These suppressive effects may be manifested by the inhibition of granulosa cell aromatase which would reduce the estrogen background that is essential for the preovulatory gonadotropin surge.[259] They may also act at the level of the pituitary by inhibiting estrogen action and estrogen retention.[260,261]

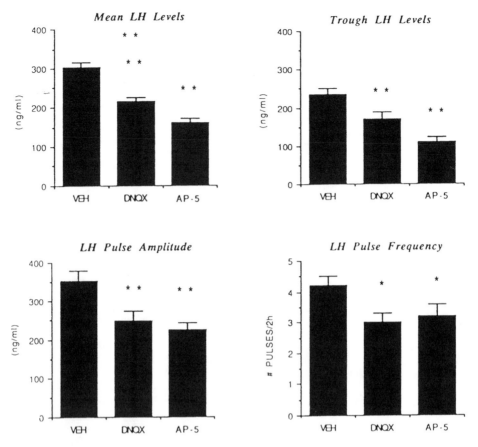

Figure 13 This four-panel graph summarizes the effects of intracerebroventricular administration of DNQX (30 n*M*) and AP-5 (10 mg) on the trough level, mean level, pulse amplitude, and frequency of LH secretion over the 2 h sampling period in ovariectomized adult rats; n = 6 rats per group. Data are expressed as the mean ± SE. Asterisks denote significance by Fisher's tests vs. the respective treatment of LH levels at confidence levels of 95% (*) and 99% or above (**). Trough LH levels after treatment with saline [vehicle (VEH)], DNQX, and AP-5 were 236.1 ± 14.6, 169.8 ± 17.6, and 108.8 ± 12.6. Mean LH levels were 303.6 ± 12,215.7 ± 10.3, and 162.8 ± 9.2. LH pulse amplitudes were 353.8 ± 24.7, 248.8 ± 24.6, and 224.1 ± 19.6. The number of pulses after each treatment were 4.1 ± 0.3, 3 ± 0.3, and 3.2 ± 0.4. Treatment with both DNQX and AP-5 significantly suppressed pulsatile LH secretion. (From Ping et al., *Endocrinology,* 135, 113, 1994. With permission.)

IX. SUMMARY AND CONCLUSIONS

The negative feedback effects of steroid hormones on the hypothalamic-pituitary complex are well recognized. During the ovulatory cycle estrogens and progesterone manifest a positive feedback effect as well, which is responsible for the preovulatory gonadotropin surge. The positive feedback effects of estrogens include release of GnRH from the hypothalamus, synthesis of GnRH receptors in the anterior pituitary in concert with hypothalamic GnRH, and increasing the sensitivity of the pituitary in the release of LH and FSH. Estrogens are also important for the synthesis of hypothalamic and pituitary progesterone receptors.

In spite of the well-documented positive feedback effects of estrogens, experimental evidence shows that estradiol by itself is not responsible for the preovulatory gonadotropin surge. The estrogen-induced gonadotropin surge in the ovariectomized animal is of considerably lesser magnitude and duration and does not have the prolonged FSH secretion observed during the night of proestrus and morning of estrous in the cycling animal. In the presence of adequate estrogen priming, progesterone can induce a preovulatory-type gonadotropin surge. Such a surge is induced by progesterone by increasing GnRH synthesis, bringing about a rapid release of GnRH, suppressing GnRH degradation in the hypothalamus, and reducing anterior pituitary occupied nuclear estrogen receptors. The physiological role of progesterone

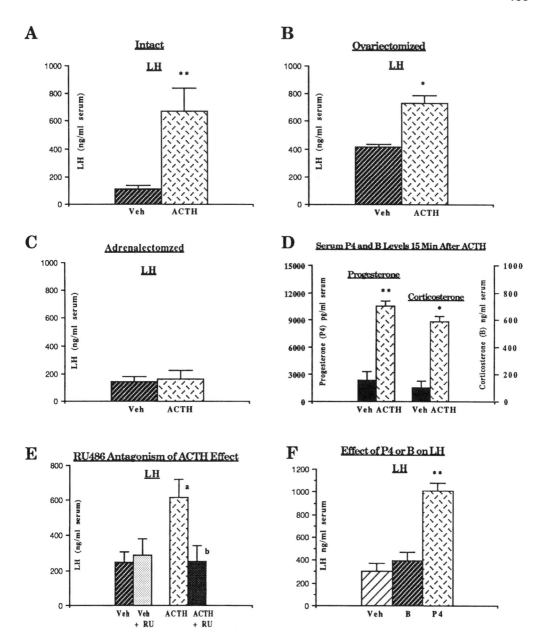

Figure 14 (A–F) Effect of ACTH and the role of progesterone in LH secretion in estrogen-primed intact, ovariectomized, and adrenalectomized immature rats. 26-day-old female rats were treated with 2 μg of estradiol in 0.2 ml 20% ethanol: saline for 2 days at 1700 h. On the third day ACTH or steroids were injected at 0900 h and the animals killed at 1500 h to measure serum LH levels. Results of the effect of injection of 100 μg ACTH on serum LH in intact rats is shown in Panel A, in ovariectomized rats in Panel B, and in adrenalectomized rats in Panel C. Serum levels of progesterone (P_4) and corticosterone (B) 15 min after the ACTH injection are shown in Panel D. The administration of 0.75 mg of RU486 at 0800 h blocked the effect of ACTH on LH release (Panel E). Progesterone (P_4: 0.8 mg/kg BW) but not corticosterone (B; 0.8 mg/kg BW) was able to induce an LH surge in ovariectomized, estrogen-primed immature rats (Panel F). *p <0.05 vs. Veh; **p <0.01 vs. Veh; a = p <0.01 vs. Veh; b = p <0.05 vs. ACTH.[104,251] (From Mahesh, V.B. and Brann, D.W., *J. Steroid Biochem. Mol. Biol.*, 41, 495, 1992. With permission.)

in inducing the preovulatory gonadotropin surge has been demonstrated by the attenuation of the surge by the progesterone receptor antagonist RU486 and progesterone synthesis inhibitors trilostane and epostane. The progesterone required for the preovulatory gonadotropin surge can come from the ovary as well as the adrenal. Prolonged administration of progesterone or administration of progesterone in the presence of inadequate estrogen priming results in the suppression of the gonadotropin surge.

The GnRH neuron has been demonstrated not to possess estrogen and progesterone receptors. Therefore, the GnRH neuron appears to be regulated by neurotransmitters which are in turn regulated by steroids. These neurotransmitters could have inhibitory as well as stimulatory influences on gonadotropin secretion. The major inhibitory neurotransmitters identified to date are the opioids, the tachykinins, and GABA. A short list of the major neurotransmitters stimulatory to GnRH secretion include catecholamines, NPY, galanin, and EEAs. Recent studies have focused on the importance of glutamate and aspartate because they represent the major EEAs present in the brain and have numerous receptors. The most studied receptors are the NMDA and non-NMDA (kainate and AMPA) receptors. NMDA receptors mediate the preovulatory- and progesterone-induced gonadotropin surge which is attenuated by the administration of the NMDA antagonist MK801. Non-NMDA receptors mediate the LH but not the FSH surge. Both receptor types are important in pulsatile discharges of LH and FSH. Steroids do not appear to regulate NMDA receptors as shown by castration and steroid replacement studies, but they may increase AMPA receptors. During the progesterone-induced preovulatory type LH surge, it is an increase of the ligand glutamate and aspartate that is responsible for the surge.

In the presence of adequate estrogen priming acute stress brings about gonadotropin release by the release of adrenal progesterone and corticosteroids that interact with the progesterone receptor such as deoxycorticosterone. Glucocorticoids also are responsible for FSH secretion and direct effects on the ovary. Prolonged stress suppresses gonadotropin secretion by increasing CRH, prolactin, progesterone, and glucocorticoids for a prolonged period. Each one of these have been shown to suppress gonadotropin secretion.

ACKNOWLEDGMENTS

The research of Dr. V. B. Mahesh was supported by grant #R01-HD-16688 and of Dr. D. W. Brann by grant #R29-HD-28964 from the National Institute of Child Health and Human Development, National Institutes of Health, Bethesda, MD.

REFERENCES

1. Kalra, S. P. and Kalra, P. S., Neural regulation of LH secretion in the rat, *Endocrinol. Rev.*, 4, 311, 1983.
2. Mahesh, V. B., The dynamic interaction between steroids and gonadotropins in the mammalian ovulatory cycle, *Neurosci. Biobehav. Rev.*, 9, 245, 1985.
3. Fink, G., Gonadotropin secretion and its control, in *The Physiology of Reproduction*, Vol. I, Knobil, E. and Neill, J., Eds., Raven Press, New York, 1988, 1349.
4. Brann, D. W. and Mahesh, V. B., Regulation of gonadotropin secretion by steroid hormones, *Front. Neuroendocrinol.*, 12, 165, 1991.
5. Mahesh, V. B. and Brann, D. W., Interaction between ovarian and adrenal steroids in the regulation of gonadotropin secretion, *J. Steroid Biochem. Mol. Biol.*, 41, 495, 1992.
6. Brann, D. W. and Mahesh, V. B., Excitatory amino acids: functions and significance in reproduction and neuroendocrine regulation, *Front. Neuroendocrinol.*, 15, 3, 1994.
7. McPherson, J. C., Eldridge, J. C., Costoff, A., and Mahesh, V. B., The pituitary-gonadal axis before puberty: effects of various estrogenic steroids in the ovariectomized rat, *Steroids*, 24, 41, 1974.
8. Yen, S. S. C. and Tasi, C. C., The biphasic pattern in the feedback actions of ethinyl estradiol and the release of pituitary FSH and LH, *J. Clin. Endocrinol. Metab.*, 33, 882, 1971.
9. Ramirez, V. D. and Sawyer, C. H., Advancement of puberty in the female rat by estrogens, *Endocrinology*, 76, 1158, 1965.
10. Smith, E. R. and Davidson, J. M., Role of estrogens in the cerebral control of puberty in female rats, *Endocrinology*, 82, 100, 1968.
11. Knudsen, J. F. and Mahesh, V. B., Initiation of precocious sexual maturation in the immature rat treated with dehydroepiandrosterone, *Endocrinology*, 97, 458, 1975.
12. Knudsen, J. F., Costoff, A., and Mahesh, V. B., Correlation of serum gonadotropins, ovarian and uterine histology in immature and pubertal rats, *Anat. Rec.*, 180, 497, 1974.
13. Parker, C. R. and Mahesh, V. B., Hormonal events surrounding the natural onset of puberty in female rats, *Biol. Reprod.*, 14, 347, 1976.
14. Sizonenko, P. C. and Paunier, L., Hormonal changes in puberty III: correlation of plasma dehydroepiandrosterone, testosterone, FSH and LH with stages of puberty and bone age in normal boys and girls and in patients with Addison's disease or hypogonadism or with premature or late adrenarche, *J. Clin. Endocrinol. Metab.*, 41, 894, 1975.

15. Brom, G. M. and Schwartz, N. B., Acute changes in the estrous cycle following ovariectomy in the golden hamster, *Neuroendocrinology*, 3, 366, 1968.
16. Brown-Grant, K., The induction of ovulation by ovarian steroids in the adult rat, *J. Endocrinol.*, 43, 553, 1969.
17. Goding, J. R., Catt, K. J., Brown, J. M., Kaltenbach, C. C., Cumming, I. A., and Mole, B. J., Radioimmunoassay for ovine LH. Secretion of LH during estrus and following estrogen administration in the sheep, *Endocrinology*, 85, 113, 1969.
18. Yoshinaga, K., Hawkins, R. A., and Stocker, J. F., Estrogen secretion by the rat ovary *in vivo* during the estrous cycle and pregnancy, *Endocrinology*, 85, 103, 1969.
19. Goldman, B. D., Mahesh, V. B., and Porter, J. C., The role of the ovary in control of cyclic LH release in the hamster, *Biol. Reprod.*, 4, 57, 1971.
20. Mann, D. R., and Barraclough, C. A., Role of estrogen and progesterone in facilitating LH release in 4-day cyclic rats, *Endocrinology*, 93, 694, 1973.
21. Norman, R. L., Blake, C. A., and Sawyer, C. H., Estrogen-dependent twenty-four hour periodicity in pituitary LH release in the female hamster, *Endocrinology*, 93, 965, 1973.
22. Legan, S. J., Coon, G. A., and Karsch, F. J., Role of estrogen as initiator of daily LH surges in the ovariectomized rat, *Endocrinology*, 96, 50, 1975.
23. Shirley, B., Wolinsky, J., and Schwartz, N. B., Effects of a single injection of an estrogen antagonist on the estrous cycle of the rat, *Endocrinology*, 82, 959, 1968.
24. Ferin, M., Tempone, A., Zimmering, P. E., and Vande Wiele, R. L., Effect of antibodies to 17β-estradiol and progesterone on the estrous cycle of the rat, *Endocrinology*, 85, 1070, 1969.
25. Labhsetwar, A. P., Role of estrogens in ovulation: a study using the estrogen-antagonist ICI 46,474, *Endocrinology*, 87, 542, 1970.
26. Yamaji, T., Dierschke, D. J., Hotchkiss, J., Bhattacharya, A. M., Surve, A. H., and Knobil, E., Estrogen induction of LH release in the rhesus monkey, *Endocrinology*, 89, 1034, 1971.
27. Monroe, S. E., Jaffe, R. B., and Midgley, A. R., Jr., Increase in serum gonadotropins in response to estradiol, *J. Clin. Endocrinol. Metab.*, 34, 342, 1972.
28. Yen, S. S., Lasley, B. L., Wang, C. F., Leblac, H., and Siler, T. M., The operating characteristics of the hypothalamic-pituitary system during the menstrual cycle and observations of biological actions of somatostatin, *Rec. Prog. Horm. Res.*, 31, 321, 1975.
29. March, C. M., Goebelsmann, U., Nakamura, R. M., and Mishell, D. R., Roles of estradiol and progesterone in eliciting the midcycle LH and FSH surges, *J. Clin. Endocrinol. Metab.*, 49, 507, 1979.
30. Greeley, G. H., Muldoon, T. G., and Mahesh, V. B., Correlative aspects of LHRH sensitivity and cytoplasmic estrogen receptor concentration in the anterior pituitary and hypothalamus of the cycling rat, *Biol. Reprod.*, 13, 505, 1975.
31. Nillius, S. J. and Wide, L., Variations in LH and FSH response to LH-releasing hormone during the menstrual cycle, *J. Obstet. Gynecol. Br. Commonw.*, 79, 865, 1972.
32. Mahesh, V. B., Muldoon, T. G., Eldridge, J. C., and Korach, K. S., The role of steroid hormones in the regulation of gonadotropin secretion, *J. Steroid Biochem.*, 6, 1025, 1975.
33. Parker, C. R., Jr., Costoff, A., Muldoon, T. G., and Mahesh, V. B., Actions of PMSG in the immature female rat: correlative changes in blood steroids, gonadotropins and cytoplasmic estradiol receptors of the anterior pituitary and hypothalamus, *Endocrinology*, 98, 129, 1976.
34. Ashiru, O. A. and Blake, C. A., Effect of ovariectomy, estrogen and LHRH on periovulatory increases in plasma gonadotropins in the cycling rat, *Biol. Reprod.*, 22, 533, 1980.
35. Turgeon, J. L. and Barraclough, C. A., Regulatory role of estradiol in pituitary responsiveness to LHRH on proestrus in the rat, *Endocrinology*, 101, 548, 1977.
36. Turgeon, J. L., Estradiol-luteinizing hormone relationship during the proestrous gonadotropin surge, *Endocrinology*, 105, 731, 1979.
37. Stumpf, W. E., Cellular and subcellular [³H]estradiol localization in the pituitary by autoradiography, *Z. Zellforsch.*, 92, 23, 1968.
38. Anderson, C. H. and Greenwald, G. S., Autoradiographic analysis of estradiol uptake in the brain and pituitary of the female rat, *Endocrinology*, 85, 1160, 1969.
39. Korach, K. S. and Muldoon, T. G., Comparison of specific 17β-estradiol receptor interactions in the anterior pituitary of male and female rats, *Endocrinology*, 92, 322, 1973.
40. Muldoon, T. G., Regulation of steroid hormone receptor activity, *Endocrine Rev.*, 1, 339, 1980.
41. McEwen, B. S., Hormones, receptors and brain functions, *Adv. Pathobiol.*, 1, 56, 1975.
42. Kato, J., Steroid hormone receptors in brain, hypothalamus and hypophysis, in *Receptors and Mechanism of Action of Steroid Hormones*, Part II, Pasqualini, J. R., Ed., Marcel Dekker, New York, 1977, 603.
43. Sarkar, D. K., Chippa, S. A., Fink, G., and Sherwood, N. M., Gonadotropin releasing hormone surge in proestrous rats, *Nature*, 264, 461, 1976.
44. Schillo, K. K., Leshin, L. S., Kuehl, D., and Jackson, G. L., Simultaneous measurement of luteinizing hormone releasing hormone and luteinizing hormone during estradiol induced luteinizing hormone surges in the ovariectomized ewe, *Biol. Reprod.*, 33, 644, 1985.

45. Moenter, S. M., Brand, R. C., and Karsch, F. J., Dynamics of GnRH secretion during the GnRH surge: insights into the mechanism of GnRH surge induction, *Endocrinology*, 130, 2978, 1992.

46. Levine, J. E., Norman, R. L., Gliessman, P. M., Oyama, T. T., Bangsberg, D. R., and Spies, H. G., *In vivo* gonadotropin releasing hormone release and serum luteinizing hormone measurements in ovariectomized estrogen treated rhesus macaques, *Endocrinology*, 117, 711, 1985.

47. Xia, L., Van Vugt, D., Alston, E. J., Lerckhaus, J., and Ferin, M., A surge of GnRH accompanies the estradiol-induced gonadotropin surge in the rhesus monkey, *Endocrinology*, 131, 2812, 1992.

48. Leadem, C. A. and Kalra, S. P., Stimulation with estrogen and progesterone of luteinizing hormone releasing hormone release from perifused adult female rat hypothalami: correlation with the LH surge, *Endocrinology*, 114, 51, 1984.

49. Kim, K. and Ramirez, V. D., *In vitro* luteinizing hormone releasing hormone release from superfused rat hypothalami: site of action of progesterone and effect of estrogen priming, *Endocrinology*, 116, 252, 1985.

50. Levine, J. E., Bethea, C. L., and Spies, H. G., *In vitro* gonadotropin releasing hormone release from hypothalamic tissue of ovariectomized estrogen treated cynomolgus macques, *Endocrinology*, 116, 431, 1985.

51. Libertun, C., Orias, R., and McCann, S. M., Biphasic effect of estrogen on the sensitivity of the pituitary to luteinizing hormone releasing factor, *Endocrinology*, 94, 1094, 1974.

52. Martin, J. E., Tyrey, L., Everett, J. W., and Fellows, R. E., Variation in responsiveness of synthetic LH-releasing factor in proestrous and diestrous-3 rats, *Endocrinology*, 94, 556, 1974.

53. Greeley, G. H., Allen, M. B., and Mahesh, V. B., Potentiation of LH release by estradiol at the level of the pituitary, *Neuroendocrinology*, 18, 233, 1975.

54. Hsueh, A. J. W., Erickson, G. F., and Yen, S. S. C., The sensitizing effect of estrogens and catechol estrogens on cultured rat pituitary cells to LHRH: its antagonism by progesterone, *Endocrinology*, 104, 807, 1979.

55. Drouin, J. and Labrie, F., Interaction between 17β-estradiol and progesterone in the control of LH and FSH release in rat anterior pituitary cells in culture, *Endocrinology*, 108, 52, 1981.

56. Lesoon, L. A. and Mahesh, V. B., Stimulatory and inhibitory effects of progesterone on FSH secretion by the anterior pituitary, *J. Steroid Biochem. Mol. Biol.*, 42, 479, 1992.

57. Moll, G. W. and Rosenfield, R. L., Direct inhibitory effects of estradiol on pituitary luteinizing hormone responsiveness to luteinizing hormone releasing hormone is specific and of rapid onset, *Biol. Reprod.*, 30, 59, 1984.

58. Frawley, L. S. and Neill, J. D., Biphasic effects of estrogen on gonadotropin-releasing hormone induced luteinizing hormone release in monolayer cultures of rat and monkey pituitary cells, *Endocrinology*, 114, 659, 1984.

59. Yen, S. S. C., Vandenberg, G., and Siler, T. M., Modulation of the pituitary responsiveness to LRF by estrogens, *J. Clin. Endocrinol. Metab.*, 39, 170, 1974.

60. Aiyer, M. S., Chiappa, S. A., and Fink, G., A priming effect of LHRF on the anterior pituitary gland in the female rat, *J. Endocrinol.*, 62, 573, 1974.

61. Fink, G., Chiappa, A., and Aiyer, M. S., Priming effects of luteinizing hormone releasing factor elicited by preoptic stimulation and by intravenous infusion and multiple injections of the synthetic decapeptide, *J. Endocrinol.*, 69, 359, 1976.

62. Waring, D. W. and Turgeon, J. L., Self-priming of gonadotropin secretion: time course of development, *Am. J. Physiol.*, 244, C410, 1983.

63. Pieper, D. R., Gala, R. R., Regiani, S. R., and Marshall, J. C., Dependence of pituitary gonadotropin releasing hormone (GnRH) receptors on GnRH secretion from the hypothalamus, *Endocrinology*, 110, 749, 1982.

64. Clayton, R. H., Channabasaviah, K., Stewart, J. M., and Catt, K. J., Hypothalamic regulation of pituitary gonadotropin releasing hormone receptors, *Endocrinology*, 110, 1108, 1982.

65. Duncan, J. A., Barkan, A., Herbon, L., and Marshall, J. C., Regulation of pituitary gonadotropin releasing hormone (GnRH) receptors by pulsatile GnRH in female rats: effects of estradiol and prolactin, *Endocrinology*, 118, 320, 1986.

66. Kaiser, V. B., Jakubowiak, A., Steinberger, A., and Chin, W. W., Regulation of rat pituitary GnRH receptor mRNA levels *in vivo* and *in vitro*, *Endocrinology*, 133, 931, 1993.

67. Clayton, R. N., Solano, A. R., Garcia-Vela, A., Dufau, M. L., and Catt, K. J., Regulation of pituitary receptors for GnRH during the rat estrous cycle, *Endocrinology*, 107, 699, 1980.

68. Savoy-Moore, R. T., Scwartz, N. B., Duncan, J. A., and Marshall, J. C., Pituitary GnRH receptors on proestrus: effect of pentobarbital blockage of ovulation in the rat, *Endocrinology*, 109, 1360, 1981.

69. Bauer-Dantoin, A. C., Hollenberg, A. N., and Jameson, J. L., dynamic regulation of GnRH receptor mRNA levels in the anterior pituitary gland during the rat estrous cycle, *Endocrinology*, 133, 1911, 1993.

70. Gregg, D. W. and Nett, T. M., Direct effects of estradiol-17β on the number of GnRH receptors in the ovine pituitary, *Biol. Reprod.*, 40, 288, 1989.

71. Gregg, D. W., Allen, M. C., and Nett, T. M., Estradiol induced increase in number of GnRH receptors in cultured ovine pituitary cells, *Biol. Reprod.*, 43, 1032, 1990.

72. Emons, G., Hoffman, H. G., Brack, C., Ortmann, O., Sturm, R., Ball, P., and Knuppen, R., Modulation of GnRH receptor concentration in cultured female rat pituitary cells by estradiol treatment, *J. Steroid Biochem.*, 31, 751, 1988.

73. Wagner, T. O. F., Adams, T. E., and Nett, T. M., GnRH interaction with anterior pituitary, I. Determination of the affinity and number of receptors for GnRH in ovine anterior pituitary, *Biol. Reprod.*, 20, 140, 1979.

74. Ferland, L., Marchetti, B., Seguin, C., Lefebvre, F. A., Reeves, J. J., and Labrie, F., Dissociated changes of pituitary LHRH receptors and responsiveness to neurohormone induced by 17β-estradiol and LHRH *in vitro* in the rat, *Endocrinology*, 109, 87, 1981.

75. Ramaley, J. A., Minireview: adrenal-gonadal interactions at puberty, *Life Sci.*, 14, 1623, 1974.
76. MacFarland, L. A. and Mann, D. R., The inhibitory effects of ACTH and adrenalectomy on reproductive maturation in female rats, *Biol. Reprod.*, 16, 306, 1977.
77. Lawton, E. E., Facilitory feedback effects of adrenal and ovarian hormones on LH secretion, *Endocrinology*, 90, 575, 1972.
78. Peppler, R. and Jacobs, J., The effect of adrenalectomy on ovulation and follicular development in the rat, *Biol. Reprod.*, 15, 173, 1976.
79. Mann, D. R., Korowitz, C., and Barraclough, C., Adrenal gland involvement in synchronizing the preovulatory release of LH in rats, *Proc. Soc. Exp. Biol. Med.*, 150, 115, 1975.
80. Meijs-Roelogs, H. and Kramer, P., Effects of adrenalectomy on the release of follicle-stimulating hormone and the onset of puberty in female rats, *J. Endocrinol.*, 75, 419, 1977.
81. Hilliard, J., Pernardi, R., and Sawyer, C., A functional role of 20α-hydroxy-pregn-4-en-3-one in the rabbit, *Endocrinology*, 80, 901, 1967.
82. Krey, L. C., Tyrey, L., and Everett, J. W., The estrogen-induced advance in the cyclic LH surge in the rat: dependency on ovarian progesterone secretion, *Endocrinology*, 93, 385, 1973.
83. Aiyer, M. S. and Fink, G., The role of sex steroid hormones in modulating the responsiveness of the anterior pituitary gland to LHRH in the female rat, *J. Endocrinol.*, 62, 533, 1974.
84. Everett, J. W., Progesterone and estrogens in the experimental control of ovulation time and other features of the estrous cycle in the rat, *Endocrinology*, 43, 389, 1948.
85. Naller, R., Antunes-Rodrigues, J., and McCann, S., Effects of progesterone on the level of plasma luteinizing hormone (LH) in normal female rats, *Endocrinology*, 79, 907, 1966.
86. Caligaris, L., Astrada, J. J., and Taleisnik, S., Stimulating and inhibiting effects of progesterone on the release of luteinizing hormone, *Acta Endocrinol. (Copenhagen)*, 59, 177, 1968.
87. Krey, L. C., Tyrey, L., and Everett, J. W., The estrogen-induced advance in the cyclic LH surge in the rat: dependency on ovarian progesterone secretion, *Endocrinology*, 93, 385, 1973.
88. Martin, J. E., Tyrey, L., and Everett, J. W., Estrogen and progesterone modulation of the pituitary response of LRF in the cyclic rat, *Endocrinology*, 95, 1664, 1974.
89. McPherson, J. C., Costoff, A., and Mahesh, V. B., Influence of estrogen-progesterone combinations on gonadotropins secretion in castrate female rats, *Endocrinology*, 91, 771, 1975.
90. McPherson, J. C. and Mahesh, V. B., Dose-related effect of a single injection of progesterone on gonadotropin secretion and pituitary sensitivity to LHRH in estrogen-primed castrated female rats, *Biol. Reprod.*, 20, 763, 1979.
91. Freeman, M. E., Duke, K. C., and Croteau, C. M., Extinction of the estrogen-induced daily signal for LH release in the rat: a role for the proestrus surge of progesterone, *Endocrinology*, 99, 223, 1976.
92. Banks, J. A. and Freeman, M. E., The temporal requirement of progesterone on proestrus for the extinction of the estrogen induced daily signal controlling luteinizing hormone release in the rat, *Endocrinology*, 102, 426, 1978.
93. Banks, J. A., Mick, C., and Freeman, M. E., A possible cause for the differing responses of the LH surge mechanisms of ovariectomized rats to short-term exposure to estradiol, *Endocrinology*, 106, 1677, 1980.
94. Odell, W. D. and Swerdloff, R. S., Progesterone-induced luteinizing and follicle-stimulating hormone surge in post-menopausal women: a simulated ovulatory peak, *Proc. Natl. Acad. Sci. U.S.A.*, 61, 529, 1968.
95. Nillius, S. J. and Wide, L., Effects of progesterone on the serum levels of FSH and LH in postmenopausal women treated with estrogen, *Acta Endocrinol.*, 67, 362, 1971.
96. Wise, A. J., Gross, M. A., and Schalach, D. A., Quantitative relationships of the pituitary-gonadal axis in postmenopausal women, *J. Lab. Clin. Med.*, 81, 28, 1973.
97. McCann, S. M., Effect of progesterone on plasma luteinizing hormone activity, *Am. J. Physiol.*, 202, 601, 1962.
98. McPherson, J. C., Costoff, A., Eldridge, J. C., and Mahesh, V. B., Effects of various progestational preparations on gonadotropin secretion in ovariectomized immature female rats, *Fertil. Steril.*, 25, 1063.
99. Smanik, E. J., Young, H. K., Muldoon, T. G., and Mahesh, V. B., Analysis of the effect of progesterone *in vivo* on estrogen receptor distribution in the rat anterior pituitary and hypothalamus, *Endocrinology*, 113, 15, 1983.
100. Peduto, J. C. and Mahesh, V. B., Effects of progesterone on hypothalamic and plasma LHRH, *Neuroendocrinology*, 40, 238, 1985.
101. Smith, M. S., Freeman, M. E., and Neill, J. D., The control of progesterone secretion during the estrous cycle and early pseudopregnancy in the rat: prolactin, gonadotropin and steroid levels associated with rescue of the corpus luteum of pseudopregnancy, *Endocrinology*, 96, 219, 1975.
102. Calderon, J. J., Muldoon, T. G., and Mahesh, V. B., Receptor-mediated interrelationships between progesterone and estradiol action on the anterior pituitary-hypothalamic axis of the ovariectomized immature rat, *Endocrinology*, 120, 2428, 1987.
103. Brann, D. W., Putnam, C. D., and Mahesh, V. B., Corticosteroid regulation of gonadotropin and prolactin secretion in the rat, *Endocrinology*, 126, 159, 1990.
104. Brann, D. W., Putnam, C. D., and Mahesh, V. B., Validation of the mechanisms proposed for the stimulatory and inhibitory effects of progesterone on gonadotropin secretion in the estrogen-primed rat. A possible role for adrenal steroids, *Steroids*, 56, 103, 1991a.

105. Brann, D. W., O'Conner, J., Wade, M., and Mahesh, V. B., LH and FSH subunit mRNA concentrations during the progesterone-induced gonadotropin surge in ovariectomized estrogen-primed immature rats, *Mol. Cell. Neurosci.*, 3, 171, 1992.

106. Brann, D. W., O'Conner, J. L., Wade, M. F., Zamorano, P. L., and Mahesh, V. B., Regulation of anterior pituitary gonadotropin subunit mRNA levels during the preovulatory gonadotropins surge: a physiological role of progesterone in regulating LH-β and FSH-β mRNA levels, *J. Steroid Biochem. Mol. Biol.*, 46, 427, 1993.

107. Zmeili, S. M., Papavasiliou, S. S., Thorner, M. O., Evans, W. S., Marshall, J. C., and Landefeld, T. D., Alpha and luteinizing hormone beta subunit messenger ribonucleic acids during the rat estrous cycle, *Endocrinology*, 119, 1867, 1986.

108. Ortolano, G. A., Haisenleder, D. J., Iliff-Sizemore, S. A., Landefeld, T. D., Maurer, R. A., and Marshall, J. C., Follicle-stimulating hormone beta subunit messenger ribonucleic concentrations during the rat estrous cycle, *Endocrinology*, 123, 2149, 1988.

109. Rao, I. M. and Mahesh, V. B., Role of progesterone in the modulation of the preovulatory surge of gonadotropins and ovulation in the PMSG primed immature rat and the adult rat, *Biol. Reprod.*, 35, 1154, 1986.

110. DePaolo, L. V., Attenuation of preovulatory gonadotropin surges by epostane: a new inhibitor of 3β-hydroxysteroid dehydrogenase, *J. Endocrinol.*, 118, 59, 1988.

111. Kalra, S. P., Observations on facilitation of the preovulatory rise of LH by estrogen, *Endocrinology*, 96, 23, 1975.

112. Lagace, L., Massicotte, J., and Labrie, F., Acute stimulatory effects of progesterone on LH and FSH release in rat anterior pituitary cell culture, *Endocrinology*, 106, 684, 1980.

113. Ortmann, O., Weise, H., Knuppen, R., and Emons, G., Acute facilitory action of progesterone on gonadotropin secretion of perifused rat pituitary cells, *Acta Endocrinol. (Copenhagen)*, 121, 426, 1989.

114. Krey, L. C. and Kamel, F., Progesterone modulation of gonadotropin secretion by dispersed rat pituitary cells in culture. I. Basal and gonadotropin-releasing hormone-stimulated luteinizing hormone release, *Mol. Cell. Endocrinol.*, 68, 85, 1990.

115. Turgeon, J. L. and Waring, J., Acute progesterone and 17β-estradiol modulation of LH secretion by pituitaries of cycling rats superfused *in vitro*, *Endocrinology*, 108, 413, 1981.

116. Witcher, J. A., Nearhoof, K. F., and Freeman, M. C., Secretion of luteinizing hormone and pituitary receptors for LH-releasing hormone as modified by the proestrous surge of progesterone, *Endocrinology*, 115, 2189, 1984.

117. Attardi, B. and Happe, H. K., Modulation of the estradiol-induced LH surge by progesterone or antiestrogens: effects on pituitary gonadotropin-releasing hormone receptors, *Endocrinology*, 119, 274, 1986.

118. Emons, G., Nill, J., Sturm, R., and Ortmann, O., Effects of progesterone on GnRH receptor concentration in cultured estrogen-primed female rat pituitary cells, *J. Steroid Biochem. Mol. Biol.*, 42, 831, 1992.

119. Krey, L. and Kamel, F., Progesterone modulation of gonadotropin secretion by dispersed rat pituitary cells in culture. III. A23187, cAMP, phorbol ester and DiC8-stimulated luteinizing hormone release, *Mol. Cell. Endocrinol.*, 70, 21, 1990.

120. Kim, K. and Ramirez, V. D., *In vitro* progesterone stimulates the release of luteinizing hormone-releasing hormone from superfused hypothalamic tissue from ovariectomized estradiol-primed prepubertal rats, *Endocrinology*, 111, 750, 1982.

121. Levine, J. E. and Ramirez, V. D., *In vivo* release of mediobasal hypothalami of ovariectomized steroid-primed rats, *Endocrinology*, 107, 1782, 1982.

122. Ke, F.-C. and Ramirez, V. D., Membrane mechanism mediates progesterone stimulatory effect on LHRH release from superfused rat hypothalami *in vitro*, *Neuroendocrinology*, 45, 514, 1987.

123. Brann, D. W., Putnam, C. D., and Mahesh, V. B., γ-Aminobutyric acid$_A$ receptors mediate 3α-hydroxy-5α-pregnan-20-one-induced gonadotropin secretion, *Endocrinology*, 126, 1854, 1990.

124. Advis, J. P., Krause, J. E., and McKelvy, J. F., Luteinizing hormone-releasing hormone peptidase activities in discrete hypothalamic regions and anterior pituitary of the rat: apparent regulation during the pre-pubertal period and first estrous cycle of puberty, *Endocrinology*, 110, 1238, 1982.

125. O'Conner, J. L., Lapp, C. A., and Mahesh, V. B., Peptidase activity in the hypothalamus and pituitary of the rat: fluctuations and possible regulatory role of luteinizing hormone releasing hormone degrading activity during the estrous cycle, *Biol. Reprod.*, 30, 855, 1985.

126. O'Conner, J. L. and Mahesh, V. B., A possible role for progesterone in the preovulatory gonadotropin surge through modulation of LHRH degrading activity, *J. Steroid Biochem.*, 29, 257, 1988.

127. Okulicz, W., Evans, R., and Leavitt, W., Progesterone regulation of the occupied form of nuclear estrogen receptor, *Science, N.Y.*, 213, 1503, 1981.

128. Mauvais-Jarvis, P., Kittenn, F., and Gompel, A., Antiestrogen action of progesterone in breast tissue, *Horm. Res.*, 28, 212, 1987.

129. Debold, J. F., Martin, J. V., and Whalen, R. E., The excitation and inhibition of sexual receptivity in female hamsters by progesterone: time and dose relationships, neural localization and mechanisms of action, *Endocrinology*, 99, 1519, 1976.

130. Marrone, B. L. and Feder, H. H., Characteristics of [³H]estrogen and [³H]progestin uptake and effects of progesterone on [³H]estrogen uptake in brain, anterior pituitary and peripheral tissues of male and female guinea-pigs, *Biol. Reprod.*, 17, 391, 1977.

131. Schwartz, S. M., Blaustein, J. D., and Wade, G. N., Inhibition of estrous behavior by progesterone in rats: role of neural estrogen and progesterone receptors, *Endocrinology*, 105, 1078, 1979.

132. Attardi, B., Facilitation and inhibition of the estrogen induced luteinizing hormone surge in the rat by progesterone: effects on cytoplasmic and nuclear estrogen receptors in the hypothalamus-preoptic area, pituitary and uterus, *Endocrinology*, 108, 1487, 1981.

133. Brann, D. W., Rao, I. M., and Mahesh, V. B., Antagonism of estrogen-induced prolactin release by progesterone, *Biol. Reprod.*, 39, 1067, 1988.

134. El Ayat, A. A. B. and Mahesh, V. B., Stimulation of 17β-hydroxysteroid dehydrogenase in the rat anterior pituitary gland by progesterone, *J. Steroid Biochem.*, 20, 1141, 1984.

135. Fuentes, M. A., Muldoon, T. G., and Mahesh, V. B., Role of 17β-hydroxysteroid dehydrogenase in the modulation of nuclear estradiol receptor binding by progesterone in the rat anterior pituitary gland and the uterus, *J. Steroid Biochem.*, 37, 57, 1990.

136. Hoff, J. D., Quigley, M. E., and Yen, S. S. C., Hormonal dynamics at midcycle: a reevaluation, *J. Clin. Endocrinol. Metab.*, 57, 792, 1983.

137. Fuentes, M., Muldoon, T., and Mahesh, V., Inhibitory effect of progesterone on occupied estrogen receptors of anterior pituitary and uterus in adult rats, *J. Neuroendocrinol.*, 2, 517, 1990.

138. Brann, D. W., Putnam, C. D., and Mahesh, V. B., Dose-related effects of progesterone and dihydroprogesterone upon estrogen-induced prolactin release, *J. Neuroendocrinol.*, 2, 341, 1990.

139. Mahesh, V. B. and Muldoon, T. G., Integration of the effects of estradiol and progesterone in the modulation of gonadotropin secretion, *J. Steroid Biochem.*, 27, 665, 1987.

140. Brann, D. W. and Mahesh, V. B., Detailed examination of the mechanism and site of action of progesterone and corticosteroids in the regulation of gonadotropin secretion: hypothalamic gonadotropin-releasing hormone and catecholamine involvement, *Biol. Reprod.*, 44, 1005, 1991.

141. Levine, J. and Ramirez, V., Luteinizing hormone-releasing hormone release during the rat estrous cycle and after ovariectomy, as estimated with push-pull cannulae, *Endocrinology*, 111, 1439, 1982.

142. Kalra, P. S. and Kalra, S. P., Temporal changes in the hypothalamic and serum luteinizing hormone-releasing hormone (LH-RH) levels and the circulating ovarian steroids during the rat oestrous cycle, *Acta Endocrinol.*, 85, 449, 1977.

143. Kim, K., Lee, B., Park, Y., and Cho, W., Progesterone increases mRNA encoding LHRH level in the hypothalamus of ovariectomized estradiol-primed prepubertal rats, *Mol. Brain Res.*, 6, 151, 1989.

144. O'Conner, J. L., Wade, M., Brann, D. W., and Mahesh, V. B., Evidence that progesterone modulates anterior pituitary neuropeptide Y levels during the progesterone-induced gonadotropin surge in the estrogen-primed intact immature female rat, *J. Steroid Biochem. Mol. Biol.*, 52, 497, 1995.

145. Lee, W., Smith, M., and Hoffman, G., Progesterone enhances the surge of luteinizing hormone by increasing the activation of luteinizing hormone-releasing hormone neurons, *Endocrinology*, 127, 2604, 1990.

146. Hoffman, G., Lee, W., Attardi, B., Yann, V., and Fitzsimmons, M., Luteinizing hormone-releasing hormone neurons express c-fos antigen after steroid activation, *Endocrinology*, 126, 1736, 1990.

147. Fox, S. R., Harlan, R., Shivers, B., and Pfaff, D. W., Chemical characterization of neuroendocrine targets for progesterone in the female rat brain and pituitary, *Neuroendocrinology*, 51, 276, 1990.

148. Leranth, C., Shanabrough, M., and Naftolin, F., Estrogen induces ultrastructural changes in progesterone receptor-containing GABA neurons of the primate hypothalamus, *Neuroendocrinology*, 54, 571, 1991.

149. Flugge, G., Oertel, W. H., and Wuttke, W., Evidence for estrogen-receptive GABAergic neurons in the preoptic anterior area of the rat brain, *Neuroendocrinology*, 43, 1, 1986.

150. Munaro, N. I., Dotti, C., and Taleisnik, S., Glutamic acid decarboxylase activity in the hypothalamus of the rat: effect of estrogens, *Adv. Biochem. Psychopharmacol.*, 42, 201, 1986.

151. Gabriel, S. M., Simpkins, J. W., and Kalra, S. P., Modulation of endogenous opioid influence on luteinizing hormone secretion by progesterone and estradiol, *Endocrinology*, 113, 1806, 1983.

152. Piva, F., Maggi, R., Limonta, P., Motta, M., and Martini, L., Effect of naloxone on luteinizing hormone, follicle stimulating hormone, and prolactin secretion in the different phases of the estrous cycle, *Endocrinology*, 117, 766, 1985.

153. Kerdelhue, B., Parnet, P., Lenoir, V., Schirar, A., Gaudox, F., Levasseur, M., Palkovits, M., Blocker, C., and Scholler, R., Interactions between 17β-estradiol and the hypothalamo-pituitary β-endorphin system in the regulation of the cycle LH secretion, *J. Steroid Biochem.*, 30, 161, 1988.

154. Bedran de Castro, J., Khorram, O., Petrovic, S., and McCann, S. M., Role of opioid peptides in pulsatile release of gonadotropins and prolactin in the rat, *Brain Res. Bull.*, 19, 539, 1987.

155. Lustig, R. H., Pfaff, D. W., and Fishman, J., Opioidergic modulation of the estradiol-induced LH surge in the rat: role of ovarian steroids, *Endocrinology*, 116, 55, 1988.

156. Limonta, P., Rovati, E., Dondi, D., and Maggi, R., Effect of estrous cyclicity on the density of mu opioid receptors in the rat hypothalamus [Abstract], *Neuroendocrinology*, Suppl. 1, 52, 82, 1990.

157. Brann, D. W., Putnam-Roberts, C. D., and Mahesh, V. B., Progesterone and corticosteroid regulation of hypothalamic and pituitary opioid content during LH surge induction, *Mol. Cell. Neurosci.*, 3, 191, 1992.

158. Wise, P. M., Scarbrough, K., Weiland, N. G., and Larson, G. H., Diurnal pattern of expression of proopiomelanocortin gene expression in the arcuate nucleus of proestrous, ovariectomized, and steroid-treated rats: a possible role in cyclic LH secretion, *Mol. Endocrinol.*, 4, 886, 1990.

159. Jirikouski, G. F., Merchenthaler, I., Rieger, G. E., and Stumpf, W. E., Estradiol target sites immunoreactive for β-endorphin in the arcuate nucleus of the rat and mouse hypothalamus, *Neurosci. Lett.*, 65, 121, 1986.

160. Leranth, C., MacLusky, N. J., Sakamoto, H., Shanabrough, M., and Naftolin, F., Glutamic acid decarboxylase-containing axons synapse on LHRH neurons in the rat medial preoptic area, *Neuroendocrinology*, 40, 536, 1985.

161. Akema, T. and Kimura, F., The mode of GABA$_B$ receptor-mediated inhibition of the preovulatory LH surge in female rats, *Brain Res.*, 562, 169, 1991.

162. Masotto, C. and Negro-Vilar, A., Activation of gamma-aminobutyric acid β-receptors abolishes naloxone-stimulated luteinizing hormone release, *Endocrinology*, 121, 2251, 1987.

163. Masotto, C., Wisniewski, G., and Negro-Vilar, A., Different gamma-aminobutyric acid receptor subtypes are involved in the regulation of opiate-dependent and independent luteinizing hormone-releasing hormone secretion, *Endocrinology*, 125, 548, 1989.

164. Morello, H., Caligaris, L., Haymal, B., and Taleisnik, S., Inhibition of proestrous LH surge and ovulation in rats evoked by stimulation of the medial raphe nucleus involves a GABA-mediated mechanism, *Neuroendocrinology*, 50, 81, 1989.

165. Wilson, C. A., James, M. D., and Leigh, A. J., Role of GABA$_A$ in the Zona incerta in the control of LH release and ovulation, *Neuroendocrinology*, 52, 354, 1990.

166. Brann, D. W., Zamorano, P. L., Putnam-Roberts, C. D., and Mahesh, V. B., Gamma aminobutyric acid-opioid interactions in the regulation of gonadotropin secretion in the immature female rat, *Neuroendocrinology*, 56, 445, 1992.

167. Vijayan, E. and McCann, S. M., The effects of intraventricular injection of gamma-aminobutyric acid (GABA) on prolactin and gonadotropin release in conscious female rats, *Brain Res.*, 155, 35, 1978.

168. Ondo, J. G., Gamma aminobutyric acid effects on pituitary gonadotropin secretion, *Science*, 186, 738, 1974.

169. McCann, S. M., Vijayan, E., Negro-Vilar, A., Mizumuma, H., and Mangat, H., GABA, a modulator of anterior pituitary hormone secretion by hypothalamic pituitary action, *Psychoneuroendocrinology*, 9, 97, 1984.

170. Virmani, M. A., Stojilkovic, S. S., and Catt, K. J., Stimulation of luteinizing hormone release by gamma-aminobutyric acid (GABA) agonists: mediation by GABA$_A$-type receptors and activation of chloride and voltage-sensitive calcium channels, *Endocrinology*, 126, 2499, 1990.

171. Racagni, G., Apud, J. A., Cocchi, D., Loatelli, V., and Muller, E. E., GABAergic control of anterior pituitary hormone secretion, *Life Sci.*, 31, 823, 1982.

172. Moguilewsky, J. A., Carbone, S., Szwarcfarb, B., and Rondina, D., Sexual maturation modifies the GABAergic control of gonadotropin secretion in female rats, *Brain Res.*, 563, 12, 1991.

173. Jarry, H., Perschl, A., and Wuttke, W., Further evidence that preoptic anterior hypothalamic GABAergic neurons are part of the GnRH pulse an surge generator, *Acta Endocrinol.*, 118, 573, 1988.

174. Jarry, H., Leonhardt, S., and Wuttke, W., Gamma-aminobutyric acid neurons in the preoptic/anterior hypothalamic area synchronize the phasic activity of the gonadotropin-releasing hormone pulse generator in ovariectomized rats, *Neuroendocrinology*, 53, 261, 1991.

175. Jarry, H., Hirsch, B., Leonhardt, S., and Wuttke, W., Aminoacid neurotransmitter release in the preoptic area of rats during the positive feedback of estradiol on LH release, *Neuroendocrinology*, 56, 133, 1992.

176. Ljungdahl, A., Hokfelt, T., and Nilsson, G., Distribution of substance P-like immunoreactivity in the central nervous system of the rat-cell bodies and nerve terminals, *Neuroscience*, 3, 861, 1978.

177. Palkovits, M., Kakucska, T., and Makara, G. B., Substance P-like immunoreactive neurons in the arcuate nucleus project to the median eminence in rat, *Brain Res.*, 486, 364, 1989.

178. Tsuro, Y., Kawano, H., Nishiyama, T., Hisano, S., and Diakokur, S., Substance P-like immunoreactive neurons in the tubero-infundibular area of rat hypothalamus. Light and electron microscopy, *Brain Res.*, 289, 1, 1983.

179. Hokfelt, T., Pernow, B., Nilsson, G., Wetterburg, L., Goldstein, M., and Jeffcoate, S., Dense plexus of substance P immunoreactive nerve terminals in eminentia medialis of the primate hypothalamus, *Proc. Natl. Acad. Sci. U.S.A.*, 75, 1013, 1978.

180. Mantyh P., Gates, T., Mantyh, P., and Maggio, J., Autoradiographic localization and characterization of tachykinin receptor binding sites in the rat brain and peripheral tissues, *J. Neurosci.*, 9, 258, 1989.

181. Akesson, T. and Micevych, P., Estrogen concentration by substance P-immunoreactive neurons in the medial basal hypothalamus of the female rat, *J. Neurosci. Res.*, 19, 412, 1988.

182. Brown, E. R., Harlan, R. E., and Krause, J. E., Gonadal steroid regulation of substance P and SP-encoding messenger ribonucleic acids in the rat anterior pituitary and the hypothalamus, *Endocrinology*, 126, 330, 1990.

183. Vijayan, E. and McCann, S. M., *In vivo* and *in vitro* effects of substance P and neurotensin on gonadotropin and prolactin release, *Endocrinology*, 105, 64, 1979.

184. Dees, W. L., Skelley, C. W., and Kozlowski, G. P., Central effects of an antagonist and an antiserum to substance P on serum gonadotropin and prolactin secretion, *Life Sci.*, 37, 1627, 1985.

185. Debeljuk, L., Lasaga, M., Horvath, J., Duvilanski, B. H., Seilicovich, A., and Diaz, M. C., Effect of an anti-substance P serum on prolactin and gonadotropins in hyperprolactinemic rats, *Regul. Pept.*, 19, 91, 1987.

186. Sahu, A., Crowley, W. R., Tatemoto, K., Balasubramanian, A., and Kalra, S. P., Effects of neuropeptide Y, NPY analog, galanin, and neuropeptide K on LH release in ovariectomized and ovx estrogen, progesterone-treated rats, *Peptides*, 8, 921, 1987.

187. Sahu, A. and Kalra, S. P., Effects of tachykinins in luteinizing hormone release in female rats: potent inhibitory action of neuropeptide K, *Endocrinology*, 130, 1571, 1992.
188. Kalra, P. S., Sahu, A., Bonavera, J., and Kalra, S., Diverse effects of tachykinins on luteinizing hormone release in male rats: mechanism of action, *Endocrinology*, 131, 1195, 1992.
189. Shamgochian, M. D. and Leeman, S. E., Substance P stimulates LH secretion from anterior pituitary cells in culture, *Endocrinology*, 131, 871, 1992.
190. Grant, L. D. and Stumpf, W. E., Localization of ^{3}H-estradiol and catecholamines in identical neurons in the hypothalamus, *J. Histochem. Cytochem.*, 21, 404, 1973.
191. Blaustein, J. D. and Turcott, J. C., A small population of tyrosine hydroxylase-immunoreactive neurons in the guinea pig arcuate nucleus contains progestin receptor-immunoreactivity, *J. Neuroendocrinol.*, 1, 333, 1989b.
192. Leranth, C., MacLusky, N., Shanabrough, M., and Naftolin, F., Catecholaminergic innervation of luteinizing hormone-releasing hormone and glutamic acid decarboxylase immunopositive neurons in the rat medial preoptic area, *Neuroendocrinology*, 48, 591, 1988.
193. Barraclough, C. A. and Wise, P. M., The role of catecholamines in the regulation of pituitary luteinizing hormone and follicle-stimulating hormone secretion, *Endocrine Rev.*, 3, 91, 1982.
194. Wise, P. M., Rance, N., and Barraclough, C. A., Effects of estradiol and progesterone on catecholamine turnover rates in discrete hypothalamic regions in ovariectomized rats, *Endocrinology*, 108, 2186, 1981.
195. Nagle, C. A. and Rosner, J. M., Plasma norepinephrine during the rat estrous cycle and after progesterone treatment to the ovariectomized estrogen-primed rat, *Neuroendocrinology*, 22, 89, 1976.
196. Coen, C. W. and Coombs, M. C., Effects of manipulating catecholamines on the incidence of the preovulatory surge of luteinizing hormone and ovulation in the rat: evidence for a necessary involvement of hypothalamic adrenaline in the normal or "midnight" surge, *Neuroscience*, 10, 187, 1983.
197. McDonald, J. K., NPY and related substances, *Crit. Rev. Neurobiol.*, 4, 97, 1988.
198. Kalra, S. P., Allen, G., Sahu, A., Kalra, P., and Crowley, W. R., Gonadal steroids and neuropeptide Y-opioid-LHRH axis: interactions and diversities, *J. Steroid Biochem.*, 30, 185, 1988.
199. Sahu, A., Crowley, W. R., and Kalra, S. P., Dynamic changes in neuropeptide Y concentrations in the median eminence in association with preovulatory LH release in the rat, *J. Neuroendocrinol.*, 126, 876.
200. Sutton, S. W., Toyama, T., Otto, S., and Plotsky, P., Evidence that neuropeptide Y (NPY) released into the hypophysial-portal circulation participates in priming gonadotropes to the effects of gonadotropin-releasing hormone (GnRH), *Endocrinology*, 123, 1208, 1988.
201. Brann, D. W., McDonald, J. K., Putnam, C. D., and Mahesh, V. B., Regulation of hypothalamic gonadotropin-releasing hormone and neuropeptide Y concentrations by progesterone and corticosteroids in immature rats: correlation with luteinizing hormone and follicle-stimulating hormone release, *Neuroendocrinology*, 54, 425, 1991.
202. O'Conner, J., Wade, M., Brann, D., and Mahesh, V. B., Direct anterior pituitary modulation of gonadotropin secretion by NPY: role of gonadal steroids, *Neuroendocrinology*, 58, 129, 1993.
203. Parker, S. L., Kalra, S. P., and Crowley, W. R., Neuropeptide Y modulates the binding of a GnRH analog to anterior pituitary GnRH receptor sites, *Endocrinology*, 125, 2309, 1991.
204. Wehrenberg, W., Corder, R., and Gaillard, R. C., A physiological role for neuropeptide Y in regulating the estrogen/progesterone induced LH surge in ovariectomized rats, *Neuroendocrinology*, 49, 680, 1989.
205. Melander, T., Hokfelt, T., Rokaeus, A., Cuello, A. C., Oertel, W. H., Verhofstad, A., and Goldstein, M., Coexistence of galanin-like immunoreactivity with catecholamines, 5-hydroxy-tryptamine, GABA and neuropeptides in the rat CNS, *J. Neurosci.*, 6, 3640, 1986.
206. Merchenthaler, I., Lopez, F. J., and Negro-Vilar, A., Colocalization of galanin and luteinizing hormone releasing hormone in a subset of preoptic hypothalamic neurons: anatomical and functional correlates, *Proc. Natl. Acad. Sci. U.S.A.*, 87, 6326, 1990.
207. Lopez, F. J. and Negro-Vilar, A., Galanin stimulates luteinizing hormone-releasing hormone secretion from arcuate nucleus-median eminence fragments *in vitro*: an involvement of an α-adrenergic mechanism, *Endocrinology*, 127, 2431, 1990.
208. Sahu, A., Crowley, W. R., Tatemoto, K., Balasubramaniam, A., and Kalra, S. P., Effect of neuropeptide Y, NPY analog (norleueine^4-NPY), galanin and neuropeptide K on LH release in ovariectomized (ovx) and ovx estrogens, progesterone-treated rats, *Peptides*, 8, 921, 1987.
209. Lopez, F. J., Merchenthaler, I., Ching, M., Wisniewski, M. G., and Negro-Vilar, A., Galanin: a hypothalamic-hypophysiotropic hormone modulating reproductive functions, *Proc. Natl. Acad. Sci. U.S.A.*, 88, 4508, 1991.
210. Lopez, F. J., Maede, E. H., and Negro-Vilar, A., Endogenous galanin modulates the gonadotropin and prolactin proestrous surges in the rat, *Endocrinology*, 132, 795, 1993.
211. Brann, D. W., Chorich, L. P., and Mahesh, V. B., Effect of progesterone on galanin mRNA levels in the hypothalamus and the pituitary: correlation with the gonadotropin surge, *Neuroendocrinology*, 58, 531, 1993.
212. van den Pol, A., Glutamate and aspartate immunoreactivity in hypothalamic presynaptic axons, *J. Neurosci.*, 11, 2087, 1991.
213. van den Pol, A., Waurin, J., and Dudek, F., Glutamate, the dominant excitatory transmitter in neuroendocrine regulation, *Science*, 250, 1276, 1990.

214. Abbud, R. and Smith, M. S., Differences in the luteinizing hormone and prolactin responses to multiple injections of kainate as compared to *N*-methyl-D,L-aspartate, in cycling rats, *Endocrinology*, 129, 3254, 1991.

215. McGeer, E. G. and McGeer, P. L., Duplication of biochemical changes of Huntington's chorea by intrastriatal injections of glutamate and kainic acids, *Nature*, 263, 517, 1976.

216. Medhamurthy, R., Gay, V. L., and Plant, T. M., Repetitive injections of L-glutamic acid, in contrast to those of *N*-methyl-D,L-aspartic acid, fail to elicity sustained hypothalamic GnRH release in the prepubertal male rhesus monkey, *Neuroendocrinology*, 55, 660, 1992.

217. Bourguignon, J., Gerard, A., and Franchimont, P., Direct activation of gonadotropin-releasing hormone secretion through different receptors to neuroexcitatory amino acids, *Neuroendocrinology* 49, 402, 1989.

218. Bourguignon, J., Gerard, A., Mathieu, J., Simons, J., and Franchimont, P., Pulsatile release of gonadotropin-releasing hormone from hypothalamic explants is restrained by blockade of *N*-methyl-D,L-aspartate receptors, *Endocrinology* 125, 1090, 1989.

219. Brann, D. W. and Mahesh, V. B., Endogenous excitatory amino acid involvement in the preovulatory and steroid-induced surge of gonadotropins in the female rat, *Endocrinology*, 128, 1541, 1991.

220. Donoso, A., Lopez, F., and Negro-Vilar, A., Glutamate receptors of the non-*N*-methyl-D-aspartic acid type mediate the increase in luteinizing hormone-releasing hormone release by excitatory amino acids, *Endocrinology*, 126, 414, 1990.

221. Lopez, F. Donozo, A., and Negro-Vilar, A., Endogenous excitatory amino acids and glutamate receptor subtypes involved in the control of hypothalamic luteinizing hormone-releasing hormone secretion, *Endocrinology*, 130, 1986, 1992.

222. Brann, D. W., Ping, L., and Mahesh, V. B., Role of non-NMDA receptor neurotransmission in steroid and preovulatory gonadotropin surge expression in the female rat, *Mol. Cell. Neurosci.*, 4, 292, 1993.

223. Brann, D. W. and Mahesh, V. B., Endogenous excitatory amino acid regulation of the progesterone-induced LH and FSH surge in estrogen-primed ovariectomized rats, *Neuroendocrinology*, 53, 107, 1991.

224. Urbanski, H. D. and Ojeda, S. R., Activation of luteinizing hormone-releasing hormone release advances the onset of female puberty, *Neuroendocrinology*, 46, 273.

225. Lopez, F., Donoso, A., and Negro-Vilar, A., Endogenous excitatory amino regulates the estradiol-induced LH surge in ovariectomized rats, *Endocrinology*, 126, 1771, 1990.

226. Seong, J. Y., Lee, Y. K., Lee, C. C., and Kim, K., NMDA receptor antagonist decreases the progesterone-induced increase in GnRH gene expression in the rat hypothalamus, *Neuroendocrinology*, 58, 234, 1993.

227. Brann, D. W. and Mahesh, V. B., Excitatory amino acid regulation of gonadotropin secretion: modulation by steroid hormones, *J. Steroid Biochem. Mol. Biol.*, 41, 847, 1992.

228. Luderer, U., Strobl, F., Levine, J., and Schwartz, N., Differential gonadotropin responses to *N*-methyl-D-aspartate (NMDA) in metestrous, proestrus, and ovariectomized rats, *Biol. Reprod.*,48, 857, 1993.

229. Reyes, A., Luckhaus, J., and Ferin, M., Unexpected inhibitory action of *N*-methyl-D,L-aspartate on luteinizing hormone release in adult ovariectomized rhesus monkeys: a role of the hypothalamic-adrenal axis, *Endocrinology*, 127, 724, 1990.

230. Carbone, S., Szwarcfarb, B., Losada, M., and Moguilevsky, J., Effects of ovarian steroids on the gonadotropin response to *N*-methyl-D-aspartate and on hypothalamic excitatory amino acid levels during sexual maturation in female rats, *Endocrinology*, 130, 1365, 1992.

231. Reyes, A., Xia, L., and Ferin, M., Modulation of the effects of *N*-methyl-D,L-aspartate on luteinizing hormone by the ovarian steroids in the adult rhesus monkey, *Neuroendocrinology*, 54, 405, 1991.

232. Brann, D. W., Zamorano, P. L., Chorich, L. P., and Mahesh, V. B., Steroid hormone effects on NMDA receptor binding and NMDA receptor mRNA levels in the hypothalamus and cerebral cortex of the adult rat, *Neuroendocrinology*, 58, 666, 1993.

233. Brann, D. W., Zamorano, P. L., Ping, L., and Mahesh, V. B., Role of excitatory amino acid neurotransmission during puberty in the female rat, *Mol. Cell. Neurosci.*, 4, 107, 1993.

234. Kus, L., Beitz, A. J., Kerr, J. E., and Handa, R. J., NMDA R1 receptor mRNA expression in the hypothalamus of intact, castrate and DHT-treated male rats, *Soc. Neurosci.*, 19, 919 (Abstract 379.5), 1993.

235. Weiland, N. G., Estradiol selectively regulates agonist binding sites on the *N*-methyl-D-aspartate receptor complex in the CA1 region of the hypothalamus, *Endocrinology*, 131, 662, 1992.

236. Ping, L., Mahesh, V. B., and Brann, D. W., Release of glutamate and aspartate from the preoptic area during the progesterone-induced LH surge: *in vivo* microdialysis studies, *Neuroendocrinology*, 59, 318, 1994.

237. Jarry, H., Hirsch, B., Leonhardt, S., and Wuttke, W., Amino acid neurotransmitter release in the preoptic area of rats during the positive feedback actions of estradiol on LH release, *Neuroendocrinology*, 56, 133, 1992.

238. Goroll, D., Arias, P., and Wuttke, W., Preoptic release of amino acid neurotransmitters evaluated in peripubertal and young adult female rats by push-pull perfusion, *Neuroendocrinology*, 58, 11, 1993.

239. Ping, L., Mahesh, V. B., and Brann, D. W., A physiological role for NMDA and non-NMDA receptors in pulsatile gonadotropin secretion in the adult female rat, *Endocrinology*, 135, 113, 1994.

240. Ping, L., Mahesh, B. B., and Brann, D. W., Effect of NMDA and non-NMDA receptor antagonists on pulsatile luteinizing hormone secretion in the adult male rat, *Neuroendocrinology*, 1994, in press.

241. Arslan, M., Pohl, C., and Plant, T., DL-2-amino-5-phosphonopentanoic acid, a specific *N*-methyl-D-aspartic acid receptor antagonist, suppresses pulsatile LH release in the rat, *Neuroendocrinology*, 47, 465, 1988.
242. Brann, D. W. and Mahesh, V. B., Role of corticosteroids in female reproduction, *FASEB J.*, 5, 2691, 1991.
243. Higuchi, T., Honda, K., and Negoro, H., Influence of estrogen and noradrenergic afferent neurones on the response of LH and oxytocin to immobilization stress, *J. Endocrinol.*, 110, 245, 1986.
244. Briski, K. P. and Sylvester, P. W., Effects of specific acute stressors on luteinizing hormone release in ovariectomized and ovariectomized estrogen-treated female rats, *Neuroendocrinology*, 47, 194, 1988.
245. Armario, A., Restrepo, C., Hildalgo, J., and Copez-Calderon, A., Differences in prolactin and LH responses to acute stress between peripubertal and adult male rats, *J. Endocrinol.*, 112, 9, 1987.
246. Paris, A., Kelly, P., and Ramaley, J., Effects of short-term stress upon fertility II. After puberty, *Fertil. Steril.*, 24, 546, 1973.
247. Bohler, H., Zoeller, T., King, J., Rubin, B., Weber, R., and Merriam, G., Corticotropin releasing hormone mRNA is elevated on the afternoon of proestrus in parvocellular paraventricular nuclei of the female rat, *Mol. Brain Res.*, 8, 259, 1990.
248. Raps, S., Bartke, P.L., and Desaulles, P.A., Plasma and adrenal corticosterone levels during the different phases of the sexual cycle in normal female rats, *Experientia*, 21, 239, 1970.
249. Buckingham, J.C., Dohler, K., and Wilson, C., Activity of the pituitary-adrenocortical system and thyroid gland during the oestrus cycle of the rat, *J. Endocrinol.*, 78, 359, 1978.
250. Shaikh, A. and Shaikh, S., Adrenal and ovarian steroid secretion in the rat estrous cycle temporally related to gonadotropins and steroid levels found in peripheral plasma, *Endocrinology*, 96, 37, 1975.
251. Putnam, C.D., Brann, D.W., and Mahesh, V.B., Acute activation of the ACTH-adrenal axis: effect upon gonadotropin secretion in the female rat, *Endocrinology*, 128, 2558, 1991.
252. Brann, D. W., Putnam, C. D., and Mahesh, V. B., Corticosteroid regulation of gonadotropin secretion and induction of ovulation in the rat, *Proc. Soc. Exp. Biol. Med.*, 190, 176, 1990.
253. Brann, D.W., Putnam, C. D., and Mahesh, V. B., Regulation of FSH secretion by natural and synthetic corticosteroids, in *Regulation and Action of Follicle Stimulating Hormone*, Hunzicker-Dunn, M. and Schwartz, N. B., Eds., Springer-Verlag, New York, 1991, 303.
254. Ringstrom, S.J., Sutter, D.E., Hostetler, J.P., and Schwartz, N.B., Cortisol regulates secretion and pituitary content of the two gonadotropins differently in female rats: effects of GnRH antagonists, *Endocrinology*, 130, 3122, 1992.
255. McAndrews, J.M., Ringstrom, S.J., Dahl, K.D., and Schwartz, N.B. Corticosterone *in vivo* increased pituitary FSH-β mRNA content and serum bioactivity selectively in female rats, *Endocrinology*, 134, 158, 1994.
256. Wang, C. and Leung, A., Glucocorticoids stimulate plasminogen activator production by granulosa cells, *Endocrinology*, 124, 1595, 1989.
257. Smith, M. S., Role of prolactin in regulating gonadotropin secretion and gonadal function in female rats, *Fed. Proc.*, 39, 2571, 1980.
258. Petraglia, F., Suttons, V. W., and Plotsky, P. M., CRF decreases plasma LH levels in female rats by inhibiting GnRH release into hypophysial-portal circulation, *Endocrinology*, 120, 1361, 1987.
259. Hsueh, A. J. and Erickson, G. F., Glucocorticoid inhibition of FSH-induced estrogen production in cultured rat granulosa cells, *Steroids*, 32, 636.
260. Terakawa, N., Shimizu, I., Toshihiro, A., Tanizawa, O., and Matsumoto, K., Dexamethasone inhibits the effects of estrogen on the pituitary gland in rats, *Acta Endocrinol.*, 112, 64, 1986.
261. Lisk, R. D. and Reuter, L. A., Dexamethasone: increased weights and deceased ³H-estradiol retention of uterus, vagina and pituitary in ovariectomized rat, *Endocrinology*, 99, 1063, 1976.

Chapter 11

CNS Steroids and the Control of Feeding

Peter C. Butera

CONTENTS

I. INTRODUCTION

It is apparent that eating is one of the most important things an organism does, and that eating is also a behavior from which animals, humans included, obtain a good deal of pleasure. Because survival is ultimately dependent upon the organism's ability to successfully procure, ingest, and utilize food, it is not surprising that the physiological and behavioral mechanisms that control eating reflect processes that can be seen as adaptive when viewed in the context of the species' natural history and ecological niche. The modulation of feeding behavior by CNS steroids and gonadal steroids in particular is a good case in point. The interactions between gonadal steroids and neural systems involved in the control of food intake examined in this chapter, when viewed in a broader sense, are a reflection of the relationship between energy balance and reproduction in female mammals. During pregnancy and throughout ovulatory cycles, ovarian steroids induce systematic changes in the consumption, storage, and expenditure of metabolic fuels. Reproductive success, especially for female mammals, depends on the coordinated interactions between neural systems that control reproductive cycles (e.g., GnRH neurons) and those involved in energy balance. It is quite possible that hypothalamic GnRH neurons detect availability of metabolic fuels by receiving information from hormone-sensitive brain areas involved in feeding and metabolism.[1] In fact, there are a number of interconnected brain areas that contain receptors for gonadal steroids, receive and integrate visceral inputs, and modulate behavioral and autonomic responses to the incoming sensory information. Thus, the data to be presented in this chapter represent attempts to decipher the interactions between gonadal steroids and this neural circuitry.

A chapter on CNS steroids and feeding raises the question: What is meant by the term CNS steroid or neurosteroid? A general definition that has recently emerged is that neurosteroids include all steroids that are synthesized within the brain.[2] Within this general category are the *neuroactive steroids*, which are conceptualized to be steroids with CNS effects that may be synthesized by the brain or by classic endocrine tissues. The focus of this chapter will be on neuroactive steroids that are synthesized by traditional endocrine glands but exert actions on the brain. Research on the control of food intake by neuroactive steroids typically focuses on the influence of: (1) gonadal steroids (e.g., estrogens, progestins, androgens), (2) adrenal steroids (e.g., corticosterone), or (3) gonadal steroids in gender differences in eating. The goal of this chapter is to offer a detailed examination of the CNS actions of gonadal steroids and their role in feeding behavior. The data presented will also be relevant to an analysis of gender differences in food intake. Thus, the pages that follow are not intended to reflect all the work in the area of steroids and ingestive behavior, but to show the progress that has been made in our understanding of how gonadal steroids interact with neural systems involved in the control of food intake. Readers interested in the effects of adrenal steroids on feeding are referred to the work of Castonguay, Dallman, and Stern,[3] Dallman,[4] King,[5] and Tempel and Leibowitz.[6]

The research on the effects of CNS steroids on feeding that will be presented and summarized in this chapter will be organized along the following lines: gender differences in food intake and organizational effects of hormones, general effects of androgens on food intake, general effects of estrogens and progestins on food intake, neural sites of action of estradiol and progesterone, potential mediating mechanisms, and estrogen as an indirect control of meal size.

II. GENDER DIFFERENCES IN FOOD INTAKE AND ORGANIZATIONAL EFFECTS OF HORMONES

In most mammalian species, males are larger and heavier than females.[7] In the rat, male pups weigh slightly but significantly more than female pups at birth, and this gender difference in body weight is greatly exaggerated during adulthood.[8] This sexual dimorphism in body weight can be accounted for in part by gender differences in food intake and activity, as male rats typically consume more food and exercise less than female rats.[9] A notable exception to this sexual dimorphism in food intake and body weight can be seen in the golden hamster. In this rodent species, adult females eat and weigh more than adult males.[10,11] In humans, gender differences in food intake and body composition begin to surface in adolescence. Prior to puberty, lean body mass, skeletal mass, and body fat are relatively similar among boys and girls.[12] Unpublished observations by L.L. Birch, cited in an article by Rolls, Fedoroff, and Guthrie,[13] also suggest that the eating behavior of boys and girls is similar throughout childhood. However, following puberty, males have 1.5 times the lean body mass and skeletal mass as females, whereas women have about 2 times as much body fat as men.[14] Gender differences in eating also surface during puberty and continue throughout adulthood, so that the daily energy intake of men is significantly greater than that of women.[15]

Although the gender difference in body weight appears to be caused by the organizing effects of neonatal androgens, gender differences in food intake appear to be less responsive to neonatal hormone manipulations. For example, gonadectomies carried out on newborn male and female rat pups abolished the sexual dimorphism in body weight normally seen in adulthood. Gonadectomies performed on 21-day-old rats (when the body weights of males and females are relatively similar) did not prevent the sexual dimorphism in adult body weight.[16] In addition, treating newborn female rat pups with a single injection of testosterone caused those animals to have higher body weights than control females during adulthood.[16,17] In this same experiment,[17] the food intake of gonadectomized adults was not significantly influenced by neonatal hormone manipulations (e.g., gonadectomy or testosterone treatment).

Interestingly, although neonatal hormone treatment alone does not significantly affect eating in gonadectomized adults, neonatal testosterone does influence responsiveness to exogenous steroids during adulthood. Ovariectomized adult rats who were treated neonatally with testosterone are less responsive to the effects of estradiol on food intake than ovariectomized females treated neonatally with a control injection.[18] The results of these experiments indicate that neonatal androgens can have organizational effects on adult body weight in the absence of activating hormones, and can also alter adult responsiveness to the activational effects of gonadal steroids. As of yet, little is known about the neural mechanisms that underlie the effects of neonatal hormone treatments on sexual dimorphism in body weight, or on adult responsiveness to the effects of activating hormones like estradiol on food intake.

III. GENERAL EFFECTS OF ANDROGENS ON FOOD INTAKE

Although early research conducted in the 1920s revealed a possible connection between gonadal steroids, feeding, and body weight, it was the isolation, identification, and synthesis of the sex hormones that allowed researchers to evaluate causal connections between gonadal steroids and ingestive behavior. In adult, male rats castration causes significant decreases in food intake and body weight.[19-21] In adult castrates, peripheral treatment with *low* doses of testosterone (<1 mg/day) increases food intake and ultimately restores body weight to the level of intact animals.[16,19,22] Although low doses of testosterone increase eating and weight gain in castrated rats, prolonged treatment with *high* doses of testosterone (≥ 1 mg/day) actually decreases food intake and body weight in these animals.[19,23,24] In males, the presence of the aromatase enzyme within specific neuronal populations accounts for estrogen formation in the brain via metabolism of testosterone to estradiol. Therefore, the data showing that higher doses of testosterone decrease eating and body weight suggests that estradiol is the hormone responsible for the changes in ingestive behavior because relatively larger amounts of the androgen are being aromatized to estradiol within the brain. Evidence indicates that this is the case. In castrated rats, the decrease in food intake and body weight that is produced by a high dose of testosterone (1 mg/day) can be blocked by treatment with the aromatase inhibitor androsta-1,4,6-triene-3,17-dione (ATD).[23] In this experiment, the food intake of males given a high dose of testosterone plus ATD was virtually identical to that of animals given a low dose of testosterone (0.2 mg). The fact that blocking aromatization prevents the anorectic effects of high doses of testosterone indicates that these effects of testosterone are being caused by estradiol. In males, estradiol appears to be acting as a classic neurosteroid on food intake because it is being synthesized from testosterone by various brain regions. The results of experiments examining changes in feeding following direct stimulation of the brain by androgens support this hypothesis. In castrated male rats, bilateral implants of testosterone or estradiol in the ventromedial hypothalamus (VMH) were shown to decrease food intake over a three-day period. Cannula placements in the preoptic area were without effect.[25] In addition, central implants of the nonaromatizable androgen dihydrotestosterone (DHT) had no significant effect on feeding in this experiment. The fact that testosterone but not DHT implants suppressed food intake suggests that estrogenic metabolites of testosterone can act on neural tissue to influence feeding behavior in males. However, the neural mechanism involved in the effects of testosterone or its metabolites on food intake remains to be identified.

IV. GENERAL EFFECTS OF ESTROGENS AND PROGESTINS ON FOOD INTAKE

A. CHANGES IN FOOD INTAKE DURING OVARIAN CYCLES

The effects of gonadal steroids on feeding behavior are more complex and often more robust in female mammals than in male mammals. In contrast to male rats, female rats display noncircadian fluctuations in the secretion of gonadal steroids that are reliable and systematic. As a result, female rats show rhythmic changes in food intake beyond the circadian fluctuations in eating that are characteristic of both males and females. For example, the female rat shows cyclic changes in eating during its 4- to 5-day estrous cycle, with reduced food intake occurring at proestrus, around the time of ovulation.[26,27] Similar relations between food intake and ovarian cycles have also been observed in female mammals that have long ovarian cycles with a prolonged luteal phase, such as the goat,[28] pig,[29] sheep,[30] cow, (cited in Reference 30) guinea pig,[31] rhesus monkey,[32] and human female.[33,34] This periovulatory decrease in food intake appears to be accomplished by a decrease in meal size without a compensatory increase in meal frequency, at least in the rat.[26] Comparable analyses of changes in meal patterns across ovarian cycles of the other species cited above have not been conducted.

B. EFFECTS OF OVARIECTOMY AND HORMONE REPLACEMENT

Withdrawal of ovarian hormones by ovariectomy produces dramatic changes in food intake and body weight. In addition, the results of hormone replacement experiments indicate that withdrawal of estradiol is responsible for the changes in ingestive behavior that accompany spaying. Ovariectomy of adult female rats causes a significant increase in food intake and a concomitant increase in body weight.[35-37] Similar effects of ovariectomy on eating and body weight are also observed in guinea pigs.[31,38] In rats, the hyperphagia induced by ovariectomy appears to result from an increase in meal size and a decrease in meal frequency.[26] Treating ovariectomized rats and guinea pigs with physiological doses of estradiol decreases food intake and body weight.[31,36,37] Similar effects of estradiol replacement on food intake have also been reported in ovariectomized rhesus monkeys.[39] In ovariectomized rats, estradiol decreases eating by causing a decrease in meal size.[26] Although it is quite likely that estradiol decreases food intake in guinea pigs and rhesus monkeys by influencing meal size, experimental data bearing on this hypothesis have yet to be collected. The fact that the anorectic effects of estradiol (and the hyperphagia following ovariectomy) appear to result from changes in meal size suggests that estradiol may influence feeding by advancing the onset of satiety. The idea that estradiol affects food intake through its actions on neural systems involved in satiety will be developed in subsequent sections.

In ovariectomized rats, peripheral treatment with progesterone alone (0.5 to 10 mg/day) has no significant effects on feeding or body weight.[37,40-42] Similar findings have also been reported in ovariectomized guinea pigs and rhesus monkeys treated with progesterone alone.[38] Although progesterone given alone has no effect on food intake, progesterone given to estradiol-treated, ovariectomized rats reverses the effects of estradiol on feeding behavior and leads to increases in food intake and body weight.[37,43,44] In contrast to the data obtained in the rat, progesterone treatment fails to antagonize the effects of estradiol on food intake in ovariectomized rhesus monkeys.[39] The fact that progesterone alone has no effect on ingestive behavior in ovariectomized rats and guinea pigs makes sense given what is known about the induction of progestin receptors in target tissue by estradiol. In the absence of estradiol, the number of progestin receptors in target tissues like the brain, uterus, and adipose tissue is very low. Following estradiol treatment, the number of progestin receptors in the brain, uterus, and adipose tissue increases dramatically.[45-47] Therefore, progesterone is capable of influencing ingestive behavior in the presence of estradiol (e.g., in gonadally intact or in ovariectomized females given estradiol), and its effects on food intake appear to stem from an attenuation of the actions of estradiol on eating.

C. NEURAL SITES OF ACTION OF ESTRADIOL AND PROGESTERONE

Nearly fifty years ago, Brobeck, Wheatland, and Strominger[48] hypothesized that gonadal hormones act directly on the brain to influence food intake, activity, and other variables that show rhythmic fluctuations during estrous cycles. According to these researchers:

> ...one might consider the possibility that one or more of the hormones in question act upon the hypothalamic cells responsible for this regulation in such a way as to direct the overall energy exchange now towards the side of energy storage, now towards the side of energy expenditure. Whether this is indeed the case cannot be decided at the present time, but the problem appears to be one which lends itself to further experimental study.

The hypothesis of Brobeck et al.[48] clearly anticipated the work done in the 1960s by Kennedy and Mitra,[49-51] whose studies were geared toward developing a model of gonadal steroids and brain function that would integrate findings suggesting that estrogen could affect a variety of physiological and behavioral processes by acting at distinct loci within the hypothalamus. Although these investigators did not present data on where estradiol could act to affect feeding, Kennedy[52] developed the idea that the rostral hypothalamus contained estrogen-sensitive neurons that were involved in the effects of estradiol on activity.

Research on the brain areas involved in the effects of estradiol on food intake was forthcoming. Wade and Zucker[53] were the first to report that direct stimulation of the VMH by estrogen, a brain area known to contain estrogen receptors,[54,55] influences feeding behavior in female rats. In their experiment, unilateral implants (27-gauge cannulae) of undiluted, crystalline estradiol benzoate (EB) decreased food intake in ovariectomized rats during the three-day period in which the cannulae remained in the brain. Control implants containing crystalline cholesterol were without effect. This phenomenon was replicated by other investigators[56,57] and was also shown to occur in gonadectomized male rats.[56] These findings were ultimately incorporated into the prevailing theories on the control of food intake which emphasized the role of the VMH as a satiety center.[58-60] An explanation of estradiol's effects on feeding that emerged was that estradiol suppresses eating by stimulating VMH neurons that normally inhibit food intake. The proposed mechanism maintained that estradiol decreases food intake as a result of estrogen-induced changes in a body weight set point or lipostat controlled by the VMH.[57,61,62]

Although VMH implants of estradiol have been shown to suppress feeding in female rats, the hypothesis that the VMH is the principal site at which estradiol acts to decrease food intake is called into question by several research findings. Experiments examining the effects of ovariectomy and hormone replacement in VMH-lesioned rats have shown that these females are still capable of lowering food intake in response to peripheral injections of estradiol.[63,64] These findings suggest that estrogen-sensitive regions within the brain spared by VMH lesions, or estrogen-sensitive mechanisms in the periphery, mediate the effects of estradiol on feeding behavior seen in these animals. It is interesting to note that although most published studies report that the effects of estradiol on food intake are not significantly altered by VMH lesions, there are suggestions that large VMH lesions do alter responsiveness to estradiol. For example, Wade[65] discusses unpublished data which indicate that ovariectomized rats with large VMH lesions are substantially less responsive to estradiol than are rats with smaller VMH lesions. It is possible that large VMH lesions attenuate the effects of estradiol on feeding not because the hormone's site of action has been removed, but because large VMH lesions damage neural connections between the medial hypothalamus (e.g., paraventricular nucleus of the hypothalamus or PVN) and the brainstem.[66,67]

Data from our lab and from other investigators indicate that estrogenic stimulation of the PVN, a brain area rich in estrogen receptors[54,55,68] and involved in the control of feeding behavior,[69,70] is also capable of suppressing food intake. In ovariectomized guinea pigs and rats, unilateral implants (28-gauge cannulae) of undiluted, crystalline estradiol placed in the PVN significantly decreased food intake.[71,72] Control implants containing crystalline cholesterol were without effect in these studies.

The fact that PVN implants of estradiol suppress eating in the absence of estrogen-induced alterations in lipoprotein lipase activity[72] suggests that peripheral metabolic changes are not responsible for the changes in feeding seen after estrogenic stimulation of the PVN. However, given the proximity of the PVN to the third ventricle, it is possible that steroid diffusion from the PVN to the VMH produced the effects on feeding reported in the studies cited above. These experiments did not report changes in sexual behavior following estrogenic stimulation of the PVN, which would provide a behavioral marker for hormonal leakage into the VMH, a critical brain area controlling the expression of female sexual behavior.[73] To control for the problem of hormone leakage, we conducted an experiment utilizing PVN implants of *dilute* estradiol (either a 3:1 or 10:1 mixture of cholesterol and estradiol), a technique that restricts steroid diffusion within the brain.[74] In addition, we measured changes in food intake and sexual behavior in order to more accurately assess hormonal leakage from the PVN implants into the VMH. In this experiment, as in the studies cited above, unilateral PVN implants (28-gauge cannulae) of undiluted estradiol decreased food intake during the 3-day period the cannulae were in the brain. However, the PVN implants of undiluted estradiol also produced levels of female sexual behavior that were not different from levels seen in animals with hormone implants in the VMH, indicating substantial hormone diffusion from the PVN to the VMH. More importantly, the PVN implants of dilute estradiol also decreased food intake but did not induce significant levels of lordosis. As shown in the figures below, larger dilutions of estradiol produced the greatest decreases in lordosis quotients. This effect probably stems from the fact that as the concentration of estradiol in the cannulae was decreased, the amount of steroid diffusing from the PVN to the VMH was reduced and the induction of female sexual behavior was diminished. Moreover, the suppression of food intake following estrogenic stimulation of the PVN remained robust even with the dilute implants, indicating that steroid diffusion from the PVN to the VMH did not underlie the observed effects on feeding.

The results of this experiment indicate that hypothalamic implants of undiluted estradiol can produce hormone leakage into nearby brain regions with behavioral consequences. Along these lines, it is quite likely that the VMH implants of undiluted estradiol which suppressed feeding in the studies cited above[53,56,57] resulted in significant steroid leakage from the VMH to the PVN. Consistent with this hypothesis is the report that estradiol implants produce a pattern of hormone diffusion that extends dorsally along the cannula track.[75] Given the anatomical relationship between the VMH and the PVN, it is conceivable that undiluted estradiol implants in the VMH provide significant amounts of estradiol to the PVN via diffusion up the cannula track. Therefore, the changes in food intake following undiluted estradiol implants in the VMH that have been previously documented could actually be due to estrogenic stimulation of the PVN.

Further evidence in support of the hypothesis that estradiol normally decreases eating via its actions in the PVN comes from lesion studies and from experiments utilizing PVN implants of anisomycin. As previously discussed, ovariectomized rats with VMH lesions are still capable of responding to the effects of estradiol on food intake;[63,64] however, lesions of the PVN can attenuate the effects of estradiol on feeding behavior.[76] In this experiment, female rats received bilateral, electrolytic lesions of the PVN or sham lesions. Two weeks later, all animals were ovariectomized and, two weeks after ovariectomy, females received daily injections of EB (2.0 μg) or the oil control over a three-day period. Although estradiol decreased food intake in animals with sham lesions or lesions of the dorsomedial nucleus of the hypothalamus (DMN), the anorectic effects of estradiol were abolished in females with bilateral PVN lesions. In contrast to the effects on feeding behavior, EB injections significantly lowered body weight across all groups. These results suggest that the effects of estradiol on food intake and body

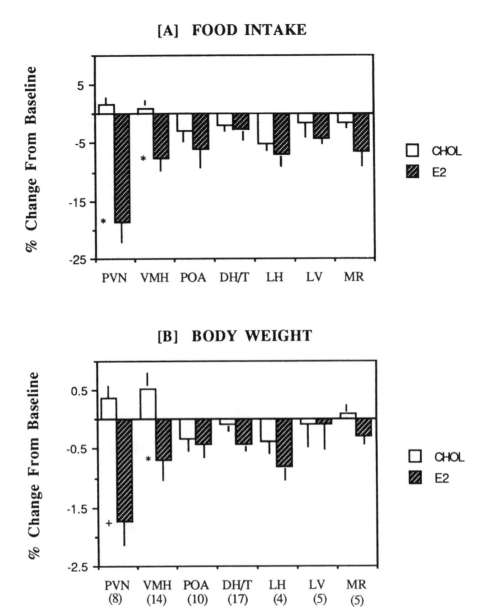

Figure 1 Effects of intracranial stimulation with cholesterol (open bars) and estradiol 17-β (shaded bars) placed into particular brain areas of ovariectomized guinea pigs. Implants were histologically verified and categorized as being in the ventromedial arcuate hypothalamic region (VMH), the paraventricular nucleus (PVN), the preoptic area (POA), the dorsal hypothalamus-thalamus (DH/T), the lateral hypothalamic area (LH), the lateral ventricles (LV), and the mammillary region of the hypothalamus (MR). Bars represent mean ± S.E. percent change from baseline for (A) food intake and (B) body weight. The number of cases per region are indicated in parentheses; *p <0.05; **p <0.01. (From Butera, P. C. and Czaja, J. A., *Brain Res.*, 322, 41, 1984. With permission.)

weight, while correlated, may be mediated by different mechanisms. For example, estrogenic effects on eating may require hormonal actions within the PVN, whereas alterations in body weight could be produced by direct actions of estradiol on enzymes in adipose tissue controlling lipolysis (e.g., lipoprotein lipase; see Reference 77 for an informative review of the metabolic effects of ovarian hormones).

Research examining the effects of estradiol in females given central implants of the protein synthesis inhibitor anisomycin also supports the hypothesis that the PVN is an important site of action for estrogenic effects on eating.[78] In this experiment, ovariectomized rats were implanted with bilateral cannulae in the PVN and assigned to a central anisomycin group or a central control group. Three weeks after surgery,

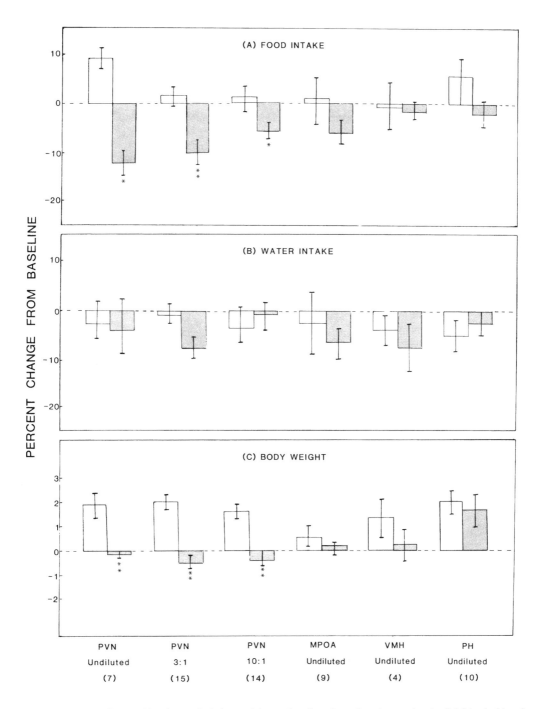

Figure 2 Effects of central implants of cholesterol (open bars) and varying doses of estradiol (shaded bars) placed into particular brain areas of ovariectomized rats. Implants were histologically verified and categorized as being in the ventromedial hypothalamus (VMH), the paraventricular nucleus (PVN), the medial preoptic area (MPOA), and the posterior hypothalamus (PH). Bars represent mean ± S.E. percent change from baseline for (A) food intake, (B) water intake, and (C) body weight. Number of cases per region indicated in parentheses; *p <0.05; **p <0.01. (From Butera, P. C. and Beikirch, R. J., *Brain Res.*, 491, 266, 1989. With permission.)

bilateral cannulae (28-gauge) containing anisomycin or no drug (control) were lowered into the brain. Thirty minutes later, females were injected with 2.0 mg of EB or the oil vehicle. The inner cannulae were removed 2 h after the hormone injections. This procedure was repeated for 3 days. Compared with

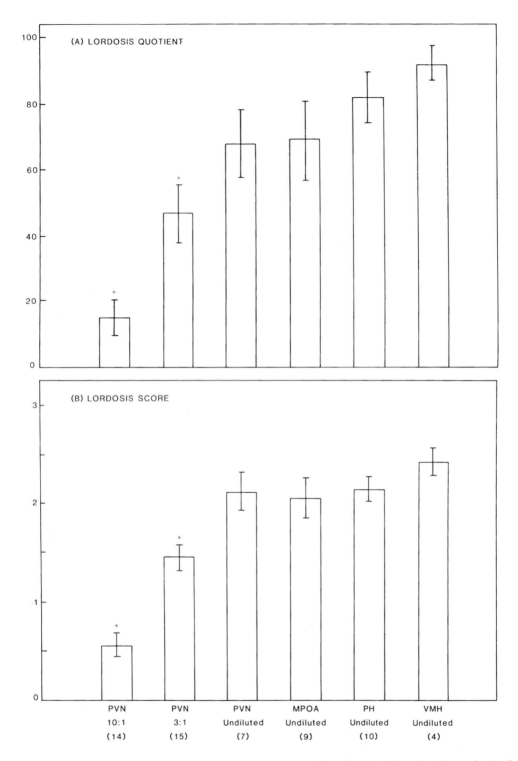

Figure 3 Changes in female sexual behavior following intracranial application of varying doses of estradiol placed into particular brain areas of ovariectomized rats. Measures of sexual behavior were collected on the third day of estradiol treatment, 5 h after an injection of progesterone (0.5 mg). Bars represent mean ± S.E. for (A) lordosis quotient and (B) lordosis score. Based on Newman-Keuls tests; *p <0.05 compared to VMH group. Number of cases per region indicated in parentheses. (From Butera, P. C. and Beikirch, R. J., *Brain Res.*, 491, 266, 1989. With permission.)

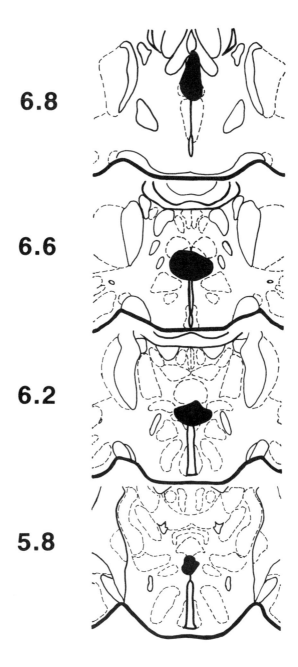

6.8

6.6

6.2

5.8

Figure 4 Schematic representation of the PVN lesions produced in this experiment. The drawings are based on the Pellegrino, Pellegrino, and Cushman atlas of the rat brain. (From Butera, P. C., Willard, D. M., and Raymond, S. A., *Brain Res.*, 576, 304, 1992. With permission.)

oil treatment, EB injections significantly lowered food intake and body weight gain of animals in the central control group. In contrast, EB treatment failed to reduce the food intake of animals given PVN implants of anisomycin. Peripheral injections of EB did lower body weight gain for these females, however. Additionally, EB treatment induced comparable levels of female sexual behavior (assessed 4 h after an injection of 0.5 mg of progesterone) in both groups, demonstrating that anisomycin implants did not interfere with the ability of estradiol to stimulate lordosis. The fact that lordosis was unaffected by PVN implants of anisomycin indicates that protein synthesis inhibition did not spread from the PVN to the VMH, since the application of anisomycin to the VMH blocks the induction of female sexual behavior by EB and progesterone.[73,79]

The data obtained from the anisomycin experiment, in conjunction with the results of the hormone implant and lesion studies discussed above, suggest that estrogenic stimulation of the PVN is necessary and sufficient for the effects of estradiol on feeding. These findings also support the hypothesis that the

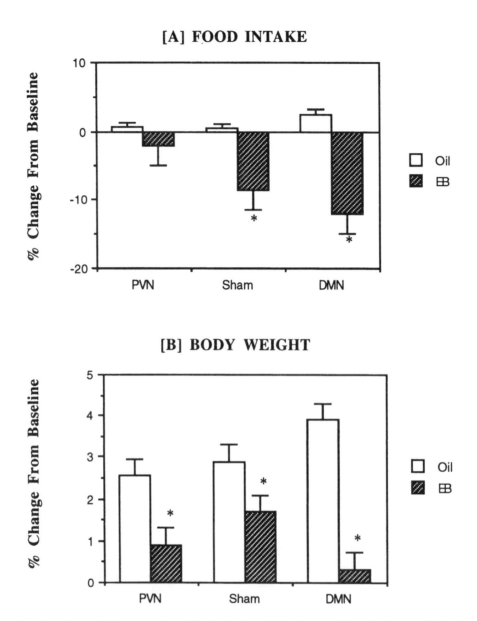

Figure 5 The effects of oil (open bars) and EB (shaded bars) injections on (A) food intake and (B) body weight in ovariectomized rats with PVN (n = 9), DMN (n = 4), or sham (n = 11) lesions. Bars represent mean ± S.E. percent change from baseline during 3 days of treatment; *p <0.01 relative to oil injections. (From Butera, P. C., Willard, D. M., and Raymond, S. A., *Brain Res.*, 576, 304, 1992. With permission.)

PVN is one of the critical brain sites at which estradiol normally acts to suppress food intake. The work disclosing the neural site of action for estradiol's effect on feeding does not simultaneously uncover the mechanism by which estradiol exerts its anorectic effects. However, research on the effects of estradiol on meal patterns and neural sites of action, together with our knowledge of gut-hindbrain-hypothalamic connections involved in the control of food intake, does offer some putative mechanisms. Specifically, it is hypothesized that the estrogenic suppression of food intake involves an interaction between estradiol and neural systems involved in meal termination (satiety).

Thus far, there has been no discussion about the neural site(s) at which progesterone acts to influence food intake. This is not an oversight. Compared with the research that has examined estrogens and feeding, very little is known about the sites and mechanisms involved in progesterone's effects on food intake. Aside from one published report showing that unilateral implants of undiluted progesterone in

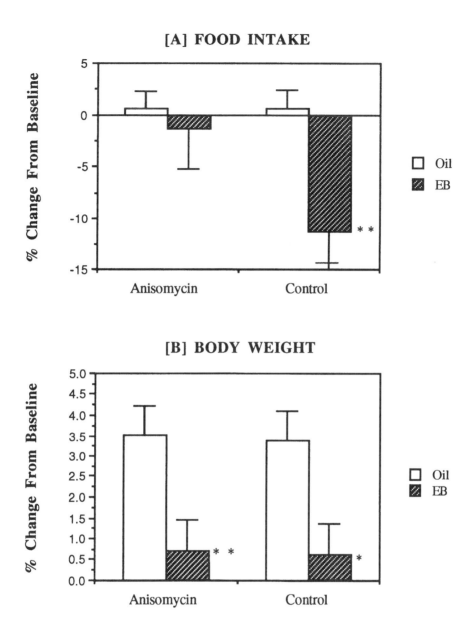

Figure 6 The effects of oil (open bars) and EB (shaded bars) injections on (A) food intake and (B) body weight in ovariectomized rats with PVN implants of anisomycin (n = 14) or blank cannulae (control, n = 10). Bars represent mean ± S.E. percent change from baseline during 3 days of treatment; *$p <0.05$ and **$p <0.01$ relative to oil injections. (From Butera, P. C., Campbell, R. B., and Bradway, D. M., *Brain Res.*, 624, 354, 1993. With permission.)

the DMN increase eating in ovariectomized rats,[57] little is known about the role of progesterone. Although the research discussed previously indicates that progesterone reverses the effects of estradiol on food intake, we do not yet know where or how progesterone exerts these effects (e.g., is it a CNS effect?). Given the dramatic changes in food intake and energy balance that accompany pregnancy, which are dependent upon the actions of progesterone,[1] it is important that we arrive at a more thorough understanding of the neural mechanisms that underlie the effects of progesterone on feeding. An analysis of the neural mechanisms involved in the effects of both progesterone and androgens on feeding would provide a more thorough understanding of the interactions that gonadal steroids have with the neural circuitry involved in the control of food intake.

V. POTENTIAL MEDIATING MECHANISMS

A. MODULATION OF A BODY WEIGHT SET-POINT

As discussed in the previous section, the research findings of the 1970s showing that estrogenic stimulation of the VMH causes a decrease in food intake were ultimately incorporated into prevailing theories on the hypothalamic control of feeding which emphasized the VMH as a critical brain site for the regulation of a body weight set-point. Simply put, the set-point hypothesis has as its central tenet that an organism defends and maintains its body weight or levels of adipose tissue around some ideal, neurologically predetermined value. According to this model, changes in food intake produced by manipulations of the hypothalamus (e.g., VMH or LH lesions) are secondary to the alterations in this predetermined value, the hypothetical set-point. It has been proposed that the estrogenic modulation of food intake results from alterations in the set-point, rather than from changes in the neural mechanisms directly involved in feeding behavior.[61,62,65] Thus, gonadal steroids were assumed to act on the VMH to alter the set-point around which levels of body fat are regulated. Changes in feeding behavior were seen as attempts to bring levels of adipose tissue into line with the new set-point. Viewed in this way, estrogen withdrawal increases the set-point and produces hyperphagia (because of set-point elevation), whereas estrogen replacement decreases the set-point and thus food intake (because of a decrease in the set-point). Support for this hypothesis includes the observations that estradiol has less effect on food intake in lighter vs. heavier age mates, and that continuous estrogenic treatment produces prolonged suppression of body weight but only a transient suppression of eating.[80,81]

Among the problems with the set-point hypothesis is the circular logic, which makes it impossible to generate experimental findings to disprove the model. The set-point concept may be a useful *description* of the relationships between estrogen, eating, and body weight, but it should not be construed as an *explanation* for these phenomena (see Reference 82 for an excellent treatment of the use of the set-point concept). Indeed, use of the set-point concept as an *explanation* could divert attention away from the research problems that must be explored in order to understand the physiological mechanisms that underlie the effects of estradiol on food intake and body weight, because researchers would be confident that an adequate explanation already exists. There are other problems that stem from a set-point explanation of food intake and body weight regulation which the interested reader can find in an excellent chapter by VanItallie and Kissileff.[83] Because of these shortcomings, the set-point hypothesis is now generally viewed as an unsatisfactory explanation for the effects of estradiol on feeding. George Wade, who articulated the set-point hypothesis in his earlier work, has recently written that a set-point explanation for the effects of estradiol on food intake is unsatisfactory and counterproductive.[1]

An alternative to the set-point hypothesis is a model that views the effects of estradiol on feeding and body weight as independent. Over the years, data have accumulated which suggest that estradiol's effects on food intake and body weight are not directly related. For example, estrogen-induced decreases in eating are not sufficient to account for changes in body weight, and estradiol can suppress food intake without affecting body weight. These observations suggest that estradiol's effects on food intake and body weight are controlled by independent mechanisms. Data supporting this hypothesis come from several sources.

In the weanling rat, estradiol lowers body weight before any effects on food intake are seen, and this decrease in body weight appears to result from metabolic effects of estradiol.[84] In the estradiol-treated ovariectomized adult, the suppression of food intake disappears within two weeks of daily hormone treatment whereas body weight remains at a low level as long as estradiol replacement continues. In the untreated ovariectomized female, the hyperphagia induced by ovariectomy subsides within a few weeks but body weight remains at an elevated level.[36,40] The results of a pair-feeding experiment indicate that restricting the food intake of untreated, ovariectomized rats to that of estradiol-treated females does not prevent the increase in body weight caused by ovariectomy.[85] In this experiment, ovariectomized animals given control injections increased their body weight by 18% within a month after surgery. Although animals given EB injections showed a 4% gain in body weight during this time, subjects that received control injections but were pair-fed with the EB-treated females increased their body weight by 18%.[85] Roy and Wade[85] also found that estradiol decreased food intake but did not lower body weight in ovariectomized females whose body weight gain after ovariectomy was prevented by food restriction. Thus, the changes in food intake caused by estradiol are neither necessary nor sufficient to account for the observed changes in body weight, and estradiol can suppress food intake without producing a concomitant decrease in body weight.

Other investigators have shown that the elevation in plasma triglycerides that accompanies estradiol treatment (which may reflect a decrease in adiposity and could be involved in the estrogenic suppression of food intake), does not occur until after 15 days of hormone treatment, a time when the anorectic effects of estradiol have subsided.[86,87] As discussed in an earlier section, PVN lesions attenuate the effects of estradiol on food intake but do not interfere with the ability of estradiol to decrease body weight.[76] Taken together, these results support the idea that estradiol's effects on food intake and body weight are independent and are probably mediated by different physiological mechanisms. Viewed in this way, estrogenic effects on body weight and adiposity may reflect direct actions of estradiol on mechanisms controlling lipid metabolism (alterations in set-point?), whereas effects on feeding involve interactions between estradiol and brain mechanisms involved in satiety. Recently, Wade and his colleagues have hypothesized that estradiol affects food intake by acting on brain areas that receive visceral inputs and by modulating the hormonal and behavioral responses to this input.[1] The research to be discussed in the next section fits nicely with this idea, and provides a testable theoretical framework for the effects of estradiol on feeding behavior.

B. INTERACTIONS WITH NEURAL SYSTEMS INVOLVED IN SATIETY

The role of neuropeptides in the control of food intake has been intensively investigated during the last 15 years. One peptide hormone that has received much research attention is cholecystokinin (CCK). CCK is a peptide originally isolated from the duodenum as a 33 amino acid chain that was found to be associated with gall bladder contractions. CCK is also present in different forms in several organ systems including the brain.[88,89] Peripheral injections of CCK octapeptide (CCK-8) have been shown to suppress food intake in a number of species,[90,91] and the consensus today is that CCK is an important physiological signal for meal termination.[92]

Although the specific pathway involved in the effects of exogenous CCK on feeding remains to be identified, it appears that the actions of CCK in the periphery are important for CCK-induced satiety. In rats, bilateral, subdiaphragmatic vagotomy abolishes the decrease in food intake normally produced by CCK injections.[93,94] The gastric branch of the vagus appears to be critical for CCK-induced satiety because gastric, but not celiac or hepatic vagotomies, block satiety elicited by CCK injections.[93,94] The small-diameter sensory axons in the afferent vagus are the important fibers for these effects, because capsaicin (a relatively selective neurotoxin for sensory nerves) applied directly to vagal trunks markedly attenuates CCK-induced satiety.[95] Thus, afferent fibers in the abdominal vagus appear to be an important component of a neural pathway involved in the effects of CCK on food intake.

In an attempt to trace the sensory pathway by which CCK reduces food intake, Crawley and colleagues[96,97] identified two central lesions that abolish CCK-induced satiety. In one experiment, bilateral midbrain transections that interrupted pathways from the nucleus of the solitary tract to the hypothalamus blocked the effects of systemic CCK injections on food intake.[96] In a second experiment, bilateral lesions of the PVN also abolished satiety normally produced by i.p. injections of CCK.[97] Although one may conclude that the neural pathway for CCK-induced satiety involves afferent vagal fibers that send information to the nucleus of the solitary tract from which efferent fibers project rostrally to the PVN, the picture is far from complete. For example, peripheral injections of CCK suppress food intake in rats with chronic decerebration produced by knife cuts from the supracollicular level to the brainstem.[98] The fact that CCK is behaviorally effective in animals where neural communication between the nucleus of the solitary tract and the hypothalamus is abolished indicates that more work is needed to clarify the brain mechanisms involved in CCK-induced satiety.

One potential mechanism by which estradiol could influence feeding is through interactions with CCK systems described above that participate in the control of food intake. For example, it is possible that estradiol decreases food intake by potentiating the afferent message that arises from the stimulation of CCK receptors in the periphery. Thus, estrogenic actions in the central nervous system (e.g., PVN, brainstem) may augment the sensory message traveling to the brain following the interaction between CCK and its peripheral receptors. Interestingly, both estradiol and CCK produce identical effects on meal size. As discussed previously, peripheral injections of EB (2.0 μg/day) decrease eating by causing a decrease in the meal size of ovariectomized rats. This suppression in meal size persists even after the total food intake of estradiol-treated females returns to the level of untreated controls.[26] Similarly, continuous i.p. infusion of CCK-8 has also been shown to suppress meal size, but not total food intake, in male rats.[99] In addition, McLaughlin, Baile, and Peiken[100] have shown that during lactation, a time when females are unresponsive to the effects of estradiol on feeding,[65] female rats are insensitive to doses of CCK that are effective in males. However, other investigators have reported somewhat different findings using different experimental paradigms.[101,102] The fact that estradiol and CCK produce similar effects on meal size, and that their effects on food intake appear to covary with the reproductive state, suggests that estradiol and CCK affect feeding behavior through a common mechanism of action. It is therefore possible that the suppression of food intake caused by estradiol is mediated by a stimulatory effect of estradiol on CCK pathways involved in the regulation of feeding behavior. The remainder of this chapter will examine the research that we and others have conducted that bears on this hypothesis.

1. Effects of Peripheral Estradiol Treatment on the Satiety Action of CCK

The following experiment was conducted to determine whether the effects of CCK on food intake could be enhanced by estradiol. To evaluate the ability of estradiol to modulate the satiety effect of exogenous CCK-8, ovariectomized rats were subcutaneously implanted with estradiol-filled Silastic capsules (5% mixture of estradiol + cholesterol) or empty capsules on the day of surgery.[103] Three weeks later, animals were food deprived for 24 h and were given i.p. injections of physiological saline or CCK-8 (5 or 10 µg/kg). Food intake (powdered rat chow) was measured 60 min after the injections. As shown in the figure below, the effects of CCK on food intake were markedly enhanced in the estradiol-treated animals. This same phenomenon was also documented independently by Linden and colleagues using a similar route of estradiol administration (Silastic capsules) and test diet.[104] Most recently, the potentiation of the satiety action of CCK by estradiol has been reported in a series of experiments by Geary and colleagues[105] using a cyclic paradigm of estradiol replacement (acute injections vs. continuous exposure), lower doses of CCK (0.5 to 4 µg/kg), and a different test diet (intake of a sucrose solution).

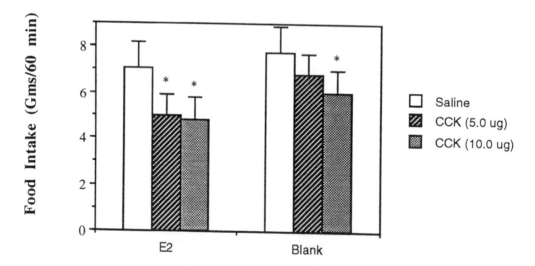

Figure 7 Effects of i.p. injections of CCK-8 or physiological saline on food intake of ovariectomized rats treated with 5% estradiol Silastic capsules (E2, n = 12) or empty capsules (blank, n = 12). Data are expressed as mean ± S.E. intake during a 60-min feeding test following 24-h food deprivation; *p <0.01 compared with saline control. (From Butera, P. C., Bradway, D. M., and Cataldo, N. J., *Physiol. Behav.*, 53, 1235, 1993. With kind permission from Elsevier Science Ltd., The Boulevard, Langford Lane, Kidlington 0X5 1Gb, UK.)

To further examine the apparent interaction between estradiol and CCK, the satiety effects of estradiol and CCK-8 administered singly and in combination were expressed as a percent suppression of intake of control animals which received neither estradiol nor CCK-8. The results indicate that the effect of combined administration of estradiol and CCK-8 is greater than that expected by addition of the suppressive effects of either estradiol or CCK-8 administered alone.

Although we and others have found that estradiol potentiates the effects of exogenous CCK on food intake,[103-105] the results of an earlier study appear to contradict these findings.[106] In this experiment, i.p. injections of CCK-8 (5 and 10 µg/kg) failed to suppress food intake in estrogen-treated, ovariectomized rats.[106] There are, however, some important methodological discrepancies between these experiments that could account for the disparate findings. In the study by Wager-Srdar et al.,[106] subjects were food

Table 1 Suppression of Food Intake after Peripheral Treatment with Estradiol, CCK-8, or Estradiol + CCK-8.

Treatment	Percent suppression
Estradiol	9.4*
CCK-8 (5 µg/kg)	12.3
CCK-8 (10 µg/kg)	22.2*
Estradiol + CCK-8 (5 µg/kg)	35.6*
Estradiol + CCK-8 (10 µg/kg)	37.9*

Note: Data are expressed as percent suppression of intake compared with untreated ovariectomized females given saline control and n = 12 for each treatment condition. *p <0.01 based on t-tests for repeated measures (CCK alone comparisons) or between-group t-tests (estradiol alone comparison; estradiol + CCK comparisons).

Reprinted from Butera, P. C., Bradway, D. M., and Cataldo, N. J., *Physiol. Behav.*, 53, 1235, 1993. With kind permission from Elsevier Science Ltd., The Boulevard, Langford Lane, Kidlington 0X5 1Gb, U.K.

deprived for only 2 h during the light phase of the L:D cycle prior to the 60-min feeding test. This mild level of food deprivation appeared to produce very low baseline levels of food intake during the feeding test. Therefore, an inability of CCK to suppress eating in this study may represent a floor effect rather than an attenuation of CCK's effects by estradiol as the authors suggest. Indeed, subjects that showed an inhibition of food intake during CCK treatment (e.g., progesterone-treated females) had baseline levels of food intake that were 1.0–1.5 g higher than those of estrogen-treated animals. Thus, an effect of estradiol alone on feeding appears to have gone undetected in this experiment. It is quite likely that these methodological differences can account for the discrepancies in CCK responsiveness that were found in these experiments.

2. Effects of Central Estradiol on the Satiety Action of CCK

Although peripheral treatment with estradiol enhances the satiety effect of CCK, the site at which such an interaction occurs remains to be specified. In the next sequence of experiments, we evaluated whether direct placement of estradiol into different brain regions could increase the satiating potency of exogenous CCK.[107] In these experiments, ovariectomized rats received unilateral implants of dilute estradiol (1 part estradiol to 10 parts cholesterol) and the cholesterol control in the PVN, VMN, preoptic area (POA), or third ventricle. In the first experiment, females were injected with physiological saline or i.p. CCK-8 at a dose of 5 µg/kg following 2 days of central steroid treatment. As in the previous study examining estradiol-CCK interactions, animals were food deprived for 24 h prior to CCK injections, and food intake was measured 60 min after the injections. In this experiment, as in the other studies previously discussed, we found that estradiol implants in the PVN significantly lowered food intake during the two-day period of central steroid treatment. In contrast, estradiol implants in the VMN or the third ventricle had no significant effects on food intake.

The effects of central implants of estradiol and i.p. injections of CCK on food intake during the 60-min feeding tests are shown in the figure below. In this experiment, there was a significant interaction between estradiol and CCK for subjects with implants in the PVN, but not for subjects with cannulae placed in the VMN or third ventricle.

When the experiment was replicated using a lower dose of CCK (0.5 µg/kg), the results were generally consistent with the data discussed above. In this study, PVN implants of dilute estradiol, but not implants in the VMN or POA, significantly decreased food intake during the two-day period of central steroid treatment.

Data obtained from the feeding tests indicated that CCK suppressed food intake in females with estradiol implants in the PVN but had no effect in animals with implants in the VMN or POA.

The data showing that dilute estradiol implants in the PVN but not the VMN suppress food intake over a two-day period confirm and extend previous findings, and lend additional support to the hypothesis that the effects of estradiol on feeding normally involve actions of the hormone in the PVN. The fact that estrogenic stimulation of the PVN also decreased food intake during the feeding tests suggests that

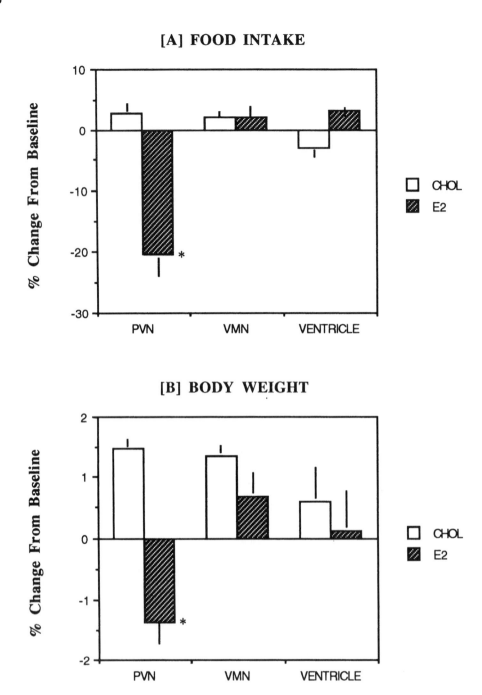

Figure 8 Effects of central implants of dilute estradiol (E2) or cholesterol (CHOL) on (A) food intake and (B) body weight in ovariectomized rats. Implants were histologically verified and categorized as being in the ventromedial nucleus of the hypothalamus (VMN), the paraventricular nucleus of the hypothalamus (PVN), and the third ventricle (ventricle). Bars represent mean ± S.E. percent change from baseline during 2 days of treatment; *p <0.01 relative to CHOL treatment.

estradiol not only influences 24-h food intake, but may also modulate short-term controls on eating by affecting meal size. The results of these experiments also shed light on a potential biochemical mechanism underlying the estrogenic suppression of food intake. Specifically, it appears that the actions of estradiol

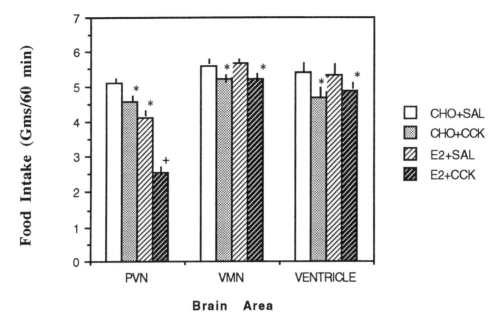

Figure 9 Effects of i.p. injections of CCK-8 (5 µg/kg) or physiological saline (SAL) on food intake of ovariectomized rats receiving central implants of dilute estradiol (E2) or cholesterol (CHO) placed into various brain
areas. Data are expressed as mean ± S.E. food intake during a 60-min feeding test following 24-h food deprivation;
*represents significant difference from control treatments (p <0.05), and + represents significant E2-CCK inter-
action (p <0.01).

in the PVN potentiate the satiety effect produced by peripheral injections of CCK. Therefore, it is possible
that the effects of estradiol on feeding involve hormonal actions in the PVN that serve to augment
sensory signals arising from the activation of CCK receptors in the periphery.

3. Effects of CCK Antagonists on the Suppression of Food Intake by Estradiol

To investigate the possibility that the anorectic effects of estradiol involve the actions of endogenous
CCK, we administered type A (devazepide) and type B (L-365,260) CCK receptor antagonists to
ovariectomized rats treated peripherally with EB.[108] In this experiment, female rats were ovariectomized
and assigned to either a devazepide or an L-365,260 group. Three weeks after ovariectomy, half of the
animals were injected with 5 µg of EB for 3 days whereas half received the oil vehicle. All subjects
were then food deprived for 24 h prior to the feeding test the following day. Food cups were presented
30 min after i.p. injections of devazepide (10 or 100 µg/kg), L-365,260 (10 or 100 µg/kg), or the propylene
glycol/ETOH vehicle, and data collected 30 min later. The procedure was repeated every five days until
all subjects received the vehicle and both doses of one antagonist. As was observed in the previous
experiments, EB-treated females consumed less food during the vehicle feeding tests than controls. In
addition, there was also a significant interaction between EB and devazepide. Specifically, although both
doses of devazepide attenuated the suppression of food intake by EB, the drug had no significant effects
on eating in oil-treated subjects. There was no interaction between EB and L-365,260 nor any effect of
L-365,260 alone on eating.

These findings are consistent with the hypothesis that estradiol's effects on eating involve an interaction with CCK satiety systems, and suggest that stimulation of type A CCK receptors is necessary for
the estrogenic suppression of food intake. A working model that emerges from these data suggests that
estradiol lowers food intake by augmenting a CCK-induced neural message that occurs during the
consumption of a meal. Based on the data from the experiments presented in the previous section, it
appears that estradiol potentiates this afferent signal through its actions in the brain. Thus, by acting on
estrogen-sensitive neurons in neural pathways involved in feeding and metabolism, estradiol may modulate behavioral responses to incoming sensory information and, as a result, produce changes in food
intake. The mechanisms by which estradiol influences the central processing of neural signals involved
in satiety remain to be investigated.

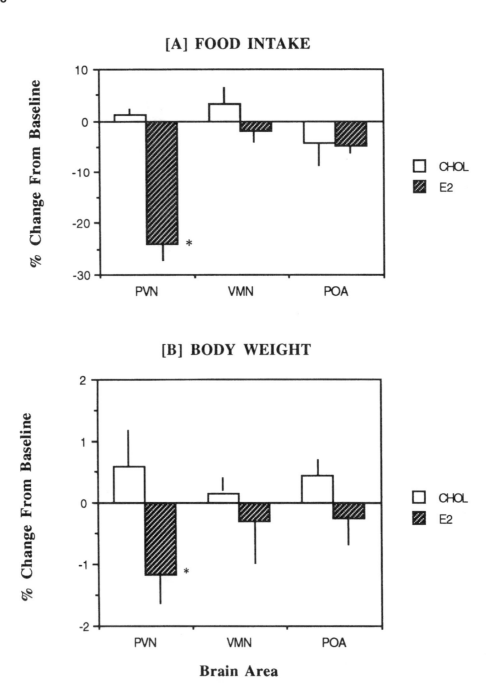

Figure 10 Effects of central implants of dilute estradiol (E2) or cholesterol (CHOL) on (A) food intake and (B) body weight in ovariectomized rats. Implants were histologically verified and categorized as being in the ventromedial nucleus of the hypothalamus (VMN), the paraventricular nucleus of the hypothalamus (PVN), and the preoptic area (POA). Bars represent mean ± S.E. percent change from baseline during 2 days of treatment; *p <0.01 relative to CHOL treatment. (From Butera, P. C., Xiong, M., Davis, R. J., and Platania, S. P., unpublished data.)

4. Estrogen as an Indirect Control of Meal Size

The research examining the effects of estradiol on food intake that has been reviewed in this chapter, and the proposed mechanism for its anorectic effects, can be placed in a broader theoretical framework of direct and indirect controls of meal size. These ideas, recently articulated by Smith,[109] conceptualize

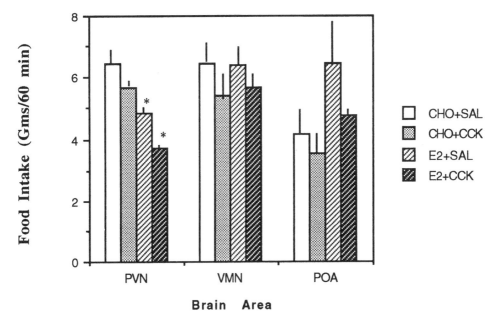

Figure 11 Effects of i.p. injections of CCK-8 (0.5 μg/kg) or physiological saline (SAL) on food intake of ova-riectomized rats receiving central implants of dilute estradiol (E2) or cholesterol (CHO) placed into various brain areas. Data are expressed as mean ± S.E. food intake during a 60-min feeding test following 24-h food deprivation; * represents significant difference from control treatments ($p < 0.01$). (From Butera, P. C., Xiong, M., Davis, R. J., and Platania, S. P., unpublished data.)

the various stimuli and conditions that change meal size as affecting one of two systems: (1) a direct sensory control system responsible for encoding, transmitting, and processing sensory stimuli that accompany ingestion (e.g., the chemical and mechanical receptors that run from the tongue to the end of the small intestine plus their afferent connections) and (2) indirect control systems that do not have direct sensory contact with food stimuli but instead encode, transmit, and process other stimuli that exert effects on meal size (e.g., ambient temperature, circadian rhythms, gonadal hormones, foraging experience, etc.).

During ingestion, food stimuli have direct contact with preabsorptive sensory receptors located in the mouth, axon terminals of visceral afferent fibers or enteric afferent neurons, and mucosal cells that manufacture and release peptides and amines. This extensive flow of sensory information that occurs during ingestion is supplemented by postingestive afferent input arising from vagal afferent fibers that emerge from the nodose and jugular ganglia and from spinal visceral afferents with cell bodies in the dorsal root ganglia. This pattern of afferent activity provides the brain with feedback about ingested food that is processed by neural networks involved in the control of food intake. Viewed in this way, meal size is determined by the relative strength and central interactions of the positive and negative sensory feedback, produced by food, that ultimately act on the central network for feeding which controls the rate and duration of eating.[109]

The indirect controls of meal size, while numerous, share some important features. According to Smith, indirect controls are not directly affected by food stimuli that activate the receptors mentioned above. In addition, the indirect controls typically have a duration of action that lasts beyond a single meal. In the theoretical system outlined by Smith, indirect controls influence meal size by modulating some features of the direct control system (e.g., central processing of the sensory input, changing metabolic or endocrine responses to food stimuli, changing the number or sensitivity of sensory receptors, etc.).

Viewing meal size in terms of direct and indirect control systems has several advantages over previous perspectives that characterize the controls of feeding in terms of central and peripheral controls, psychological and physiological controls, and short- and long-term controls. Unlike these perspectives, the model developed by Smith clearly specifies the functional relations among the different controls, facilitates the analysis of the mechanisms controlling meal size, and is heuristic for experimental research on feeding behavior.

230

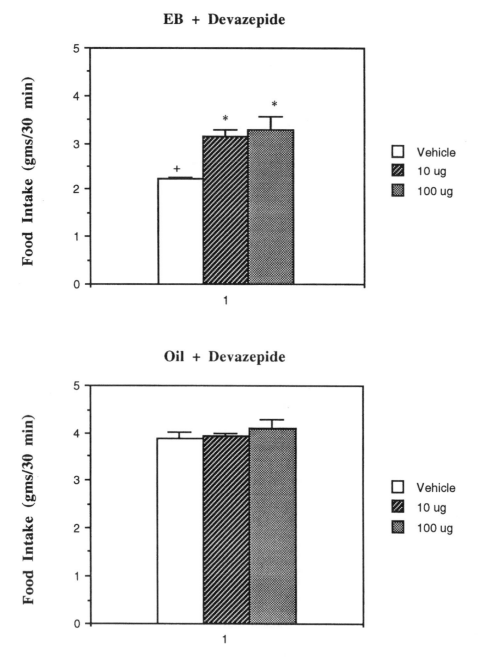

Figure 12 Effects of i.p. injections of devazepide or the propylene glycol/ETOH vehicle on food intake of ovariectomized rats receiving EB (5 μg) or oil treatments. Data are expressed as mean ± S.E. intake during a 30-min feeding test following 24-h food deprivation; * represents significant difference from vehicle treatment ($p < 0.01$), and + represents significant difference from the oil control group ($p < 0.01$).

Indeed, previous models tend to characterize the effects of estrogen in terms of hormonal modulation of long-term controls on feeding (see References 110 and 65). Viewing estradiol as one of the indirect controls on food intake has several advantages. According to Smith, indirect controls on meal size operate by modulating one or more of the direct controls, presumably by influencing the central processing of these direct sensory controls. This facilitates experimental analysis of estrogenic effects on eating because it allows us to systematically examine the interactions that estradiol has on direct sensory controls that are known (e.g., oral and postingestive input; paracrine, endocrine, and metabolic signals), instead of

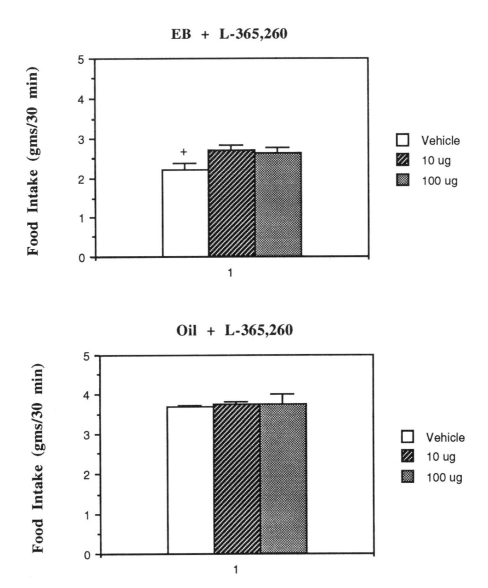

Figure 13 Effects of i.p. injections of L-365,260 or the propylene glycol/ETOH vehicle on food intake of ovariectomized rats receiving EB (5 μg) or oil treatments. Data are expressed as mean ± S.E. intake during a 30-min feeding test following 24-h food deprivation; + represents significant difference from the oil control group ($p < 0.01$).

examining hormonal modulation of a hypothetical body weight set-point. Viewing estradiol as an indirect control of meal size is also consistent with existing data on the ovarian influences on meal patterns.[26] Finally, the research done in our lab and by others on estrogen-CCK interactions is consistent with this conceptual framework because it clearly identifies a direct sensory control of meal size that is modulated by estradiol.

However, the hypothesis that estradiol suppresses food intake by interacting with CCK systems involved in satiety is not meant to be restrictive. It is quite likely that CCK is only one of the direct sensory controls of meal size with which ovarian hormones interact. For example, unpublished data from our lab suggest that peripheral treatment with estradiol also enhances the satiety action of i.p. injections of bombesin (4 μg/kg) in ovariectomized rats (Butera, Davis, and Ahmed, unpublished observations). In mammals, bombesin-like peptides appear to represent another class of putative sensory signals, stimulated by the presence of food in the gut, that exert negative sensory feedback and inhibit food intake. Therefore, the model of estrogen action being developed here maintains that estradiol lowers food intake by augmenting direct sensory controls of meal size which are stimulated by food during eating and provide negative feedback on eating and meal size. Based on the data presented in earlier sections, and in accordance with the conceptual framework outlined by Smith, it is hypothesized that estradiol influences these inhibitory sensory messages on meal size through its actions on estrogen-sensitive neurons in brain areas involved in feeding behavior. This model does not preclude interactions between estradiol and direct sensory controls other than CCK that exert inhibitory effects on meal size (e.g., bombesin-like peptides). Viewing estradiol as an indirect control of meal size would also open up avenues of experimental investigation to evaluate whether the anorectic effects of estradiol also involve mechanisms that serve to diminish the positive feedback of food stimuli derived from orosensory stimulation.

ACKNOWLEDGMENTS

Supported by Research Grants MH 42127 from the National Institute of Mental Health and NS 26020 and DK 45499 from the National Institutes of Health. I would like to thank Drs. Nori Geary and Irene Rykaszewski for their helpful comments and advice on the manuscript, and Dr. Gerard Smith for sharing his unpublished manuscripts. The CCK antagonists were generously supplied by Dr. Roger Freidinger of Merck, Sharp, and Dohme.

REFERENCES

1. Wade, G. N. and Schneider, J. E., Metabolic fuels and reproduction in female mammals, *Neurosci. Biobehav. Rev.*, 16, 235, 1992.
2. Mellon, S. H., Neurosteroids: biochemistry, modes of action, and clinical relevance, *J. Clin. Endocrinol., Metab.*, 78, 1003, 1994.
3. Castonguay, T. W., Dallman, M. F., and Stern, J. S., Some metabolic and behavioral effects of adrenalectomy on obese Zucker rats, *Am. J. Physiol.*, 251, R923, 1986.
4. Dallman, M. F., Viewing the ventromedial hypothalamus from the adrenal gland, *Am. J. Physiol.*, 246, R1, 1984.
5. King, B. M., Glucocorticoids and hypothalamic obesity, *Neurosci. Biobehav. Rev.*, 12, 29, 1988.
6. Tempel, D. L. and Leibowitz, S. F., Adrenal steroid receptors: interactions with brain neuropeptide systems in relation to nutrient intake and metabolism, *J. Neuroendocrinol.*, 6, 479, 1994.
7. Tanner, J. M., *Growth at Adolescence*, Blackwell, Oxford, 1962.
8. King, H. D., The growth and variability in the body weight of the albino rat, *Anat. Rec.*, 9, 751, 1915.
9. Wang, G. H., Age and sex differences in the daily food intake of the albino rat, *Am. J. Physiol.*, 71, 729, 1924.
10. Swanson, H. H., Effects of pre- and post-pubertal gonadectomy on sex differences in growth, adrenal and pituitary weights of hamsters, *J. Endocrinol.*, 39, 555, 1967.
11. Zucker, I., Wade, G. N., and Ziegler, R., Sexual and hormonal influences on eating, taste preferences and body weight of hamsters, *Physiol. Behav.*, 8, 101, 1972.
12. Johnston, F. E., Health implications of childhood obesity, *Ann. Int. Med.*, 103, 1068, 1985.
13. Rolls, B. J., Fedoroff, I. C., and Guthrie, J. F., Gender differences in eating behavior and body weight regulation, *Health Psychol.*, 10, 133, 1991.
14. Warren, M. P., Physical and biological aspects of puberty, in *Girls at Puberty: Biological and Psychosocial Perspectives*, Brooks-Gunn, J. and Petersen, A. C., Eds., Plenum, New York, 1983, 3.
15. Basiotis, P. P., Thomas, R. G., Kelsay, J. L., and Mertz, W., Sources of variation in energy intake by men and women as determined from one year's daily dietary records, *Am. J. Clin. Nutr.*, 50, 448, 1989.
16. Slob, A. K. and van der Werff ten Bosch, J. J., Sex differences in body growth in the rat, *Physiol. Behav.*, 14, 353, 1975.
17. Bell, D. D. and Zucker, I., Sex differences in body weight and eating: organization and activation by gonadal hormones in the rat, *Physiol. Behav.*, 7, 27, 1971.
18. Gentry, R. T. and Wade, G.N., Sex differences in sensitivity of food intake, body weight, and running wheel activity to ovarian steroids in rats, *J. Comp. Physiol. Psychol.*, 90, 747, 1976.
19. Gentry, R. T. and Wade, G.N., Androgenic control of food intake and body weight in male rats, *J. Comp. Physiol. Psychol.*, 90, 18, 1976.
20. Hoskins, R. G., Studies on vigor II: the effect of castration on voluntary activity, *Am. J. Physiol.*, 72, 324, 1925.
21. Wang, G. H., Richter, C. P., and Guttmacher, A. F., Activity studies on male castrated rats with ovarian transplants and correlation of activity with the histology of the grafts, *Am. J. Physiol.*, 73, 581, 1925.
22. Kochakian, C. D. and Webster, J. A., Effect of testosterone on the appetite, body weight, and composition of the normal rat, *Endocrinology*, 63, 737, 1958.
23. Gray, J. M., Nunez, A. A., Siegel, L. I., and Wade, G. N., Effects of testosterone on body weight and adipose tissue: role of aromatization, *Physiol. Behav.*, 23, 465, 1979.

24. Kochakian, C. D. and Endahl, B., Changes in body weight of normal and castrated rats by different doses of testosterone propionate, *Proc. Soc. Exp. Biol. Med.,* 100, 520, 1959.

25. Nunez, A. A., Siegel, L. I., and Wade, G. N., Central effects of testosterone on food intake in male rats, *Physiol. Behav.,* 24, 469, 1980.

26. Blaustein, J. D. and Wade, G. N., Ovarian influences on the meal patterns of female rats, *Physiol. Behav.,* 17, 201, 1976.

27. ter Haar, M. B., Circadian and estrual rhythms in food intake in the rat, *Horm. Behav.,* 3, 213, 1972.

28. Forbes, J. M., Effects of estradiol-17β on voluntary food intake in sheep and goats, *J. Clin. Endocrinol.,* 52, VIII, 1972.

29. Friend, D. W., Self-selection of feeds and water by swine during pregnancy and lactation, *J. Anim. Sci.,* 32, 658, 1969.

30. Tarttelin, M. F., Cyclical variations in food and water intake in ewes, *J. Physiol.,* 195, 29P, 1968.

31. Czaja, J. A., Butera, P. C., and McCaffrey, T. A., Independent effects of estradiol on water and food intake, *Behav. Neurosci.,* 97, 210, 1983.

32. Czaja, J. A., Food rejection by female rhesus monkeys during the menstrual cycle and early pregnancy, *Physiol. Behav.,* 4, 579, 1975.

33. Dalvit-McPhillips, S. P., The effect of the human menstrual cycle on nutrient intake, *Physiol. Behav.,* 31, 209, 1983.

34. Gong, E. J., Garrel, D., and Calloway, D. H., Menstrual cycle and voluntary food intake, *Am. J. Clin. Nutr.,* 49, 252, 1989.

35. Holt, H., Keeton, R. W., and Vennesland, B., The effect of gonadectomy on body structure and body weight in albino rats, *Am. J. Physiol.,* 114, 515, 1936.

36. Tarttelin, M. F. and Gorski, R. A., The effects of ovarian steroids on food and water intake and body weight in the female rat, *Acta Endocrinol. (Copenhagen),* 72, 551, 1973.

37. Wade, G. N., Some effects of ovarian hormones on food intake and body weight in female rats, *J. Comp. Physiol. Psychol.,* 88, 183, 1975.

38. Czaja, J. A. and Goy, R. W., Ovarian hormones and food intake in female guinea pigs and rhesus monkeys, *Horm. Behav.,* 6, 329, 1975.

39. Czaja, J. A., Ovarian influences on primate food intake: assessment of progesterone actions, *Physiol. Behav.,* 21, 923, 1978.

40. Galetti, F. and Klopper, A., The effect of progesterone on the quantity and distribution of body fat in the female rat, *Acta Endocrinol. (Copenhagen),* 46, 379, 1964.

41. Hervey, G. R. and Hervey, E., Interaction of the effects of estradiol and progesterone on body weight in the rat, *J. Endocrinol.,* 33, ix, 1965.

42. Zucker I., Hormonal determinants of sex differences in saccharin preference, food intake, and body weight, *Physiol. Behav.,* 4, 595, 1969.

43. Roberts, S., Kenney, N. J., and Mook, D. G., Overeating induced by progesterone in the ovariectomized, adrenalectomized rat, *Horm. Behav.,* 3, 627, 1972.

44. Ross, G. E. and Zucker, I., Progesterone and ovarian-adrenal modulation of energy balance in rats, *Horm. Behav.,* 5, 43, 1974.

45. Gray, J. M. and Wade, G. N., Cytoplasmic progestin binding in rat adipose tissues, *Endocrinology,* 104, 1377, 1979.

46. Leavitt, W. W., Chen, T. J., and Allen, T. C., Regulation of progesterone receptor formation by estrogen action, *Ann. N.Y. Acad. Sci.,* 286, 210, 1977.

47. MacLusky, N. J. and McEwen, B. S., Estrogen modulates progestin receptor concentrations in some brain regions but not in others, *Nature,* 274, 276, 1978.

48. Brobeck, J. R., Wheatland, M., and Strominger, J. L., Variations in the regulation of energy exchange associated with estrous, diestrus, and pseudopregnancy in rats, *Endocrinology,* 40, 65, 1947.

49. Kennedy, G. C. and Mitra, J., Hypothalamic control of energy balance and the reproductive cycle in the rat, *J. Physiol. (London),* 166, 395, 1963.

50. Kennedy, G. C. and Mitra, J., Body weight and food intake as initiating factors for puberty in the rat, *J. Physiol. (London),* 166, 408, 1963.

51. Kennedy, G. C. and Mitra, J., Spontaneous pseudopregnancy and obesity in the rat, *J. Physiol. (London),* 166, 419, 1963.

52. Kennedy, G. C., Hypothalamic control of the endocrine and behavioral changes associated with oestrus in the rat, *J. Physiol. (London),* 172, 383, 1964.

53. Wade, G. N. and Zucker, I., Modulation of food intake and locomotor activity in female rats by diencephalic hormone implants, *J. Comp. Physiol. Psychol.,* 72, 328, 1970.

54. Pfaff, D. W. and Keiner, M., An atlas of estradiol-concentrating cells in the central nervous system of the female rat, *J. Comp. Neurol.,* 151, 121, 1973.

55. Stumpf, W. E., Estrogen-neurons and estrogen-neuron systems in the periventricular brain, *Am. J. Anat.,* 130, 218, 1970.

56. Beatty, W. W., O'Briant, D. A., and Vilberg, T. R., Suppression of feeding by hypothalamic implants of estradiol in male and female rats, *Bull. Psychon. Soc.,* 3, 273, 1974.

57. Jankowiak, R. and Stern, J. J., Food intake and body weight modifications following medial hypothalamic hormone implants in female rats, *Physiol. Behav.,* 12, 875, 1974.

58. Hoebel, B. G., Feeding: Neural control of intake, *Ann. Rev. Physiol.*, 33, 533, 1971.

59. Stellar, E., The physiology of motivation, *Psychol. Rev.*, 61, 5, 1954.

60. Wyrwicka, W. and Dobrzecka, C., Relationship between feeding and satiation centers of the hypothalamus, *Science*, 131, 805, 1960.

61. Mook, D. G., Kenney, N. J., Roberts, S., Nussbaum, A. I., and Rodier, W. I., III., Ovarian-adrenal interactions in regulation of body weight by female rats, *J. Comp. Physiol. Psychol.*, 81, 198, 1972.

62. Wade, G. N., Gonadal hormones and the behavioral regulation of body weight, *Physiol. Behav.*, 8, 523, 1972.

63. Beatty, W. W., O'Briant, D. A., and Vilberg, T. R., Effects of ovariectomy and estradiol injections on food intake and body weight in rats with ventromedial hypothalamic lesions, *Pharmacol. Biochem. Behav.*, 3, 539, 1975.

64. King, J. M. and Cox, V. C., The effects of estrogens on food intake and body weight following ventromedial hypothalamic lesions, *Physiol. Psychol.*, 1, 262, 1973.

65. Wade, G. N., Sex hormones, regulatory behaviors, and body weight, in *Advances in the Study of Behavior*, Vol. 6, Rosenblatt, J., Hinde, R., Shaw, E., and Beer, C., Eds., Academic Press, New York, 1976, 201–279.

66. Ahlskog, J. E. and Hoebel, B. G., Overeating and obesity from damage to a noradrenergic system in the brain, *Science*, 182, 166, 1973.

67. Kapatos, G. and Gold, R. M., Evidence for ascending noradrenergic mediation of hypothalamic hyperphagia, *Pharmacol. Biochem. Behav.*, 1, 81, 1973.

68. Simerly, R. B., Chang, C., Muramatsu, M., and Swanson, L. W., Distribution of androgen and estrogen receptor mRNA-containing cells in the rat brain: an *in situ* hybridization study, *J. Comp. Neurol.*, 294, 76, 1990.

69. Kirchgessner, A. L. and Sclafani, A., PVN-hindbrain pathway involved in the hypothalamic hyperphagia-obesity syndrome, *Physiol. Behav.*, 42, 517, 1988.

70. Leibowitz, S. F. and Stanley, B. G., Neurochemical controls of appetite, in *Feeding Behavior: Neural and Humoral Controls*, Ritter, R. and Ritter, S., Eds., Academic Press, New York, 1986, 191–234.

71. Butera, P. C. and Czaja, J. A., Intracranial estradiol in ovariectomized guinea pigs: effects on ingestive behaviors and body weight, *Brain Res.*, 322, 41, 1984.

72. Palmer, K. and Gray, J. M., Central vs. peripheral effects of estrogen on food intake and lipoprotein lipase activity in ovariectomized rats, *Physiol. Behav.*, 32, 187, 1986.

73. Meisel, R. L. and Pfaff, D. W., RNA and protein synthesis inhibitors: effects on sexual behavior in female rats, *Brain Res. Bull.*, 12, 187, 1984.

74. Butera, P. C. and Beikirch, R. J., Central implants of diluted estradiol: independent effects on ingestive and reproductive behaviors of ovariectomized rats, *Brain Res.*, 491, 266, 1989.

75. Roy, E. J., Estradiol implants in the rat striatum stimulate locomotor activity in running wheels, *Soc. Neurosci. Abstr.*, Part 1, 224, 1987.

76. Butera, P. C., Willard, D. M., and Raymond, S. A., Effects of PVN lesions on the responsiveness of female rats to estradiol, *Brain Res.*, 576, 304, 1992.

77. Wade, G. N. and Gray, J. M., Gonadal effects on food intake and adiposity: a metabolic hypothesis, *Physiol. Behav.*, 220, 583, 1979.

78. Butera, P. C., Campbell, R. B., and Bradway, D. M., Antagonism of estrogenic effects on feeding behavior by central implants of anisomycin, *Brain Res.*, 624, 354, 1993.

79. Rainbow, T. C., McGinnis, M. Y., Davis, P., and McEwen, B. S., Application of anisomycin to the lateral ventromedial nucleus of the hypothalamus inhibits the activation of sexual behavior by estradiol and progesterone, *Brain Res.*, 233, 417, 1982.

80. Zucker, I., Body weight and age as factors determining estrogen responsiveness in the rat feeding system, *Behav. Biol.*, 7, 527, 1972.

81. Gray, J. M. and Wade, G. N., Food intake, body weight, and carcass adiposity in female rats: actions and interactions of progestins and antiestrogens, *Am. J. Physiol.*, 250, E474, 1981.

82. Mrosovsky, N. and Powley, T. L., Set points for body weight and fat, *Behav. Biol.*, 20, 205, 1977.

83. VanItallie, T. B. and Kissileff, H. R., Human obesity: a problem in body energy economics, in *Handbook of Behavioral Neurobiology, 10, Neurobiology of Food and Fluid Intake*, Stricker, E. M., Ed., Plenum Press, New York, 1990, 207.

84. Wade, G. N., Interaction between estradiol 17-B and growth hormone in weanling rats, *J. Comp. Physiol. Psychol.*, 86, 354, 1974.

85. Roy, E. J. and Wade, G. N., Role of food intake in estradiol-induced body weight changes in female rats, *Horm. Behav.*, 8, 265, 1977.

86. Ferreri, L. F. and Naito, H. K., Effect of estrogens on rat serum cholesterol concentrations: consideration of dose, type of estrogen, and treatment duration, *Endocrinology*, 102, 1621, 1978.

87. Ramirez, I., Relation between estrogen-induced hyperlipemia and food intake and body weight in rats, *Physiol. Behav.*, 25, 511, 1980.

88. Beinfeld, M. C., Cholecystokinin in the central nervous system: a mini-review, *Neuropeptides*, 3, 411, 1983.

89. Liddle, R., Goldfine, I. D., Rosen, M. S., Taplitz, R. A., and Williams, J. A., Molecular forms, responses to feeding and relationship to gallbladder contractions, *J. Clin. Invest.*, 75, 1144, 1985.

90. Baile, C. A. and Della-Ferra, M. A., Peptidergic control of food intake in food-producing animals, *Fed. Proc.*, 43, 2898, 1984.

91. Smith, G. P., Gibbs, J., and Young, R. C., Cholecystokinin and intestinal satiety in the rat, *Fed. Proc.*, 33, 1146, 1974.

92. Smith, G. P. and Gibbs, J., The development and proof of the cholecystokinin hypothesis of satiety, in *Multiple Cholecystokinin Receptors in the CNS*, Dourish, C. T., Cooper, S. J., Iversen, S. D., and Iversen, L. L., Eds., Oxford University Press, 1992, 166.

93. Lorenz, D. N., Kreielsheimer, G., and Smith, G. P., Effect of cholecystokinin, gastrin, secretin and GIP on sham feeding in the rat, *Physiol. Behav.*, 23, 1065, 1979.

94. Smith, G. P., Falasco, J., Moran, T. H., Joyner, K. M. S., and Gibbs, J., CCK-8 decreases food intake and gastric emptying after pylorectomy and pyloroplasty, *Am. J. Physiol.*, 255, R111, 1988.

95. South, E. H. and Ritter, R. C., Capsaicin application to central or peripheral vagal fibers attenuates CCK satiety, *Peptides*, 9, 601, 1988.

96. Crawley, J. N., Kiss, J. Z., and Mezey, E., Bilateral midbrain transections block the behavioral effects of cholecystokinin on feeding and exploration in rats, *Brain Res.*, 322, 316, 1984.

97. Crawley, J. N. and Kiss, J. Z., Paraventricular nucleus lesions abolish the inhibition of feeding induced by systemic cholecystokinin, *Peptides*, 6, 927, 1985.

98. Grill, H. J. and Smith, G. P., Cholecystokinin decreases sucrose intake in chronic decerebrate rats, *Am. J. Physiol.*, 254, R853, 1988.

99. West, D. B., Fey, D., and Woods, S. C., Cholecystokinin persistently suppresses meal size but not food intake in free-feeding rats, *Am. J. Physiol.*, 246, R777, 1984.

100. McLaughlin, C. L., Baile, C. A., and Peiken, S. R., Hyperphagia during lactation: satiety response to CCK and growth of the pancreas, *Am. J. Physiol.*, 224, E62, 1983.

101. Wager-Srdar, S. A., Morley, J. E., and Levine, A. S., The effect of cholecystokinin, bombesin and calcitonin on food intake in virgin, lactating and postweaning female rats, *Peptides*, 7, 729, 1986.

102. Linden, A., Uvnas-Moberg, K., Forsberg, G., Bednar, I., Eneroth, P., and Sodersten, P., Involvement of cholecystokinin in food intake: II. lactational hyperphagia in the rat, *J. Endocrinol.*, 2, 791, 1990.

103. Butera, P. C., Bradway, D. M., and Cataldo, N. J., Modulation of the satiety effect of cholecystokinin by estradiol, *Physiol. Behav.*, 53, 1235, 1993.

104. Linden, A., Uvnas-Moberg, K., Forsberg, G., Bednar, I., and Sodersten, P., Involvement of cholecystokinin in food intake: III. oestradiol potentiates the inhibitory effect of cholecystokinin octapeptide on food intake in ovariectomized rats, *J. Neuroendocrinol.*, 2, 797, 1990.

105. Geary, N., Trace, D., McEwen, B., and Smith, G. P., Cyclic estradiol replacement increases the satiety effect of CCK-8 in ovariectomized rats, *Physiol. Behav.*, 56, 281, 1994.

106. Wager-Srdar, S. A., Gannon, M., and Levine, A. S., The effect of cholecystokinin on food intake in gonadectomized and intact rats: the influence of sex hormones, *Physiol. Behav.*, 40, 25, 1987.

107. Butera, P. C., Xiong, M., Davis, R. J., and Platania, S. P., Potentiation of the satiety effect of CCK-8 by central implants of dilute estradiol, *Soc. Neurosci. Abstr.*, Part 3, 1823, 1993.

108. Butera, P. C., Davis, R. J., and Platania, S. P., Effects of CCK antagonists on the suppression of food intake by estradiol, *Soc. Neurosci. Abstr.*, Part 2, 587, 1994.

109. Smith, G. P., The direct and indirect controls of meal size, *Neurosci. Biobehav. Rev.*, in press.

110. Nance, D. M., The developmental and neural determinants of the effects of estrogen on feeding behavior in the rat: a theoretical perspective, *Neurosci. Biobehav. Rev.*, 7, 189, 1983.

INDEX